Intramolecular Charge Transfer

Intramolecular Charge Transfer

Theory and Applications

Ramprasad Misra and S. P. Bhattacharyya

WILEY-VCH

Authors

Dr. Ramprasad Misra
Indian Association for the Cultivation of Science
Department of Physical Chemistry
2 A&B Raja S.C. Mullick Road
700032 Kolkata
India

Prof. S. P. Bhattacharyya
Indian Association for the Cultivation of Science
Department of Physical Chemistry
2 A&B Raja S.C. Mullick Road
700032 Kolkolta
India

Library of Congress Card No.: applied for

British Library Cataloguing-in-Publication Data
A catalogue record for this book is available from the British Library.

Bibliographic information published by the Deutsche Nationalbibliothek
The Deutsche Nationalbibliothek lists this publication in the Deutsche Nationalbibliografie; detailed bibliographic data are available on the Internet at <http://dnb.d-nb.de>.

Print ISBN: 978-3-527-34156-6
ePDF ISBN: 978-3-527-80194-7
ePub ISBN: 978-3-527-80192-3
Mobi ISBN: 978-3-527-80193-0
oBook ISBN: 978-3-527-80191-6

Cover Design Formgeber, Mannheim, Germany
Typesetting SPi Global, Chennai, India
Printing and Binding C.O.S. Printers Pte Ltd Singapore

Printed on acid-free paper

10 9 8 7 6 5 4 3 2 1

Contents

Preface

Over the last two decades, intramolecular charge transfer (ICT) molecules have been receiving wide-spread attention of scientists in view of their potential for technological applications in molecular electronics, solar cells, quantum optics, sensors, and so on. Organic ICT molecules in particular have been at the primary focus as they provide better stability and flexibility in design than the standard inorganic molecules. A large number of research publications dealing with different aspects of these molecules have been added to the literature over the recent years. We felt that a wide gap already exists between information available in books and monographs accessible to the graduate and Masters students as well as beginners in research and those available in specialized journals, and the gap is increasing by the day. A researcher interested in designing new organic nonlinear optical (NLO) material or ICT based sensor or molecule based materials for organic electronics, or a general reader curious to understand the ICT phenomenon more completely would welcome a book that bridges the that gap seems to exist. This book attempts to address the issue by providing a pedagogical description of the different stages and facets of ICT phenomenon and what goes behind the designing of ICT based materials for technological applications; what problems are to be taken care of and how to leverage theory and experiments in a specific context, and so on. In addition to an overview of the recent theoretical and experimental developments relating to ICT molecules and the ICT phenomenon the book contains a brief history of past efforts in this area. Charge transfer is a rather ubiquitous process in nature and a fundamental step in many chemical and biological processes, like photosynthesis and metabolism, for example. The ICT in conjugated π-electronic systems has attracted serious attention in view of the immense technological applications that the process has for example, in organic electronics and photovoltaics. Materials based on such molecules are potential candidates for organic light-emitting diodes (OLEDs), field effect transistors, dye-sensitized solar cells, and so on. In fact the main impetus to study electron transfer (ET) process in organic molecules and materials came from the urge to understand the mechanism of photosynthesis in plants and bacteria with the hope that the knowledge gained in the process would help the scientists in designing artificial photosynthetic systems for efficient conversion and storage of solar energy. The search for such systems is still on with a long way to go. Regular stocktaking is therefore essential in such a critical field of research.

The contents of the book have been divided among seven chapters as detailed in the following. In the introductory chapter (Chapter 1) the basics of the ICT process has been discussed laying the foundation for the next five chapters (Chapters 2–6). The rather long history of evolution of the idea of ICT process and ICT molecules has been presented in Chapter 2 to place the modern developments in a proper perspective. The ICT molecules most frequently studied with steady state and time-resolved spectroscopic techniques in the UV–visible range; specially, time-correlated single photon counting (TCSPC) and fluorescence up-conversion techniques have been found widespread use for time-resolved studies on ICT. Two other useful techniques for probing ICT have been femtosecond transient absorption (TA) and resonance Raman spectroscopy. A relatively recent and novel experimental tool to follow the ICT dynamics is the terahertz (THz) spectroscopy. The idea behind the THz spectroscopy is simple. The ICT process involves movement of electronic charge from one end of the molecule to the other. If the charge is accelerated, electromagnetic (EM) radiation will be emitted. Assuming that the ICT occurs on a timescale of picoseconds, the frequency of the EM radiation will be in the THz range that can be detected and monitored. It is, as if, the moving charge "broadcasts" its own dynamics opening up a direct route to follow the ICT dynamics. We have devoted some space to this new mode of studying the ICT process – the so called THz spectroscopy.

A lot of theoretical calculations (modeling as well as calculations) on ICT molecules are now available in literature. The calculations have been mostly done at the Hartree–Fock (HF) level and of late increasingly at the level of the complete active space self-consistent field (CASSCF) method or CASSCF with second-order correction incorporated (CASPT2) along with variants of density functional theory (DFT) and time-dependent density functional theory (TDDFT) methods. Although majority of such calculations have been done at the adiabatic level, there has been a "paradigm shift" in the sense that theoreticians are increasingly looking into the very important role played by non-adiabatic factors in shaping the ICT process. The Chapter 3 is devoted to the relatively newer theoretical and experimental techniques that are being used to unravel the mystery of the ICT process and model ICT rates.

It is now firmly established that the ICT process is critically affected by the properties of the medium in which it occurs. Not only are the spectral signatures of the ICT molecules modulated by the polarity and hydrogen bonding abilities of the solvents, the ICT rate is also affected by thermal effects (temperature) and friction (solvent viscosity). Several ICT molecules have therefore been exploited as polarity and viscosity sensors. Such media effects on the ICT process have been elaborately discussed in Chapter 4. As mentioned already, the ICT molecules have technological relevance and are being explored for fabricating new molecules based NLO or halochromic materials, materials for solar cells, for OLEDs, for viscosity and polarity sensors, and so on. In view of the importance of the field and diversity of issues, we have devoted two consecutive chapters to discuss technological applications of ICT molecules (Chapters 5 and 6). The NLO phenomena with emphasis on hyperpolarizabilities that the ICT molecules are endowed with and several aspects of ICT based two-photon absorbing materials are dealt with in Chapter 5. Several issues with technological applications of ICT molecules in sensing, in OLEDs, and so on are examined in the penultimate Chapter 6 while Chapter 7 is devoted to consider future projection of research in ICT in backdrop of important developments that have already taken place.

This book is a review of studies of ICT process and related phenomena. We have reported results available in contemporary literature in good faith. Permissions for copyrighted materials have been duly obtained from the copyright holders. The book is primarily intended for the Masters and graduate (doctoral) students of chemistry and chemical physics. We hope that more specialized people too, will find the book useful. We have tried to strike a balance between experiment and theoretical aspects with the hope that it caters the need of the both theoreticians and experimentalists alike. In spite of our best efforts to make the compilation error-free, some unfortunate and unintended omissions might still be crept in. We sincerely regret any such blemishes.

We are happy to acknowledge the help and constant encouragement of colleagues and friends without which the book might not have seen the light of the day. A major part of the book was written while the authors were in the Indian Association for the Cultivation of Science (IACS). We thank all our colleagues in the Department of Physical Chemistry, IACS for providing the intellectual ambience needed for undertaking the project. We specially thank Prof. D.S. Ray and Prof. S. Adhikari for all the help and encouragements they extended to us. One of the authors (RM) wishes to thank his present mentor Prof. M. Sheves of the Weizmann Institute of Science for his constant support. RM is grateful to his parents, wife (Piyali), and other family members for their unfailing supports and encouragement. SPB wishes to express indebtedness to Bharati (wife), Rupsha (daughter), and Sayan (son) for their unfailing support and the colleagues in the Department of Chemistry, IIT Bombay for making his stay in the department (2012–2015) as Raja Ramanna fellow (DAE) fruitful.

August 21, 2017
Kolkata

Ramprasad Misra
S. P. Bhattacharyya

1

Introduction

1.1 An Overview of the ICT Process

Charge transfer is a fundamental step in many chemical and biological processes, including photosynthesis and metabolism [1–4]. The recent technological applications of charge-transfer–based materials include organic-light-emitting diodes (OLEDs), solar energy conversion, fluorescence sensing, nonlinear optical (NLO) materials, and so on [5–7]. The charge-transfer process may be divided into two broad categories. The transfer of charge from an electron-rich donor moiety to an electron-poor acceptor part located in different molecules is known as intermolecular charge-transfer process. However, if the donor and the acceptor belong to the same molecule, the phenomenon is called intramolecular charge-transfer (ICT) process. The ICT process generally occurs in the photoexcited state which a molecule reaches due to absorption of light of proper wavelength. The photoexcitation facilitates transfer of an electron from one part of a molecule/ion to its other part in the excited state, which makes the charge distribution in the excited state markedly different from that in the ground state. The through-bond ICT occurs in molecules in which the donor and the acceptor groups are connected through a π-electron bridge (Figure 1.1). In some rare cases, an intramolecular through-space charge transfer is observed, where the transfer of charge through the conjugative path is denied but donor and acceptor groups are in a favorable position for charge transfer. Although the intermolecular interaction mediated by through-space charge transfer dictates the properties of many π-stacked molecular systems, studies of intramolecular through-space charge transfer is scarce. In π-conjugated organic molecules comprising electron donor (D) and acceptor (A) subunits, the process has attracted a lot of attention due to their immense technological implications in organic electronics and photovoltaics [5–14]. Materials based on such organic molecules are potential candidates for use in OLEDs, field-effect transistors, dye-sensitized solar cells, to name a few. In this book, our focus is mainly on the excited-state ICT in stable organic molecules as well as in inorganic complexes. A few examples of the electron-transfer process in biomolecules have also been discussed. The signatures of ICT, spectroscopic techniques, and theoretical tools employed to study this process are also mentioned. It is now known that

Intramolecular Charge Transfer: Theory and Applications,
First Edition. Ramprasad Misra and S. P. Bhattacharyya.

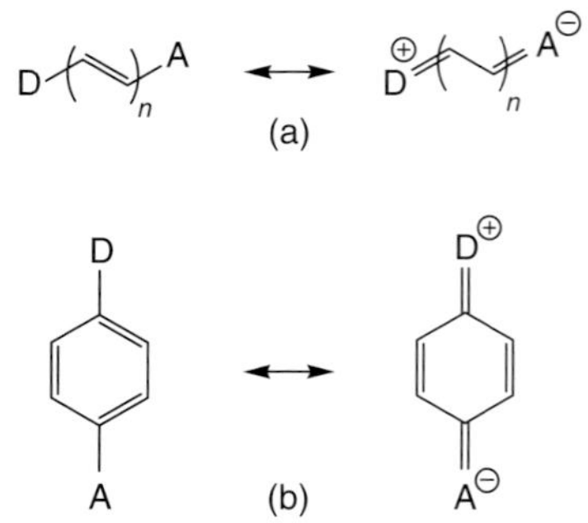

Figure 1.1 The intramolecular charge-transfer processes (a) in a (poly)ene system and (b) in an aromatic donor–acceptor molecule. Donor (D) to acceptor (A) charge transfer occurs in the excited state through the π-electron bridge to form an ICT state of higher dipole moment.

excited-state ICT in organic molecules may give rise to dual emission in its electronic spectrum. The peak seen at the blue end of the emission spectrum is generally believed to be arising from a locally excited (LE) state of the molecule, while the peak at the red end is generally assumed to bear the signature of an ICT species formed in the excited state. The ICT process generally occurs in polar solvents and the Stokes-shifted ICT fluorescence is observed due to solvent stabilization in the excited state [15]. In Figure 1.2, the potential energy surfaces (PES) for the ground state S_0 and the first two excited states (S_1, S_2) have been depicted along with the LE and ICT states. The vertical coordinate represents energy, while the horizontal coordinate (ξ) comprises all molecular changes accompanying the LE → ICT reaction, such as changes in bond lengths and bond angles. In the given example, excitation of the molecule leads it to the S_2 state, which relaxes through internal conversion to the equilibrated LE state. The ICT reaction proceeds from the LE to the ICT state that has a

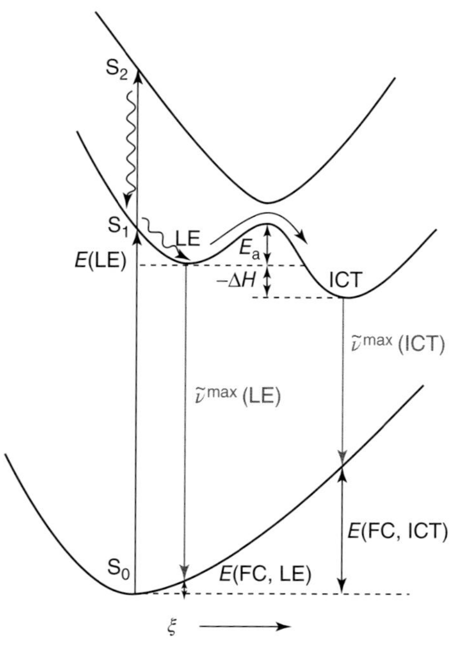

Figure 1.2 Schematic representation of intramolecular charge transfer (ICT) process in a model system. In the given scheme, excitation leads the molecule to the S_2 state from which it reaches the S_1-LE state through internal conversion. The ICT state is accessible through the S_1-LE state. Emission form both the LE and ICT states leads to dual emission from the molecule. (Druzhinin *et al.* 2010 [15]. Reproduced with permission of American Chemical Society.)

reaction barrier E_a and an enthalpy difference of ΔH. Fluorescence from the LE and ICT states reaches the corresponding Franck–Condon states E(FC, LE) and E(FC, ICT). This gives rise to dual fluorescence with emission maxima of $\tilde{\nu}_{max}$(LE) and $\tilde{\nu}_{max}$(ICT), respectively. Lippert and coworkers reported the dual fluorescence in 4-*N*,*N*-dimethylaminobenzonitrile (DMABN) for the first time in 1962 [16]. The debate continued regarding the origin of the dual emission of DMABN for some years. Most of the studies accepted that the observed dual emission from DMABN is due to excited-state ICT from the dimethylamine to the cyano group through the π-electron bridge. Later on, many congeners of DMABN were put under the scanner for deciphering or decoding the nature of the ICT process and its dynamics. Many experimental or theoretical studies or both have been devoted to investigate the charge-transfer mechanism in different organic molecules. The early experimental and theoretical results on the ICT process support a twisted intramolecular charge-transfer state (called TICT state) [16]. In the framework of the TICT model, the dual emission of an ICT probe originates from the primary excited, called LE, state as well as from the ICT state. The ICT state is accessible only by an adiabatic photoreaction from the LE state that includes rotational motion around the bond connecting the donor and acceptor moieties. If there is no energy barrier between the LE and ICT states, the excited-state relaxation can occur extremely rapidly, resulting in emission from the ICT species only. Although the TICT mechanism is till date the most popular concept in describing the structure of the excited state, this model was challenged by several groups. Later on, several other models, like planar intramolecular charge transfer (PICT), rehybridized intramolecular charge transfer (RICT), and wagging intramolecular charge transfer (WICT), were proposed by several groups to explain their experimental results. For example, Domcke and coworkers put forward a RICT model to account for the formation of the ICT state of DMABN and its analogs. Later on, the formation of a charge-transfer state from a rigid molecule, *N*-phenylpyrrole (PP) and its free analog, fluorazene phenylpyrrole (FPP) put a question mark on the validity of the TICT model. The similarity in spectral signatures of these molecules could not be explained using the TICT model. This led to the proposal of a PICT mechanism to explain the spectroscopic response of the aforesaid molecules. There are several other examples that support the fact that the ICT state of the molecule does not need to be twisted. The origin and drawbacks of some of these models have been described in Chapter 2.

Another point regarding the formation of the ICT state that is still under intense debate is the pathways of the charge-transfer process, that is, the mechanism through which the ICT process in a molecule occurs. Several high-level calculations and state-of-the-art experimental techniques have been used to settle this issue. In spite of several studies that describe the PES of the ICT process, this issue is yet to be settled amicably. A current account of this topic has been described briefly in Chapter 2. The charge-transfer studies in some other molecules, including rhodamine derivatives, coumarin, oxazine, flavin, Nile red, and so on, are discussed in Chapter 2.

1.2 Experimental and Theoretical Studies of the ICT Process

As we mentioned earlier, several studies have been devoted to explore the structure and mechanism of formation of the ICT state. Initially, steady-state UV–visible absorption and emission spectroscopy have been widely used for studying the ICT phenomenon in several organic molecules as dual fluorescence is a major observation of the ICT process. Later on, time-resolved spectral techniques, including, picosecond (ps) time-resolved fluorescence, femtosecond transient absorption (TA) and fluorescence upconversion spectroscopies are used by the scientific community to study the ICT process. Recently, Fleming and coworkers [17] have used a combined ultrafast electronic pump and infrared (IR) probe spectroscopy to study the excited-state CT. Gaffney and coworkers [18] have used polarization-resolved UV pump–mid-IR probe spectroscopy, in conjunction with time-dependent density functional theory (TDDFT) calculations to investigate the dynamics of charge-transfer–induced intramolecular rotation in julolidine malononitrile (JDMN), an ICT probe. Of late, Terahertz (THz) spectroscopy [19] is being used to measure the ICT in a molecule directly. When the ICT process occurs in a molecule, electronic charge moves from one end to another, which will undergo acceleration, resulting in radiation of electromagnetic (EM) pulses. If the charge-transfer process occurs on ps timescale, the radiated EM pulse will fall in the THz region of the spectrum. There are some phenomena similar to ICT, like energy transfer, intramolecular proton transfer (IPT), and so on that has similarity to this process but are different in nature. Dual fluorescence has also been observed in molecules undergoing excited-state intramolecular proton transfer (ESIPT). Although ESIPT is also a charge-transfer process, it is sometimes necessary to distinguish between the ICT and IPT processes occurring in a molecule. It has been seen that the IPT process occurs on a much faster timescale compared to the ICT process. On the other hand, in an ESIPT process, the hydrogen donor and acceptor groups must be present within a certain distance of each other, while ICT can occur even if the donor and the acceptor groups in a molecule are placed quite a distance apart either by a "through-bond" process mediated by a π-electron bridge or by a "through-space" interaction. It is now known that both the ICT and ESIPT processes are dependent on the solvents. One can therefore distinguish these two processes by changing the medium, which in turn will affect their spectroscopic properties.

During the ICT process, the electronic charge distribution of a molecule changes, generally resulting in an ICT species with higher dipole moment compared to its ground-state counterpart. The idea of the formation of a higher dipole moment species has also been supported by the fact that in polar solvent the ICT reaction becomes faster due to better stabilization of the polar ICT species in polar medium. The formation of the ICT process in a molecule is also indicated through the redshifted emission as one goes from a nonpolar to a polar solvent. Although many of the ICT molecules show dual emission in polar solvents, not all of them show dual emission. It is believed that due to ICT the non-radiative deactivation channels (like solvent relaxation) become active, finally decreasing the emission quantum yield. The properties of the medium play a crucial role in determining the rate of formation of the ICT state as well as its structure.

It is worth mentioning here that not all ICT molecules show redshifted emission maxima upon photoexcitation. Molecules with zwitterionic ground state gets charge transfer in the excited state, generally resulting in a species with lower dipole moment than that in the ground state. This may result in a blueshift in their emission maxima. For example, betaine-30, also known as Reichardt's dye (Figure 1.3) shows negative solvatochromism [20]. This was explained by the change in its dipole moment in the ground state to that in the excited state. The molecule remains in a charge-transferred zwitterionic (D^+-A^-) state in the ground state that turns into a D-A state upon photoexcitation. The dipole moment of the betaine-30 is reported to be about 15 D and 6.2 D in the ground and excited states, respectively.

Several theoretical calculations have been employed to study the ICT process. Initially, semiempirical methods were used by several groups to explore the structure of the ICT state. Later, Hartree–Fock (HF), density functional theory (DFT), complete active space self-consistent field method (CASSCF) and second-order correction to CASSCF (CASPT2) level of theories were employed either alone or in conjunction with experimental studies to explore the ICT process in the excited state. Many of these theories now can reproduce the experimental results on the ICT process accurately. It is now known that standard DFT functional like B3LYP in many cases failed to reproduce the spectral properties of charge-transfer–based molecules. Therefore, several studies have devoted to propose new DFT functional and basis set combinations to study the ICT process. Some of these studies are discussed in the following

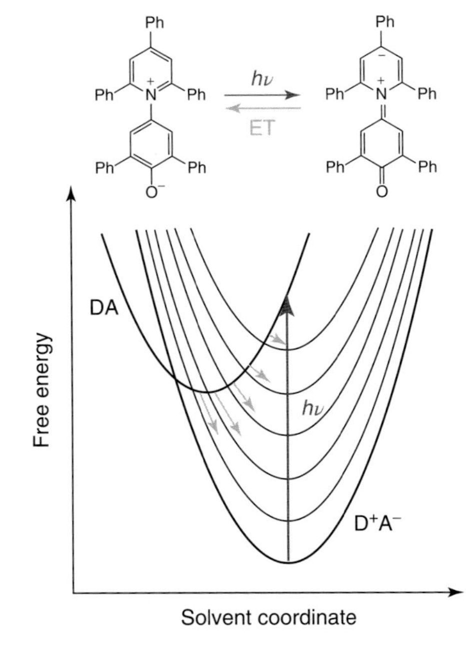

Figure 1.3 The free energy curves of the ground (S_0) and first excited (S_1) states of betaine-30. The vibrational levels and possible ET mechanism are also shown. (Kumpulainen *et al.* 2017 [20]. Reproduced with permission of American Chemical Society.)

chapters of this book. Initial theoretical studies involved optimization of the ground state, followed by calculation of spectroscopic properties using the ground-state geometry with certain approximation. Of late, optimization of the excited-state geometry using configuration interaction singles (CIS), TDDFT, and CASSCF level of theories has been undertaken to understand the physical nature of the charge transfer in specific organic molecules. It is apparent that in a molecule the formation of the ICT state from the LE state competes with the radiative deactivation process. These states occupy different regions of the PES of the relevant molecule. To understand and interpret the phenomenon unambiguously, the details of the PES of these molecules in the ground and excited states are therefore necessary. Of late, theoretical calculations at the CIS, TDDFT, CASSCF, and CASPT2 level of theories and femtosecond pump-probe spectroscopy have been essential tools in such endeavors. A detailed description on theoretical studies on ICT can be found in Chapter 3.

Electron transfer (ET) reaction can be considered as one of the simplest chemical reactions occurring in nature. Initially, the classical Marcus theory, proposed during the mid-fifties of the previous century became very popular. In the model described by Marcus theory, the free energy of the reactant (R) and product (P) goes along the reaction coordinate, comprising both solvent and intramolecular modes [21]. The horizontal displacement, as shown in Figure 1.4, accounts for the difference in equilibrium geometries of the reactant and product states and is quantized by the reorganization energy (λ). On the other hand, the vertical displacement represents the driving force of the reaction and is denoted as $-\Delta G_{ET}$. The reorganization energy can be further divided into contribution from the solvent and intramolecular modes. In classical Marcus theory, the ET reaction is considered as a thermally activated process and the rate of this reaction (k_{ET}) can be expressed in an Arrhenius-type equation, as shown in Equation 1.1.

$$k_{ET} = A \exp\left[-\frac{(\Delta G_{ET} + \lambda)^2}{4\lambda k_B T}\right] \tag{1.1}$$

where k_B is the Boltzmann constant and T is the absolute temperature.

From Equation 1.1, three conditions can arise: (i) In the normal region, $-\Delta G_{ET} < \lambda$, the values of k_{ET} increases with the driving force. (ii) In the

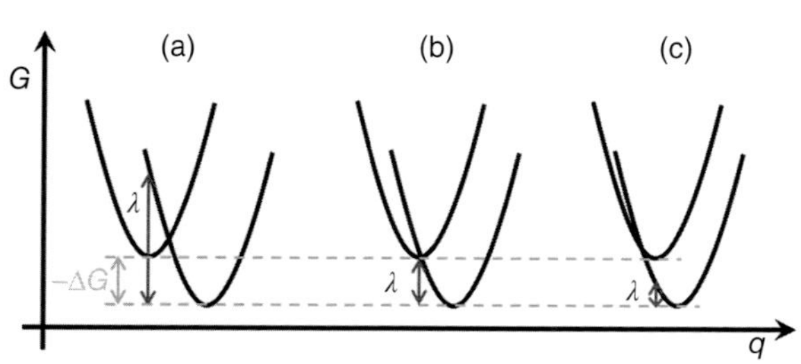

Figure 1.4 Potential energy curves along the reaction coordinate, q, for the reactant and product states. The curves a, b, and c represent the normal, barrier-less and inverted Marcus regions, respectively (see text for details). (Pelzer and Darling 2016 [22]. Reproduced with permission of Royal Society of Chemistry.)

barrier-less region $-\Delta G_{ET}$ is almost equal to λ, the k_{ET} value reaches the maxima. (iii) In the inverted region, $-\Delta G_{ET} > \lambda$, the values of k_{ET} decreases with increasing driving force.

Later, several semiclassical and quantum mechanical descriptions of the ET process were considered [20, 22]. Fermi Golden rule [23] is one of the popular models to study the ET process. Sumi–Marcus theory [24] of ET considers the possibility of the ET process taking place faster than the solvent relaxation by splitting the reaction coordinate into fast and a slow coordinates, where they are associated with the relaxing intramolecular modes and with the solvents, respectively. Some of the methods to find the rate of electron-transfer process are discussed in Chapter 3. Recently, Manna and Dunietz [25] studied the rate of photoinduced charge-transfer processes within dyads consisting of porphyrin derivatives in which one ring ligates a zinc (Zn) metal center and where the rings vary by their degree of conjugation.

The separation of charge that follows the ICT process generally enhances the dipole moment of the system in the excited state with some exceptions, like betaine-30 (discussed in this chapter). So, it is quite likely that the ICT species will be stabilized progressively with increase in the polarity of the medium as polar solvents surround the polar probe to exert stability to the probe molecule. The signatures of the ICT process are therefore expected to be more recognizable in strongly polar solvents. A solute (probe) interacts with the solvent in different ways. In suitable cases, it can form molecular clusters of specific sizes and stoichiometries through specific interaction (mostly through hydrogen bonding) with the solvent molecules. On the other hand, the solute molecules may be solvated through the dielectric property of the solvent. This process is called macroscopic or bulk solvation. Both specific or microscopic solvation and macroscopic solvation in dielectric continuum could shape the photophysics of the solute. Many experimental, quantum chemical studies and simulations have been therefore undertaken by the scientific community to unravel the microscopic details the solvent effects on the ICT process [26–38]. It is now known that investigations of structure and dynamics of hydrogen-bonded molecular clusters provide us with the opportunity to understand the effect of the bulk solvation systematically. The studies of cluster formation between solvent and solute molecules also enable one to understand many complex physical and chemical processes from a molecular point of view, which is not possible in the study of bulk solvents. The investigations on the formation of molecular clusters have been fueled due to advent of experimental methods like the mass-selective resonance two-photon ionization (R2PI), rotational coherence spectroscopy (RCS), two-photon ionization and IR ion depletion (IR/R2PI) spectroscopy, among others [39]. A large volume of research has been devoted to explore the molecular clusters in the ground state. Of late, invention of many state-of-the-art time-resolved spectroscopic techniques, like femtosecond TA spectroscopy, femtosecond time-resolved fluorescence spectroscopy, time-resolved vibrational spectroscopy, and so on, have enabled the scientific community to study hydrogen bonding in the excited states. The effect of the medium polarity and hydrogen bonding ability on the ICT process on some molecules is discussed in Chapter 4. Studies of the ICT process is generally done in solution. However, a few studies of ICT have been reported in literature either in the gas phase or in solid state. A few representative examples of ICT studies in the gaseous and solid states can be found in Chapters 2 and 4.

As mentioned earlier, several new ICT molecules are being reported in literature to understand the ICT process in detail that could be useful in using them for technological applications. Zhao *et al.* [40] have reported the photoinduced ICT process in thiophene-π–conjugated donor–acceptor compounds. As thiophene has lower delocalization energy than benzene, the authors hoped that the former could be a better spacer for π-conjugation in ICT molecules. In fact, several compounds containing thiophene ring have been investigated for possible use in optoelectronic devices, such as NLO, OLEDs, and dye-sensitized solar cells [41–43]. To test their hypothesis, the authors compared the spectral properties of QTCP with its benzene-bridged analog QBTP (Figure 1.5). They assigned the large Stokes shift and strong solvatochromism of these molecules to the formation of the ICT state. They also found that the electron-withdrawing ability of the substituents at the thienyl 2-position affected the ICT process of the molecules studied. In addition to this, they also observed emission from the S_2 state of these molecules, which they confirmed through their fluorescence excitation spectra.

Studies of the ICT process led to the proposal of several theoretical models to mostly predict the rate of the ICT process. Some of these models have been discussed in Chapter 3. Recently, Ivanov and coworkers [44] have reported their simulation studies of ultrafast charge recombination in ion pairs formed through intramolecular electron-transfer process in the light of free-energy-gap law. They simulated the kinetics of charge transfer and ensuing ultrafast charge recombination within the framework of multichannel stochastic model that considers reorganization of both the solvents and a number of intramolecular high-frequency vibrational modes. They described the solvent relaxation in terms of two relaxation modes. The authors found that for ultrafast charge recombination the free-energy-gap law strongly depended on parameters like electronic coupling, reorganization energy of the intramolecular high frequency vibrational modes, as well as the vibrational and solvent relaxation times.

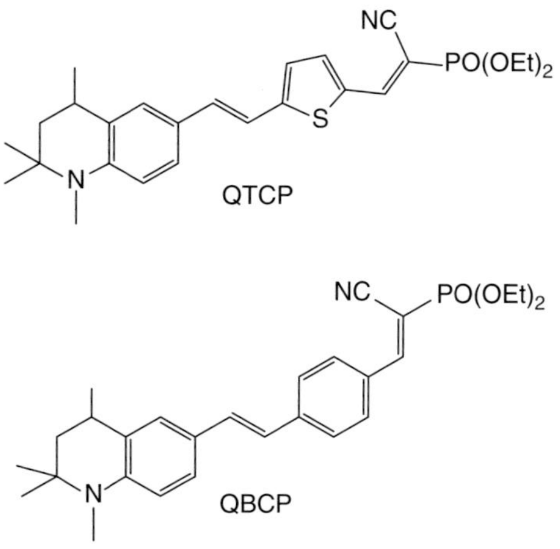

Figure 1.5 The chemical structure of QTCP and QBTP.

They also found that the semilog dependence of the charge recombination rate constant on the free-energy gap varied from a parabolic shape to an almost linear one, while the electronic coupling was increased and the vibrational relaxation time was decreased. The authors predicted that the dynamical solvent effect in charge recombination was large in the area of strong exergonicity, while the effect was small in the area of weak exergonicity.

As mentioned earlier, the properties of the medium play a crucial role in determining the properties of an ICT molecule. Dube and coworkers [45] have reported that the internal motions of hemithioindigo (HTI) photoswitches could be controlled through a change of solvent. The control of the internal motions of molecules using stimuli from outside is important for generation of responsive as well as complex molecular behavior and functionality. Photoswitches that react to light reversibly in a specific way are used as “engine” unit to trigger specific motion as well as events in functional molecular and biological systems [46–48]. HTI derivatives typically consist of a central double bond that functions as a bridge between the thioindigo fragment and a stilbene fragment via one additional single bond. After photoexcitation, unsubstituted HTI derivatives undergo de-excitation in which rotation of the double bond occurs, with additional pyramidalization of the corresponding carbon atoms. The presence of a single C—C bond in these molecules raises the possibility of rotation of this bond during the de-excitation process. Simultaneous rotation of the single and double bonds, called hula-twisting, is generally not considered to take place during de-excitation of unsubstituted HTI derivatives. To test the possible formation of TICT states, the authors considered four HTI derivatives acronymed *Z*-1 to *Z*-4 (Figure 1.6). HTI derivatives *Z*-1 to *Z*-3 possess substituents in both *ortho*-positions of the stilbene fragment. They used the planar derivative HTI

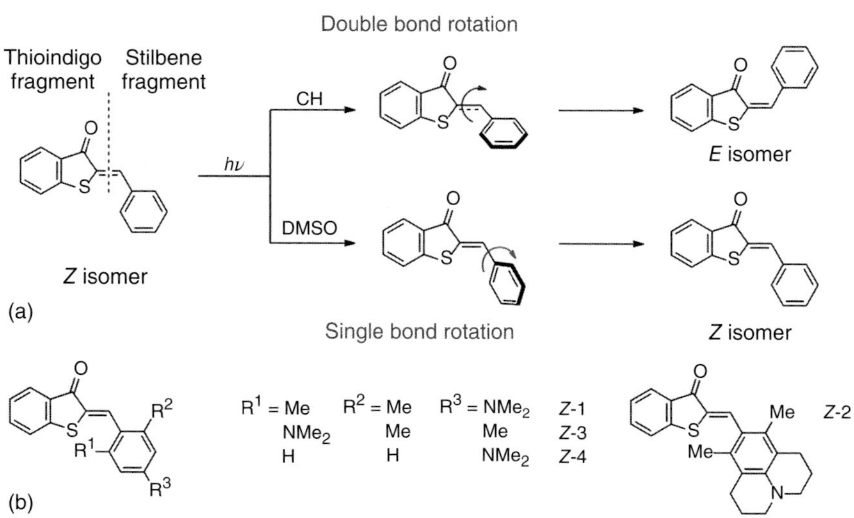

Figure 1.6 (a) Possible light-induced motions of the HTI derivatives, as proposed by Dube *et al.* (b) The chemical structures of the HTI photoswitches *Z*-1 to *Z*-4. (Wiedbrauk *et al.* 2016 [45]. Reproduced with permission of American Chemical Society.)

Z-4 as control, which does not undergo the TICT process in the excited state. All the four derivatives have strong electron-donor dialkylamino substituent (the electronic effect of additional alkyl substituents in *Z*-1 to *Z*-3 was considered to be not much). The twofold *ortho*-substitution in *Z*-1 to *Z*-3 resulted in a significant twist along the single-bond axis that could be measured using crystal structure analysis. The authors found that the dihedral angle around the rotatable single bond was highest for *Z*-1, with a value of 75°. The aforesaid dihedral angles for *Z*-2 and *Z*-3 were found to be 60° and 32°, respectively. Their planar analog *Z*-4 showed the dihedral angle to be 7° in its crystal structure analysis. The authors studied the chemical shift of indicative proton signals in ^{1}H NMR spectra of *Z*-1 to *Z*-3 that showed a similar twist of these molecules even in solution phase. As expected, the planar molecule HTI *Z*-4 showed moderate solvatochromism in both absorption and fluorescence. The absorption spectra of HTIs *Z*-1 to *Z*-3 also showed moderate solvatochromism, although somewhat broader than their planar analog *Z*-4. The fluorescence spectra of HTIs *Z*-1 and *Z*-2 were affected significantly due to the change in the polarity of the medium. In polar solvents, those molecules showed Stokes shift of more than 200 nm. Dual emission of *Z*-1 and *Z*-2 were observed in polar solvents like dimethyl formamide (DMF), dimethyl sulfoxide (DMSO) and acetonitrile. The steady-state absorption and emission behavior of HTI *Z*-3 was quite similar to HTIs *Z*-1 and *Z*-2, although the authors did not find any clear indication of dual emission of this probe. The authors undertook time-resolved absorption, time-resolved emission, and quantum yield measurements to understand the photophysics of the aforesaid molecules in detail. They followed the ultrafast kinetics taking place after photoexcitation of HTIs *Z*-1 to *Z*-4 in different media at 22 °C by TA spectroscopy with 150 fs resolution. Although the fitting of the time-resolved absorption curve of HTI *Z*-4 resulted in several timescales, the authors identified the important processes to describe the excited-state dynamics of these molecules. For this molecule, very fast nuclear motions and solvent reorganization were observed due to photoexcitation to the Franck–Condon (FC) region before reaching the relaxed first excited state (S_{1Min}). The S_{1Min} was identified in all the solvents studied through small redshifted ground-state bleaches and simulated emissions. The redshift increases with increasing solvent polarity. The authors argued that the decay of this electronic excited state to be related to the *Z*/*E* photoisomerization of this molecule (HTI *Z*-4). The rate of photoisomerization of HTI *Z*-4 was found to be solvent dependent, which decreased with increasing solvent polarity. The slowing down of photoisomerization with increase in the polarity of the solvents was attributed to the significantly polar nature of the S_{1Min}. However, no indication of formation of the TICT state was observed in HTI *Z*-4. For twisted HTI derivative *Z*-1, the initial absorption behavior in nonpolar solvents like cyclohexane was found to be quite similar to its planar counterpart HTI *Z*-4, although the decay of the excited-state features and appearance of *E* isomer absorption in this molecule was found to be faster than that in HTI-*Z*-4. The authors argued that the steric interactions imposed by the *ortho*-substituents in HTI *Z*-1 could be the reason for fast isomerization of this molecule in nonpolar solvents. In moderately polar solvents like dichloromethane (CH_2Cl_2), the time-resolved absorption profile of

HTI *Z*-1 varied significantly that of its planar counterpart. Dual emission of *Z*-1 and *Z*-2 were observed in polar solvents like DMF, DMSO, and acetonitrile. They also found that the excited electronic state had redshifted simulated emission and new excited-state absorption (ESA) features. This new excited state was found in HTI *Z*-1 in every polar solvent used for the study and the time constant of the relatively longer lived state was connected to the decay of S_{1Min}, while the amplitude is determined by the properties of the solvent as well as the specific decay time. The authors found that the new excited state (they call it T as it was only observed in the twisted HTI derivatives) got stabilized significantly with increase in solvent polarity. This indicated that the new excited state was highly polar and could possess significant charge-transfer character, a characteristic of the TICT state. Similar studies were carried out for HTIs *Z*-2 and *Z*-3, which also indicated the formation of TICT states in these molecules. The steady-state absorption and emission, time-resolved absorption and emission spectral studies, and quantum yield measurements of the probe led the authors to conclude the following. For planar HTI derivative *Z*-4, the de-excitation occurs through the photoisomerization of the double bond using the conical intersection ($CoIn_{S2}$) between the second excited state (S_2) and the ground state (S_0). The excited population leaves the initially reached S_{1Min} state and crosses a barrier to reach the S_2 hyperpotential surface. Finally, it reached the $CoIn_{S2}$ from which the transition to the S_0 state occurs. The behavior of HTI *Z*-4 is solvent independent. In twisted HTI derivative *Z*-1, the excited-state behavior is solvent dependent. In nonpolar solvent cyclohexane, the excited TICT state is not accessible and the de-excitation proceeds through the double-bond photoisomerization pathway. This led to fluorescence of HTI *Z*-1 from the S_{1Min} of this molecule in cyclohexane. In solvents with medium polarity (CH_2Cl_2), the TICT state of HTI *Z*-1 is stabilized significantly and gets populated upon photoexcitation. A strong fluorescence of this molecule from the TICT state occurs in CH_2Cl_2. Increasing solvent polarity further (e.g., DMSO) stabilizes the TICT more which leads to opening of another radiation-less de-excitation pathway through rotation around the single bond. The depopulation of the TICT state in DMSO becomes rapid and another blueshifted emission from S_{1Min} is observed in this solvent. All these results are schematically shown in Figure 1.7. The results presented by the authors showed that an unprecedented control over the photoinduced intramolecular rotations could be achieved for the HTI photoswitches through change in solvent polarity.

It is worthwhile to note here that just connecting the donor and acceptor moieties through the π-electron bridge in a molecule is not enough for ICT to occur in the exited state. Dube and coworkers [49] have studied the structural and electronic factors required for formation of excited ICT state in a series of HTI dyes (Figure 1.8). These dyes have been studied extensively for possible use as photoswitches that have advantages like high thermal bistability, fatigue-resistant switching, and visible light responsiveness. The TICT process is reported to be a second de-excitation pathway that is complementary to photoisomerization in these molecules. Dube and coworkers have considered eight HTI derivatives with varying structural and electronic properties to decipher the parameters required to facilitate the ICT process in the excited state. The

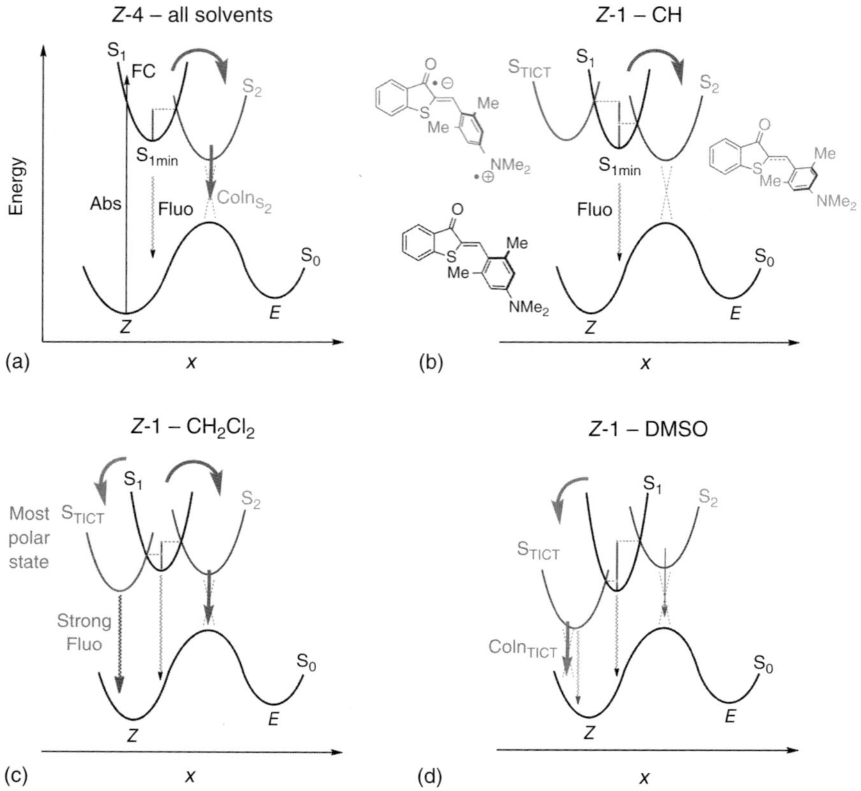

Figure 1.7 Schematic representation of excited-state deactivation mechanism of HTI derivatives: (a) *Z*-4 (in all solvents) and *Z*-1 (in cyclohexane, CH_2Cl_2 and DMSO). (Wiedbrauk *et al.* 2016 [45]. Reproduced with permission of American Chemical Society.)

HTI derivatives have different substitutions at the stilbene fragment and the twist angles around the rotatable single bond connecting the stilbene moiety with the photoisomerizable double bond are also different. HTI derivatives with acronyms *Z*-1 and *Z*-2 have strong donor units in conjugation with the photoisomerizable double bond. They also have large twisting of the stilbene fragment in the ground state. In all the HTI molecules used in this study, the thioindigo fragment with its strong electron-withdrawing carbonyl is used as the electron acceptor. HTI derivatives *Z*-3 and *Z*-4 possess strong donor groups but they are planar in the ground state. *Z*-5 and *Z*-6 do not possess strong electron-donating groups, although they are severely twisted around the rotatable single bond in their respective ground states. HTI derivatives *Z*-7 and *Z*-8 have substituents with moderate donating strength and they are twisted in the ground state. The authors argued that the varying donor strengths of the molecules studied can alter the push–pull character of the molecules, which is an important ingredient for formation of the ICT state. The steady-state electronic absorption spectra of *Z*-1 and *Z*-2 showed redshift in their absorption maxima with the increase in polarity of the media, although the effect is rather smaller than their

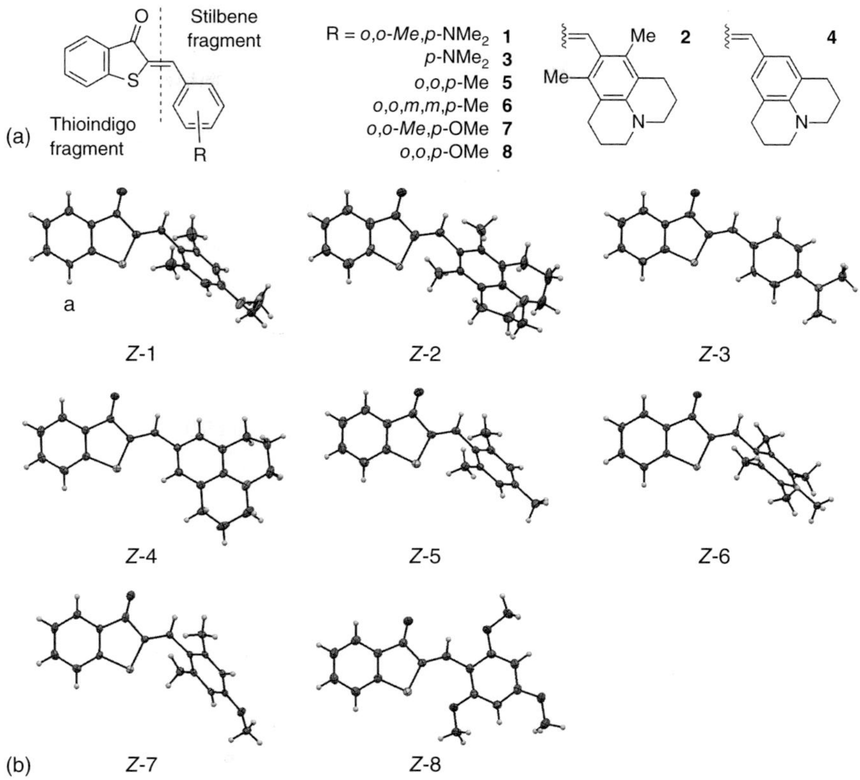

Figure 1.8 (a) Chemical structure of the HTI derivatives *Z*-1 to *Z*-8 studied by Dube and coworkers; (b) crystal structures of this *Z*-1 to *Z*-8. (Wiedbrauk *et al.* 2017 [49]. Reproduced with permission of American Chemical Society.)

corresponding emission spectra. The steady-state emission spectra of these derivatives showed strong bathochromic (red) shifts with increasing solvent polarity and occurrence of dual emission in solvents with large polarities. The planar derivatives *Z*-3 and *Z*-4 showed prominent solvent response, although much less compared to that of *Z*-1 and *Z*-2. The authors found that the effect of solvent polarity on the absorption and emission maxima of *Z*-5 and *Z*-6 are even smaller than that of *Z*-3 and *Z*-4. In spite of having twisted ground-state structures and donating groups with moderate strengths, the solvatochromic response of *Z*-7 and *Z*-8 are not very strong. Since the effect of solvent polarity on the emission energy is larger than the corresponding absorption energy of the molecules studied, the authors argued that the excited-state minima from which the fluorescence occurs are more polar than the FC region. As TICT is an alternative yet independent pathway of de-excitation along with the *Z*/*E* isomerization and the TICT process is highly dependent on the polarity of the solvents, it is expected that the quantum yield of *Z*/*E* photoisomerization will decrease significantly with increasing solvent polarity. In fact, the authors have used the quantum yield of *Z*/*E* photoisomerization ($\phi_{Z/E}$) as a tool for identification of the TICT process.

The HTI derivatives Z-1 and Z-2 show high $\phi_{Z/E}$ values in apolar solvents like cyclohexane, where no TICT is expected to occur. The values of $\phi_{Z/E}$ of these two compounds decrease with increasing solvent polarity, indicating formation of the TICT state. The $\phi_{Z/E}$ values for HTIs Z-3 and Z-4 also decrease with increases in solvent polarity, although the change is much lower than that of Z-1 and Z-2. The authors found that the $\phi_{Z/E}$ values for Z-5 to Z-8 were almost independent of the solvent polarity. The quantum yields of fluorescence of these derivatives are reported to be low irrespective of solvent polarity. The authors carried out time-resolved absorption spectroscopic measurements to explore TICT formation in the HTI derivatives. They noted that the TICT state in the HTI compounds were characterized by redshifted ESA and stimulated emissions that decay significantly slower than the initially populated excited-state minimum. In addition, the lifetime of the TICT state gets shortened with increasing solvent polarity. Time-resolved spectral data of Z-1 and Z-2 clearly indicated the formation of TICT in these compounds, while no evidence of TICT formation was observed in HTI derivatives Z-3 to Z-8. So, from their studies the authors conclude that although HTI derivatives Z-3 and Z-4 showed some indication of the TICT formation as indicated in decreasing values of $\phi_{Z/E}$ with solvent polarity, a complete study rules out formation of TICT in these molecules. No evidence of TICT formation was also observed in other HTIs (Z-5 to Z-8). Therefore, the authors conclude that considerable pretwisting of the stilbene fragment along with very strong donor group are required for the TICT to occur in the HTI derivatives reported in their study. From the nonoccurrence of TICT in HTI derivatives other than Z-1 and Z-2, although some of them possess pretwisted ground state and moderately strong donor groups, the authors infer that either of the aforesaid conditions are not enough for formation of TICT in these molecules.

The studies of ICT in 2-arylbenzotriazoles (Figure 1.9) by Turro and coworkers [50] also underscored the need for twisting of donor/acceptor groups for formation of the excited TICT states. The authors explored the ability of the aforesaid molecules as ultraviolet absorbers (UVAs) that dissipate the excited-state energy rapidly without any irreversible photochemistry. The quantum yields of decomposition of the UVAs are extremely low (in the order of 10^{-6} or even less), which renders them with unusual photostabilities. 2-(2-Hydroxy-5-methylphenyl)benzotriazole, also known as Tinuvin P, showed exceptional photostability due to efficient deactivation of excited singlet state through ESIPT process. Substitution of a methyl group at the 6′-position of 2-arylbenzotriazoles endowed the excited molecule with another deactivation mechanism. The authors undertook steady-state absorption and emission as well as time-resolved emission spectral studies to decipher the excited singlet state deactivation mechanism of 6′-methyl-2-arylbenzotriazoles that indicated the new process to be TICT. Their studies also revealed that due to steric requirement of the 6′-methyl group the TICT process was more facile in the molecules in which the phenyl and triazole rings were not in coplanar arrangements.

The authors installed two important features to the 2-arylbenzotriazole compounds to enhance the possibility of formation of excited TICT state. One of them is incorporation of methyl group in the 6′-position that provided an

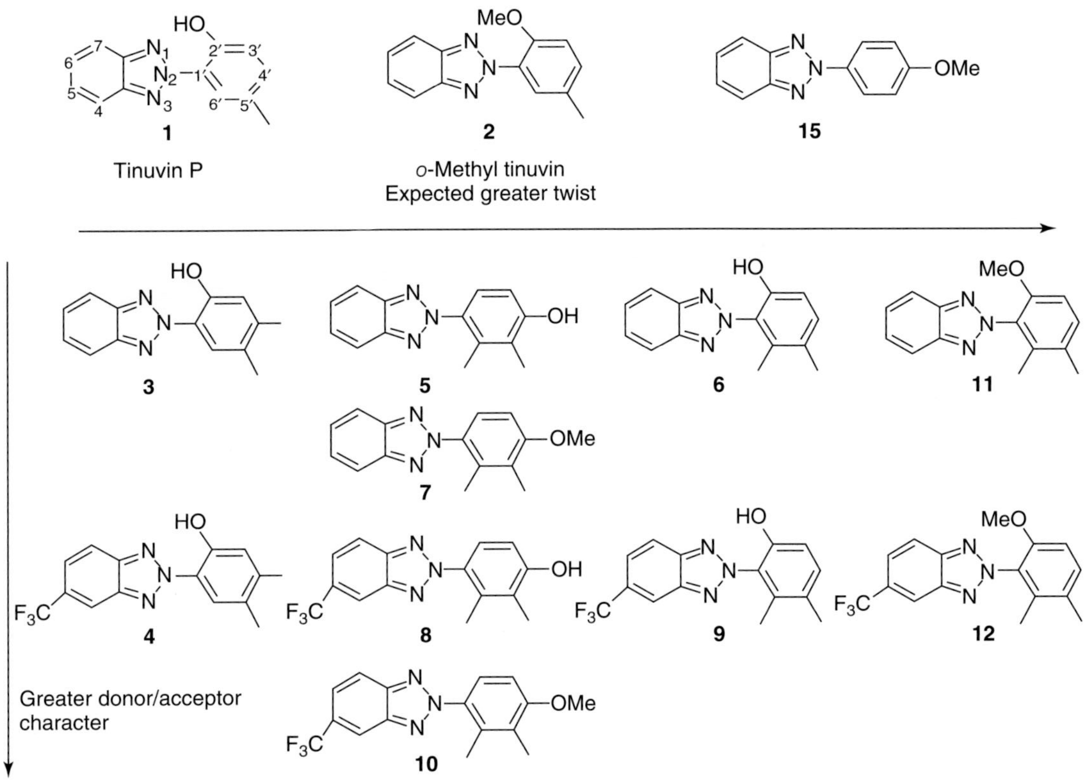

Figure 1.9 Chemical structures of the molecules studied by Turro and coworkers. The molecules were ordered on the basis of expected degree of twist along the triazole C—N bond as well as their donor/acceptor character. (Maliakal *et al*. 2002 [50]. Reproduced with permission of American Chemical Society.)

extra propensity of twisting in these molecules, while the other is addition of trifluoromethyl group to the triazole ring to increase their overall donor/acceptor character. They also explored the degree of twist on the formation of the TICT state by varying the substitution pattern. Their investigation revealed that the excitation led the molecule to the LE state (also FC state) which rapidly twisted to undergo the ICT process through which the TICT state was formed. The deactivation of the TICT state occurred through internal conversion either via a close approach or through a conical intersection with the ground state. In the framework of TICT, the rotation around the bond connecting the donor and the acceptor moieties that leads to the decoupling of the orbitals of these two groups occurs. This orbital decoupling could facilitate almost complete charge transfer in the molecule from the donor to the acceptor moiety, resulting in a highly polar twisted excited state. This polar excited state gets stabilized preferentially in solvents with high polarity, compared to its planar LE counterpart. As twisting makes the molecule energetically destabilized in the ground state, the formation of the excited TICT state brings the ground and excited states closer, which facilitates better internal conversion. The authors noted that although the donor/acceptor character was important, the degree of twist was a more important parameter for the TICT deactivation in these molecules. Several spectroscopic studies led the authors to conclude that the 2-(hydroxy or methoxy-6′-methylaryl)benzotriazoles could form the excited TICT state and the deactivation through the charge-transfer pathway accelerated the internal conversion from the TICT state to the ground state. They found that the compound with both twisting and strong donor/acceptor character (compound **12**) showed the most efficient fluorescence quenching. The compounds possessing both 6′-methyl and hydroxy groups on the phenyl ring showed diffusion-controlled quenching in DMSO solvent. The authors argued that the quenching process could be resulting from either partial or complete excited-state proton transfer to DMSO that enhances the deactivation of the singlet excited state through the TICT process.

As mentioned earlier, the debate regarding the structure of the excited state of ICT molecules is a matter of active debate in the scientific community. There are reports in the literature that the excited state can be planar as well. For example, Ghosh and coworkers [51] reported a novel concept to design planar (they call them "zero-twist" donor–acceptor molecules) ICT molecules for obtaining emission from both the solution and solid states. The authors took the help of quantum chemical calculations at the DFT (B3LYP functional/6-31G(d) basis set) level of theory. Their aim was to utilize the efficient segregation of highest occupied molecular orbital–lowest occupied molecular orbital (HOMO–LUMO) charge densities without the need for twisting of the donor and acceptor groups.

The authors [51] mentioned earlier studies that reported that planar emitters could suffer from an aggregation-caused quenching (ACQ), mostly due to face-to-face stacking as well as intramolecular interactions [52]. They also noted that the structural planarity, stacking modulator, and conformational rigidity were the major requirements for dual-state (solution and aggregate) emission from small molecules. Several organic donor–acceptor systems give rise to the TICT state, which may be weakly fluorescent or even nonfluorescent in solvents, limiting their application in the solution phase. Therefore, to avoid the formation of the "dark" TICT state, the authors have envisioned molecules with a planar excited state that would be fluorescent in the solution as well as in their aggregated state. Using their quantum chemical studies, the authors proposed an ICT molecule possessing butterfly-shaped phenothiazine as electron donor group and oxazole as electron acceptor, which are connected through a covalent bond. They called the system as D-π^0-Ψ^0-A system as the system was zero-twist ($\Psi = 0$) and there was no additional π-spacer. The authors found that the allowed orbital overlap in the compounds studied was responsible for their solution-state fluorescence properties. Their molecular-level studies using single-crystal analysis pointed out the detrimental effects of intermolecular short-range forces on the solid-state emission properties of compound **1**. Therefore, they introduced an anisole group as bulky substitution to subtle stacked packing of compound **2**. The solid-state emission properties of the aforesaid compounds were studied by fabricating green-emitting devices of those compounds in unoptimized devices for OLED applications. As mentioned earlier, the authors undertook quantum chemical calculations to predict the zero-twist molecules with dual-state emitters. Their theoretical calculations showed that the oxazole moiety shared the plane of the phenothiazine unit and there was no distortion between those two units. In compound **2**, the only distortion was due to the presence of the anisole group. The authors envisioned that the π-stacking of compound **2** in solid state could be avoided with the help of non-planarity of the anisole group and the butterfly shape of the phenothiazine moiety. They found that both the compounds showed similar distribution of HOMO and LUMO energy levels and the presence of the anisole group in compound **2** did not affect the HOMO and LUMO levels of the molecule due to the high dihedral angle (96.8°) with the phenothiazine plane. In the aforesaid compounds, the maximum contribution to the HOMO and LUMO came from the phenothiazine and oxazole units, respectively. Therefore, charge transfer from the phenothiazine unit to oxazole unit could be observed upon photoexcitation. Natural transition orbital (NTO) analyses carried out by the authors indicated similar hole and particle distribution behavior. The excited-state geometries optimized at the DFT [B3LYP/6-31G(d)] level of theory predicted the planarization of the first excited states (S_1) of compounds **1** and **2** in which the butterfly bend of the phenothiazine unit became fully planar (Figure 1.10). These results led the authors to infer that the excitation energy dissipated to attain more planar structure in the excited state, thereby making the twisting around the single bond connecting the donor and acceptor units less favorable. Their emission quantum yield measurements also supported the formation of planar excited state. The quantum yields of emission of compounds **1** and **2** were 0.7 and 0.9 in DMSO, while those were 0.5 and 0.4 in dichloromethane (DCM), respectively. They related the increase in quantum yields with increasing solvent polarity to the rigidification of the conformations that resulted in loss of vibrational motions [53]. To understand the excited-state properties of compounds **1** and **2** more extensively, the authors carried out femtosecond TA measurements in acetonitrile medium. They used a 410-nm laser source of 35 fs pulse duration and measured the TA spectra of both the compounds at different pump probe delays.

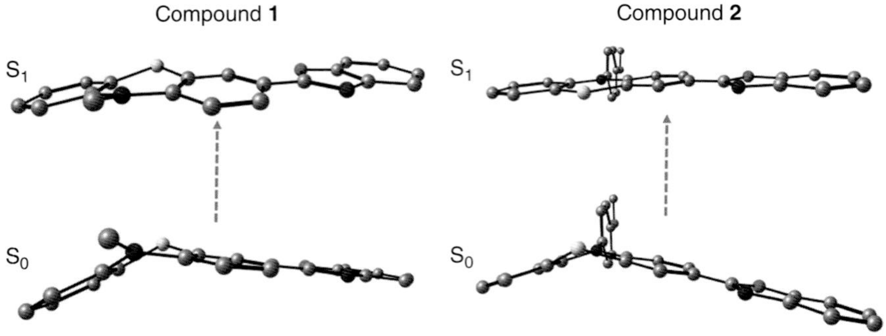

Figure 1.10 The ground- (S_0) and first-excited state (S_1) optimized structures of compounds **1** and **2**, studied by Ghosh and coworkers (Kumar *et al.* 2016 [51]. Reproduced with permission of American Chemical Society.)

The TA measurements revealed positive absorption bands in the 450–700 nm region with peaks at 550 and 525 nm for compounds **1** and **2**, respectively. They attributed these absorption bands to the ESA of those compounds. A gradual blueshift of the maximum of the absorptions bands was observed for both the compounds upon increasing delay of the pump-probe pulses up to 3 ps, while no such shift was observed thereafter. The aforesaid behavior (blueshift of TA bands) could be observed due to stabilization of the ICT excited states through solvation or geometrical relaxation of the molecules in the excited state [54, 55]. The authors also measured the TA kinetics at the ESA maxima of the compounds and fitted them with a biexponential function that consists of a rise and a decay component. They attributed the growth component with 667 fs lifetime to the solvent relaxation of the excited state, although they did not exclude the possibility of involvement of geometrical or vibrational (FC state) relaxation. The time constant (decay component) of 376 ps was assigned to the decay of the excited state back to the ground state of compound **1**. The authors argued that the relatively slower nature of the decay component could be due to the non-twisting of the excited state. Therefore, the results obtained from the TA studies supported the observation made from quantum chemical studies that the excited states of these compounds were planar in nature. The authors fabricated multilayer fluorescent OLED devices to evaluate their compounds for light-emitting applications. They used the compounds **1** and **2** as emissive layers in their unoptimized preliminary devices and compared their performances. Electroluminescence spectra and current–voltage–luminescence (*J*–*V*–*L*) characteristics of both the devices were examined. The color purity in the device using compound **2** was attributed to the relatively lesser intermolecular interaction, compared to that of compound **1**. Therefore, from the device application studies the authors inferred that the zero-twist D-π^0-Ψ^0-A emitter made of compound **2** could be used for designing organic emitters developed from the ACQ system. The molecular-level studies performed by the authors also underscores the role of steric bulkiness to avoid stacking in compound **2** that increased its emission quantum yield, compared to that of compound **1**.

1.3 Applications of ICT Molecules

The ICT-based molecules have been studied for potential applications in OLEDs, bioimaging applications through aggregation-induced emission (AIE), sensor for polarity and viscosity, and so on. As mentioned earlier, the ICT process is a fluorescence deactivation process which makes these molecules ready for potential use as UVAs. UVAs dissipate the excited-state energy quickly without the need for any irreversible photochemical reactions. Therefore, several studies that utilize ICT molecules to design UVAs have been reported in literature [56–58]. Interest in studies of OLEDs has been fueled as they provide several advantages over the conventional LEDs in terms of power consumption, big area fabrication compatibility, wide-viewing angle, fancy architecture, lightweight, higher brightness and sharper image, and so on [59]. ICT through which ultrafast deactivation of excited molecule could occur has been exploited in applications as laser dye absorbers that could be used in mode-locked lasers [60].

Teran and Reynolds [61] reported the designing strategies for using donor–acceptor conjugated systems in their neutral and oxidized states for use in high-contrast electrochromics. They mentioned that the conjugated molecules and polymers can be used for several technological applications as the structure and properties of these materials could be modulated with ease using myriad techniques of synthetic organic chemistry. One of the well-known methods is to link electron-rich and electron-deficient heterocycles covalently to prepare multi-ring molecules or copolymerized in such a way that an ICT interaction is induced. The aforesaid technique has been used by several research groups to design low-energy gap materials for light absorption in solar cells [62, 63], to develop materials with dual-band absorption [64], and so on. In electrochromics, the differences in the absorption/transmission properties of the neutral and charged species of a conjugated system is accessed through application of voltage and compound that shows sufficient contrast can be used in displays and smart windows. It is worth noting here that for several electronic applications, the absorption properties of the both the neutral and charged states should be situated within specific regions of the EM spectrum. Teran *et al.* have reported the neutral and oxidized state absorption and spin properties of a family of donor–acceptor–donor (D-A-D) molecules based on electron-rich dioxythiophenes (DOTs), electron-deficient 2,1,3-benzothiadiazole (BTD) and the heterocycles function as building blocks for electrochromic materials. They also studied the polymer with discrete D-A-D conjugated system obtained from the aforesaid units. They examined the steric and electronic structural requirements for designing high-contrast electrochromics and also proposed a method to translate the small conjugated molecules into polymeric materials that retain their well-defined properties for practical applications. The authors synthesized the three-, five- and seven-heterocycle conjugated molecules based on DOT and BTD units. They found that all the materials showed well-defined dual-band absorptions in the neutral state, arising from the charge-transfer interaction and the π–π^* high-energy excitation. The oxidized state generated through chemical and electrochemical doping took a structure in which the positive charge was localized in the donor unit and the acceptor ring served as conjugation break. In penta-heterocycle (EPBPE), this led the polaron pair and its π-dimer to be the dominant charge carriers in the low- as well as high-oxidation states. In hepta-heterocycles (EPPBPPE), the longer donor segment generated polaron pairs with significantly redshifted absorption with peak outside the visible region. The authors found that the extended structure allowed formation of bipolarons at higher doping levels. The absorption maxima of both the charge carriers were in the near-IR region. They designed polymers of the aforesaid D-A materials by linking the units through aliphatic chains. Although poly-EPBPE showed well-defined and narrow transitions in its absorption spectra, in higher oxidation states the effect of π–π interactions became significantly higher. This led the authors to propose that one needs to design the linker carefully (like sterically bulky aliphatic groups) to develop polymeric materials that are devoid of undesired interactions from discrete chromophores.

Of late, several ICT-based molecules are being used as fluorescent and colorimetric sensors as well as NLO switches. Novel ICT-based molecules are being designed to detect ions and molecules in solution as well as in living cells. Applications of ICT probes in AIE, dye-sensitized solar cells, and OLEDs are also being reported. Some of these applications are discussed in Chapters 5 and 6. We have already mentioned that the dipole moment of a charge-transfer molecule could be higher in the exited state than in the ground state, leading to a bathochromic shift in its electronic absorption spectrum. A simple two-state model, proposed by Ouder [65, 66] suggests that the ICT molecules may possess high quadratic first hyperpolarizability, a term used to quantize the first-order NLO response. The NLO response property of a material is ultimately shaped by the NLO response of the individual chromophores, sometimes referred to as the building blocks. A high first hyperpolarizability from a single chromophore is generally taken as a signature of a low-lying ICT state in the building block. Tuning of the ICT process of a molecule is therefore an essential ingredient in the designing of materials with high NLO response. It has been reported by many research groups that organic molecules with ICT can be possible chromophores for NLO materials. According to the two-state model proposed by Ouder [65, 66], the static or field-independent first hyperpolarizability (β) of the system can be calculated using Equation 1.2.

$$\beta_0 = \frac{3}{2}\left[\frac{\Delta\mu f_0}{(\Delta E)^3}\right] \tag{1.2}$$

where ΔE is the difference in energy between the ground and excited states of a molecule, while the f_0 stands for the oscillator strength of the transition. The term $\Delta\mu$ represents the difference in dipole moments between the ground and excited states of the molecule.

From Ouder's model one can therefore anticipate that the molecules having high HOMO–LUMO energy gap and large values of $\Delta\mu$ can result into high second-order NLO response. Although the NLO response of a material is measured using experimental techniques like electric-field–induced second harmonic generation (EFISH), Z-scan, hyper-Rayleigh scattering (HRS), and so on, the polarizability and hyperpolarizabilities (β_0) are the properties of a free molecule. The values of the polarizability (α) and hyperpolarizabilities (β, γ, ..) are independent of the material and can be calculated using quantum chemical techniques. Due to availability of vast literature, we have separated the application of ICT molecules for NLO response from other applications and discussed it in little more detail in Chapter 5. Two-photon absorption (TPA), another nonlinear phenomenon has drawn the interest of the scientific community for possible technological applications of TPA-based materials. Several ICT-based molecules have been examined recently for preparation of materials with high TPA activities. The basic theory behind the calculations of polarizability and hyperpolarizabilities of a molecule are also discussed in the same chapter (Chapter 5). *para*-Nitroaniline (PNA) is the one of the very first organic molecules that showed very high NLO response. Later on, several molecules with ICT were explored for designing possible NLO materials [67–78]. Till date, most of the NLO studies concern the second-order NLO response, while few reports of third-order NLO response have been reported in literature. Some representative examples of studies of second- and third-order NLO response and TPA properties of ICT-based molecules are discussed in Chapter 5. It is worth mentioning that the choice of theoretical method is important to accurately calculate the polarizability and hyperpolarizabilities of a molecule [79–85]. Electron correlation is reported to be an important parameter for calculations of linear and NLO response properties of a molecule [86, 87]. The vibrational contribution and effect of the medium are other two interesting topics for studies of NLO response [88]. As expected, several studies have been reported in literature concerning these issues. Coupled cluster singles and doubles (CCSD) and coupled cluster singles and doubles with a perturbative estimate of triples [CCSD(T)] level of theories have been used to calculate the polarizabilities and hyperpolarizabilities of several relatively small molecules, while using these theories for larger molecules are costly [85]. Therefore, several alternate methodologies have been adopted by the scientific community to reproduce the experimental values of α, β, and γ with some approximations. For ICT-based molecules, use of MP2 theory seems to be one of the popular choices. Several DFT functionals and basis sets have been designed to calculate the values of α, β, and γ of conjugated molecules [85].

Champagne and coworker [81] studied the effect of electron correlation on the first hyperpolarizability of a number of push–pull π-conjugated molecules (Figure 1.11). They found the 6-31+G(d,p) basis set gave a balance between the computational resources and accuracy for the polyene linkers. The authors reported that due to cancellation between higher order contributions the MP2 method was most accurate in predicting the values of β of push–pull polyene molecules as the values closely reproduced those obtained through CCSD(T) calculations. In case of exchange-correlation DFT functionals, the authors found the LC-BLYP was reliable while predicting the changes of β while enlarging the π-conjugation network or upon changing the polyene linker to a polyene segment, although its reliability is reported to be close to that of HF method and MP2 method was far superior.

Luthi and coworkers [86] reported their theoretical investigation of polarizability and second hyperpolarizability of polyacetylene oligomer chains with

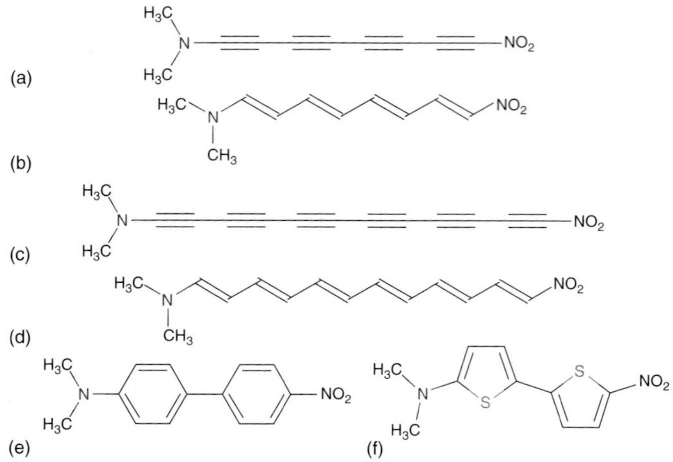

Figure 1.11 Chemical structures of π-conjugated donor–acceptor molecules studies by Champagne and coworkers. (de Wergifosse and Champagne 2011 [81]. Reproduced with permission of American Institute of Physics.)

increasing size up to $C_{24}H_{26}$. These polyacetylenes are prototypical systems for studying the π-conjugation in a linear framework and the interactions between the adjacent double bonds can be tuned through incorporation of electron donor and acceptor groups at two ends. The authors used long-range corrected CAM-B3LYP functional for calculation of α and γ of a series of polyacetylenes and compared the theoretically estimated values with that obtained experimentally. They found this functional could remove the large part of overestimation generally observed using standard DFT methods and even in some cases the values are close to coupled cluster calculations. They found that CAM-B3LYP reproduced the structure of the molecules found experimentally and was found better than the standard B3LYP functional and MP2 method. The authors reported that the experimental values of second hyperpolarizability of ethylene, butadiene, and hexatriene could be reproduced, although they found CAM-B3LYP and CCSD generally overestimated the values by 25% and 10%, respectively.

Several molecules exhibit structural changes in response to external stimuli like changing polarity and acidity/basicity (pH) of the medium, irradiating the sample with light of specific frequency, the temperature of the system, the redox potential, and so on [87–89]. If these alterations lead to spectral changes (generally in the visible region but not always), then the molecules are referred to as photochromes, thermochromes, and acidochromes, when the changes are induced by irradiation, changing temperature, and acidity of the medium, respectively. Some properties of these compounds can be tuned reversibly to achieve "on" and "off" states that could be useful in devices such as sensors, actuators, memory, and so on. Similar to change in linear absorption and emission maxima of the probe, the external stimuli can also change the NLO properties of a molecule, such as second-harmonic generation (SHG), third-harmonic generation, and TPA [90, 91]. Molecules in which the structural change triggers changes in NLO response properties are called NLO switches. A vast majority of the NLO switches exhibit changes in their first hyperpolarizability (β), which is the molecular property at the origin of SHG. Although the change in β with structural changes were observed in several molecules, including azobenzenes, nitrobenzylpyridines, diarylethenes, *N*-salicyclidene-aniline, and spiropyrans, several organometallic complexes with Fe, Ru, Zn, and Pt transition metal atoms, and, indolinooxazolidine and spiropyran derivatives were shown to behave as highly efficient NLO switches that showed large differences in their NLO responses due to structural changes induced through external stimuli [89].

OLEDs usually consist of ultrathin layers of soft and amorphous organic semiconductors [92]. OLEDs are believed to be potential candidates for next-generation displays and solid-state lighting owing to their low driving voltage, high brightness, and high efficiency. These devices can also realize the extreme wide viewing and high color quality. The compatibility with production processes has made these devices attractive for the scientific community. Solution processing is believed to be the most favorable method for fabrication of OLED devices for low-cost and large-area applications [93]. Once the fluorescence is switched on, the holes and electrons are fed into the organic semiconductor in which they bind together to form the excitons [92]. These excitons are generally localized on single molecules and carry a distinct spin character. It can be either a singlet state, which is light emitting, or a triplet state, which is non-radiative. It has been reported that due to unfavorable spin statistics, for every singlet state that is created, three triplet states will also be created. That is, the population of singlet and triplet states will be 25% and 75%, respectively. As these triplet states are non-emissive, this will lead to loss of about two-thirds of the total energy that limits use of organic semiconductors for OLED applications. The problem of unfavorable spin statistics was taken care of using phosphorescence-based molecules as emitters. In these devices, heavy metal ions such as iridium and platinum that leads to strong spin–orbit coupling (SOC) within the molecule are introduced. The SOC process leads to fast and efficient transition from singlet to triplet state (called intersystem crossing) and also enhances the transition from the triplet to the ground state (called phosphorescence). This method forces the system to luminescence from the triplet state, irrespective of the initial spin state of the excitons, offering a theoretical internal quantum efficiency of cent percent. However, the applications of phosphorescence-based materials are limited by their high cost due to use of precious metals. Adachi and coworkers have introduced the concept of thermally activated delayed fluorescence (TADF) to overcome the unfavorable spin statistics in organic materials [94]. The TADF method relies on reverse intersystem crossing from the triplet state to the radiative singlet state. They have shown that when the singlet and triplet states are close in energy, using the thermal energy at room temperature the triplets can transition to the singlet state, thereby making the internal quantum efficiency of the system to 100%. One of the key rules to design the system with very small gap between the singlet and triplet states (ΔE_{ST}) is to minimize the spatial overlap between the HOMO and LUMO of the molecule. ICT molecules, in which electron donor and acceptor groups are connected through a π-electron bridge, are used to prepare TADF molecules for OLED applications [95]. This

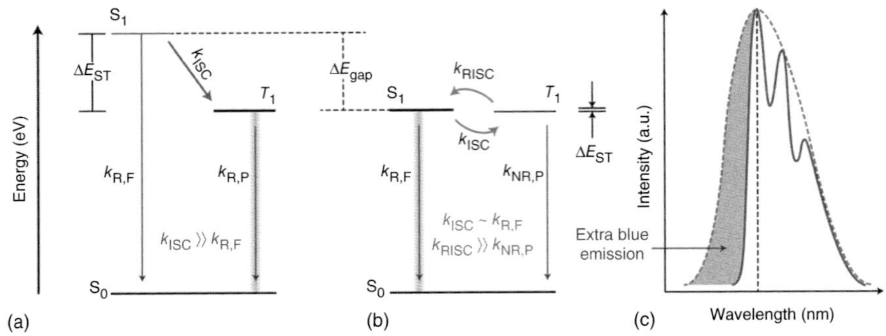

Figure 1.12 Simplified Jablonski diagrams for (a) phosphorescence- and (b) TADF-based emitters. The difference between the spectral distributions of phosphorescence (dashed curve) and TADF emissions (solid curve) are shown in (c). Important rate processes are indicated using arrows. In the figure R, NR, F, P, ISC, RISC, and ΔE_{ST} stand for radiative, non-radiative, fluorescence, phosphorescence, intersystem crossing, reverse ISC and singlet–triplet splitting energy, respectively. (Reineke 2014 [92]. Reproduced with permission of Macmillian Publishers Limited.)

is due to the fact that in ICT molecules the delocalization of π-electrons leads to separation of the HOMO and LUMO, thereby giving rise to small values of ΔE_{ST}. Adachi *et al.* have shown that in intermolecular charge-transfer complexes, the TADF process can be observed through separation of the wavefunction over a heterointerface between the molecules forming the complex. The difference between the phosphorescence-based emitters and TADF-based emitters has been shown in Figure 1.12.

From the preceding discussion it is apparent that ICT-based molecules have been used to design materials for applications in molecular switch, in nonlinear optics, in OLEDs, and so on. Some of these applications are discussed in detail in Chapter 6. Chapter 7 summarizes the studies presented in this book. Some of the unresolved issues concerning the ICT processes are also discussed.

References

1 Sedghi, G., Sawada, K., Esdaile, L.J., Hoffmann, M., Anderson, H.L., Bethell, D., Haiss, W., Higgins, S.J., and Nichols, R.J. (2008) *J. Am. Chem. Soc.*, **130**, 8582.

2 Closs, G.L. and Miller, J.R. (1988) *Science*, **240**, 440.

3 Bredas, J.L., Calbert, J.P., da Silva Filho, D.A., and Cornil, J. (2002) *Proc. Natl. Acad. Sci. U.S.A.*, **99**, 5804.

4 Barbara, P.F., Walker, G.C., and Smith, T.P. (1992) *Science*, **256**, 975.

5 Kobori, Y., Yamauchi, S., Akiyama, K., Tero-Kubota, S., Imahori, H., Fukuzumi, S., and Norris, J.R. (2005) *Proc. Natl. Acad. Sci. U.S.A.*, **102**, 10017.

6 Zhao, G.-J., Liu, J.-Y., Zhou, L.-C., and Han, K.-L. (2007) *J. Phys. Chem. B*, **111**, 8940.

7 Bulheller, B.M., Miles, A.J., Wallace, B.A., and Hirst, J.D. (2008) *J. Phys. Chem. B*, **112**, 1866.

8 Li, G., Josowicz, M., Janata, J., and Semancik, S. (2004) *Appl. Phys. Lett.*, **85**, 1187.

9 Arzhantsev, S., Zachariasse, K.A., and Maroncelli, M. (2006) *J. Phys. Chem. A*, **110**, 3454.

10 Cao, X., Tolbert, R.W., McHale, J.L., and Edwards, W.D. (1998) *J. Phys. Chem. A*, **102**, 2739.

11 Thar, J., Zahn, S., and Kirchner, B. (2008) *J. Phys. Chem. B*, **112**, 1456.

12 Akemann, W., Laage, D., Plaza, P., Martin, M.M., and Blanchard-Desce, M. (2008) *J. Phys. Chem. B*, **112**, 358.

13 Zyss, J., Ledoux, I., Volkov, S., Chernyak, V., Mukamel, S., Bartholomew, G.P., and Bazan, G.C. (2000) *J. Am. Chem. Soc.*, **122**, 11956.

14 Zachariasse, K.A. (2000) *Chem. Phys. Lett.*, **320**, 8.

15 Druzhinin, S.I. *et al.* (2010) *J. Am. Chem. Soc.*, **132**, 7730.

16 Grabowski, Z.R., Rotkiewicz, K., and Rettig, W. (2003) *Chem. Rev.*, **103**, 3899.

17 Tassel, A.J.V., Prantil, M.A., and Fleming, G.A. (2006) *J. Phys. Chem. B*, **110**, 18989.

18 Zhang, W., Lan, Z., Sun, Z., and Gaffney, K.J. (2012) *J. Phys. Chem. B*, **116**, 11527.

19 Schmuttenmaer, C.A. (2004) *Chem. Rev.*, **104**, 1759.

20 Kumpulainen, T., Lang, B., Rosspeintner, A., and Vauthey, E. (2017) *Chem. Rev.* doi: 10.1021/acs.chemrev.6b00491

21 Marcus, R.A. (1993) *Rev. Mod. Phys.*, **65**, 599.

22 Pelzer, K.M. and Darling, S.B. (2016) *Mol. Syst. Des. Eng.*, **1**, 10.

23 Paddon-Row, M.N. (2001) in *Electron Transfer in Chemistry* (ed. V. Balzani), Wiley-VCH Verlag GmbH & Co. KGaA, Weinheim.

24 Sumi, H. and Marcus, R.A. (1986) *J. Chem. Phys.*, **84**, 4894.

25 Manna, A.K. and Dunietz, B.D. (2014) *J. Chem. Phys.*, **141**, 121102.

26 Rettig, W., Bliss, B., and Dirnberger, K. (1999) *Chem. Phys. Lett.*, **305**, 8.

27 Rettig, W. (1986) *Angew. Chem. Int. Ed. Engl.*, **25**, 971.

28 Zilberg, S. and Haas, Y. (2002) *J. Phys. Chem. A*, **106**, 1.

29 Yoshihara, T., Druzhinin, S.I., and Zachariasse, K.A. (2004) *J. Am. Chem. Soc.*, **126**, 8535.

30 Gomez, I., Reguero, M., Boggio-Pasqua, M., and Robb, M.A. (2005) *J. Am. Chem. Soc.*, **127**, 7119.

31 Cogan, S., Zilberg, S., and Haas, Y. (2006) *J. Am. Chem. Soc.*, **128**, 3335.

32 Chu, G. and Yangbo, F. (1987) *J. Chem. Soc., Faraday Trans.*, **83**, 2533.

33 Sobolewski, A.J. and Domcke, W. (1996) *Chem. Phys. Lett.*, **259**, 119.

34 Barbara, P.F. and Jarzeba, W. (1988) *Acc. Chem. Res.*, **21**, 195.

35 Tavernier, H.L., Barzykin, A.V., Tachiya, M., and Fayer, M.D. (1998) *J. Phys. Chem. B*, **102**, 6078.

36 Kosower, E.M. and Huppert, D. (1983) *Chem. Phys. Lett.*, **96**, 433.

37 Li, X. and Maroncelli, M. (2011) *J. Phys. Chem. A*, **115**, 3746.

38 Samanta, A., Paul, B.K., and Guchhait, N. (2012) *J. Lumin.*, **132**, 517.

39 Brutschy, B. (2000) *Chem. Rev.*, **100**, 3891.

40 Zhao, G.-J., Chem, R.-K., Sun, M.-T., Liu, J.-Y., Li, G.-Y., Gao, Y.-L., Han, K.-L., Yang, X.-C., and Sun, L. (2008) *Chem. Eur. J.*, **14**, 6935.

41 Wasielewski, M.R. (1992) *Chem. Rev.*, **92**, 435.

42 Zhang, J., Xu, Q., Feng, Z., Li, M., and Li, C. (2008) *Angew. Chem.*, **120**, 1790.
43 Hara, K., Dan-oh, Y., Kasada, C., Yasuyo, O., Shinpo, A., Suga, S., Sayama, K., and Arakawa, H. (2004) *Langmuir*, **20**, 4205.
44 Nazarov, A.E., Malykhin, R., and Ivanov, A.I. (2017) *J. Phys. Chem. B*, **121**, 589.
45 Wiedbrauk, S., Maerz, B., Samoylova, E., Reiner, A., Trommer, F., Mayer, P., Zinth, W., and Dube, H. (2016) *J. Am. Chem. Soc.*, **138**, 12219.
46 Muraoka, T., Kinbara, K., and Aida, T. (2006) *Nature*, **440**, 512.
47 Gostl, R., Senf, A., and Hecht, S. (2014) *Chem. Soc. Rev.*, **43**, 1982.
48 Samanta, S., Qin, C., Lough, A.J., and Wolley, G.A. (2012) *Angew. Chem. Int. Ed.*, **51**, 6452.
49 Wiedbrauk, S., Maerz, B., Samoylova, E., Mayer, P., Zinth, W., and Dube, H. (2017) *J. Phys. Chem. Lett.*, **8**, 1585.
50 Maliakal, A., Lem, G., Turro, N.J., Ravichandran, R., Suhadolnik, J.C., DeBellis, A.D., Wood, M.G., and Lau, J. (2002) *J. Phys. Chem. A*, **106**, 7680.
51 Kumar, S., Singh, P., Kumar, P., Srivastava, R., Pal, S.K., and Ghosh, S. (2016) *J. Phys. Chem. C*, **120**, 12723.
52 Saigusa, H. and Lim, E.C. (1995) *J. Phys. Chem.*, **99**, 15738.
53 Chen, G., Li, W., Zhou, T. *et al* (2015) *Adv. Mater.*, **27**, 4496.
54 Fakis, M., Hrobarik, P., Yuschenko, O. *et al* (2014) *J. Phys. Chem. C*, **118**, 28509.
55 Bredas, J.L., Calbert, J.P., da Silva Filho, D.A., and Cornil, J. (2002) *Proc. Natl. Acad. Sci. U.S.A.*, **99**, 5809.
56 Paterson, M.J., Robb, M.A., Blancafort, L., and DeBellis, A.D. (2004) *J. Am. Chem. Soc.*, **126**, 2912.
57 Elbe, F., Keck, J., Fluegge, A. *et al* (2000) *J. Phys. Chem. A*, **104**, 8296.
58 Keck, J., Roesller, M., Schroeder, C. *et al* (1998) *J. Phys. Chem. B*, **102**, 6975.
59 Liang, J., Li, L., Niu, X., Yu, Z., and Pei, Q. (2013) *Nat. Photonics*, **7**, 817.
60 Vogel, M. and Rettig, W. (1985) *Ber. Bunsen Ges. Phys. Chem.*, **89**, 962.
61 Teran, N. and Reynolds, J.R. (2017) *Chem. Mater.*, **29**, 1290.
62 Zhou, H., Yang, L., and You, W. (2012) *Macromolecules*, **45**, 607.
63 Dow, L., Liu, Y., Hong, Z., Li, G., and Yang, Y. (2015) *Chem. Rev.*, **115**, 12633.
64 Baeujuge, P.M., Amb, C.M., and Reynolds, J.R. (2010) *Acc. Chem. Res.*, **43**, 1396.
65 Ouder, J.L. (1977) *J. Chem. Phys.*, **67**, 446.
66 Ouder, J.L. (1977) *J. Chem. Phys.*, **67**, 2664.
67 Albert, I.D.L., Marks, T.J., and Ratner, M.A. (1998) *J. Am. Chem. Soc.*, **120**, 11174.
68 Yang, G. and Su, Z. (2009) *Int. J. Quantum Chem.*, **109**, 1553.
69 Nandi, P.K., Panja, N., Ghanty, T.K., and Kar, T. (2009) *J. Phys. Chem. A*, **113**, 2623.
70 Ishow, E., Bellaiche, C., Bouteiller, L., Nakatani, K., and Delaire, J.A. (2003) *J. Am. Chem. Soc.*, **125**, 15744.
71 Wang, C.-K. and Yang, W.-H. (2003) *J. Chem. Phys.*, **119**, 4409.
72 Geskin, V.M., Lambert, C., and Bredas, J.L. (2003) *J. Am. Chem. Soc.*, **125**, 15651.

73 Albert, I.D.L., Marks, T.J., and Ratner, M.A. (1997) *J. Am. Chem. Soc.*, **119**, 6575.

74 Zyss, J. (1979) *J. Chem. Phys.*, **71**, 909.

75 Lacroix, P.G., Padilla-Martinez, I.I., Sandoval, H.L., and Nakatai, K.N. (2004) *New J. Chem.*, **28**, 542.

76 Marder, S.R. and Perry, J.W. (1993) *Adv. Mater.*, **5**, 804.

77 Kang, H., Facchetti, A., Jiang, H., Cariati, E., Rietto, S., Ugo, R., Zuccaccia, C., Macchioni, A., Stern, C.L., Liu, Z., Ho, S.-T., Brown, E.C., Ratner, M.A., and Marks, T.J. (2007) *J. Am. Chem. Soc.*, **127**, 3267.

78 Dehu, C., Meyers, F., Hendrickx, E., Clays, K., Persoons, A., Marder, S.R., and Bredas, J.L. (1995) *J. Am. Chem. Soc.*, **117**, 10127.

79 Champgne, B. (2009) in *Polarizabilities and Hyperpolarizabilities in Chemical Modeling*, vol. **6** (ed. M. Springborg), Royal Society of Chemistry, London.

80 Maroulis, G., Bancewicz, T., and Champagne, B. (eds) (2011) *Atomic and Molecular Nonlinear Optics: Theory, Experiment and Computation: A Homage to the Pioneering Work of Stanisław Kielich (1925–1993)*, IOS Press, Amsterdam.

81 de Wergifosse, M. and Champagne, B. (2011) *J. Chem. Phys.*, **134**, 074113.

82 Zhang, L., Qi, D., Zhao, L., Chen, C., Bian, Y., and Li, W. (2012) *J. Phys. Chem. A*, **116**, 10249.

83 Bai, Y., Zhou, Z.J., Wang, J.J., Li, Y., Wu, D., Chen, W., Li, Z.R., and Sun, C.C. (2013) *J. Phys. Chem. A*, **117**, 2835.

84 Karamanis, P. and Maroulis, G. (2011) *J. Phys. Org. Chem.*, **24**, 588.

85 Alparone, A. (2013) *Chem. Phys. Lett.*, **563**, 88.

86 Limacher, P.A., Mikkelsen, K.V., and Luthi, H.P. (2009) *J. Chem. Phys.*, **130**, 194114.

87 Berkovic, G., Krongauz, V., and Weiss, V. (2000) *Chem. Rev.*, **100**, 1741.

88 Zhang, J., Zou, Q., and Tian, H. (2013) *Adv. Mater.*, **25**, 378.

89 Beaujean, P., Bondu, F., Plaquet, A., Garcia-Amoros, J., Cusido, J., Raymo, F.M., Castet, F., Rodriguez, V., and Champagne, B. (2016) *J. Am. Chem. Soc.*, **138**, 5052.

90 Castet, F., Rodriguez, V., Pozzo, J.-L., Ducasse, L., Plaquet, A., and Champagne, B. (2013) *Acc. Chem. Res.*, **46**, 2656.

91 Delaire, J.A. and Nakatani, K. (2000) *Chem. Rev.*, **100**, 1817.

92 Reineke, S. (2014) *Nat. Photonics*, **8**, 269.

93 Burroughes, J.H., Bradley, D.D.C., Brown, A.R., Mackay, R.N., Marks, K., Friend, R.H., Burns, P.L., and Holmes, A.B. (1990) *Nature*, **347**, 539.

94 Nakanotani, H., Masui, K., Nishide, J., Shibata, T., and Adachi, C. (2013) *Sci. Rep.*, **3**, 2127.

95 Endo, A. *et al* (2009) *Adv. Mater.*, **21**, 4802.

2

Brief History of ICT Molecules

2.1 Introduction

Intramolecular charge transfer (ICT) involves transfer of electronic charge from an electron-rich donor to an electron-deficient acceptor in the same molecule. As mentioned in Chapter 1, several ICT molecules show dual emission in their steady-state emission spectra. One of these arises from a $\pi\pi^*$ state (L_b type, according to Platt's notation), often referred to as locally excited (LE) state, while the redshifted fluorescence band is attributed to the ICT state (L_a type) formed in the excited state. 4-*N*,*N*-Dimethylaminobenzonitrile (DMABN) is a prototype of electron donor–acceptor (D-A) ICT molecules and its charge-transfer (CT) properties have been studied widely using several state-of-the-art spectroscopic studies as well as using quantum chemical calculations [1]. Studies of the CT process caught the attention of the scientific community as it is one of the most fundamental processes in chemistry and biology [2–15]. CT is a key step of many important biological processes, including photosynthesis and metabolism. Recently, ICT-based organic molecules are being used for several technological applications, including organic light-emitting diodes (OLEDs), dye-sensitized solar cells, nonlinear optical (NLO) materials, and so on [16, 17]. The studies of CT in small molecules have been fueled as it helps us in understanding the mechanism of many complex biological processes. If the donor (D) and the acceptor (A) are parts of two different molecules, atoms, or clusters, the process is called intermolecular CT. The donor–acceptor substituted aromatic systems are one of the popular systems for studies of the ICT process (discussed in detail in this chapter). The other popular system for studying the ICT process and NLO response is the donor–acceptor substituted polyenes [18, 19]. In a D-A organic molecule, the ICT process can take place through two different ways [20, 21]. If the electron passes over from the donor to the acceptor moiety through a π-electron bridge, it is known as through-bond charge transfer. Unless stated otherwise, while referring to ICT in this book, we mean through-bond charge transfer. On the other hand, if the donor and acceptor are not connected through a π-electron bridge, CT can still take place by a mechanism called through-space charge transfer [22]. Although the through-space CT process is relatively less known, it can be useful in describing many CT-driven processes.

Intramolecular Charge Transfer: Theory and Applications,
First Edition. Ramprasad Misra and S. P. Bhattacharyya.

Studies of interaction of light with matter constitute a large volume of research in science [23, 24]. When light-containing radiation of a wide range of wavelength (e.g., white light) falls on a substance, some of the radiation is absorbed and the rest is either transmitted or reflected by the substance, which indicates that it does not absorb radiation of any arbitrary wavelength. It is well known that a substance absorbs light of particular wavelengths only and this phenomenon is guided by quantum mechanical principles. This principle is true as much for an atom as for a molecule. After absorption of the radiation, several photophysical and/or photochemical processes may occur [25–27]. Following photoabsorption, for example, a molecule can break up into fragments that can react to give different products. If it does not, the following photophysical processes may happen, which can be summarized using a simplified version of the Jablonski diagram [28–30], as shown in Figure 2.1.

In the example shown in Figure 2.1, absorption (Abs) of radiation of suitable wavelength could excite the molecule from zero vibrational level of the ground (S_0) to a higher vibrational level of the first excited (S_1) state. Immediately, it relaxes from higher vibrational level of S_1 state to its zero vibrational level. The molecule now can relax this energy by either radiative or non-radiative processes or by both. Fluorescence resonance energy transfer or Förster resonance energy transfer (FRET) may occur through dipole–dipole coupling. The excited molecule also can relax this energy through dynamic quenching (DQ) by collision with other molecules. It may undergo some other non-radiative processes (Qu). The molecule may relax through radiative process as it gives rise to fluorescence (Fl). Transfer of energy from S_1 state to a triplet state (T_1, in this case) through intersystem crossing (ISC) and followed by phosphorescence (Ph), which is also a radiative process, may also take place. The rates of these radiative (k_R) and non-radiative (k_{NR}) processes are related to the quantum yield (ϕ) and

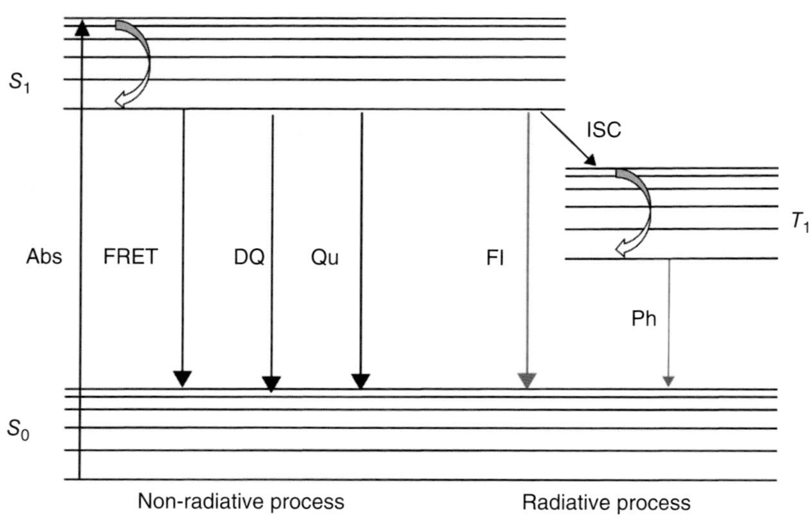

Figure 2.1 A simplified version of the Jablonski diagram showing a molecule undergoing different photophysical processes after photoexcitation. The photophysical processes are discussed in the text.

excited-state lifetime (τ) of a molecule [28].

$$\phi = \frac{k_R}{k_R + k_{NR}} \tag{2.1}$$

$$\tau = \frac{1}{k_R + k_{NR}} \tag{2.2}$$

If the ultrafast processes like ICT or intramolecular proton transfer (IPT) occur in a molecule, the photophysics becomes somewhat more complicated than what has been shown in Figure 2.1 [9, 31–35]. In these cases, a new state called ICT or IPT state is formed, the signature of which is often imprinted in dual emission that takes place as shown in Figure 2.2. Therefore, while studying the ICT process in a molecule, another important CT process, namely, IPT must be taken into account. The IPT process stands in sharp contrast to

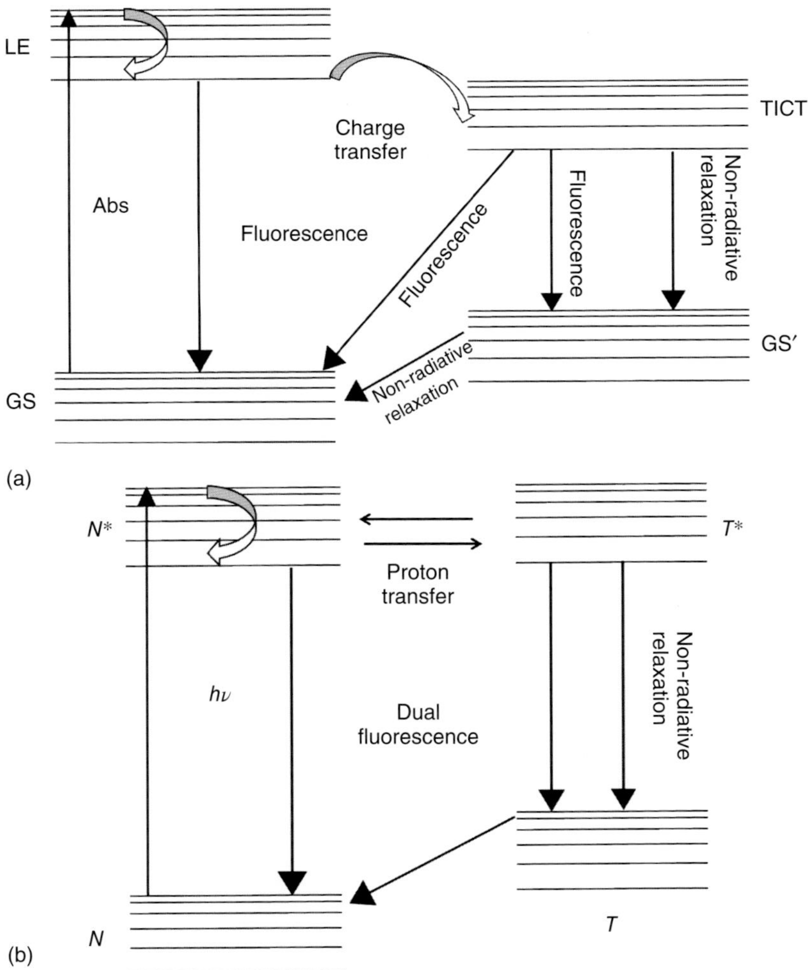

Figure 2.2 Jablonski diagram showing a molecule undergoing (a) ICT process and (b) IPT process after photoexcitation. Some other photophysical processes are also shown.

the ICT process. The proton transfer in a molecule can also give rise to dual fluorescence from its normal (N) and proton (T)-transferred states, which sometimes are referred to as keto and enol forms, respectively. In Figure 2.2, the N form in the ground state (GS) gets excited to the N* state after photoexcitation, followed by excited-state proton transfer to form the T* state. These two states, in many cases, give rise to dual emission. For ICT as well as for IPT, the emission pathways compete with non-radiative pathways. The ICT and IPT processes in a molecule can be distinguished on the basis of timescale- and solvent-dependent studies. The IPT process is much faster than the ICT process. Dual fluorescence in the ICT process generally occurs in solvents with high polarity, while in the IPT process the dual emission can be seen in nonpolar solvents due to formation of the proton-transferred state upon excitation.

On the other hand, for the IPT in a molecule to take place, a minimum distance condition between the proton donor and the proton acceptor groups is needed, while ICT processes can occur over arbitrarily long distances. Several donor–acceptor systems can undergo energy-transfer processes, like FRET and Dexter energy transfer (DET), that have some similarity in the photophysics with the ICT process [16], although a detailed discussion of these processes are out of the scope of this book.

2.2 Background of Studies on Charge Transfer

It has been reported that the dipole moment (μ) of a molecule could greatly increase in the excited state from that in the ground state due to the change in electronic charge distribution over the atoms of the molecule in the excited state [9–11, 36–39]. This change in μ is quite common for D-A–based excited-state CT molecules (also known as D-π-A), which exist in two notably different forms in the ground and excited states. The two forms can be referred to as the normal and quinonoid forms, as shown in Figure 2.3.

Let the wavefunctions of the normal and the quinonoid structure of the molecule be ψ_N and ψ_Q and the wavefunctions describing the molecule in the ground and the excited states be ψ_e and ψ_g, respectively. We can define ψ_e and ψ_g as superpositions of these two state functions using Equations 2.3 and 2.4 [1].

$$\psi_g = c_N \psi_N + c_Q \psi_Q \tag{2.3}$$

$$\psi_e = c_Q \psi_N - c_N \psi_Q \tag{2.4}$$

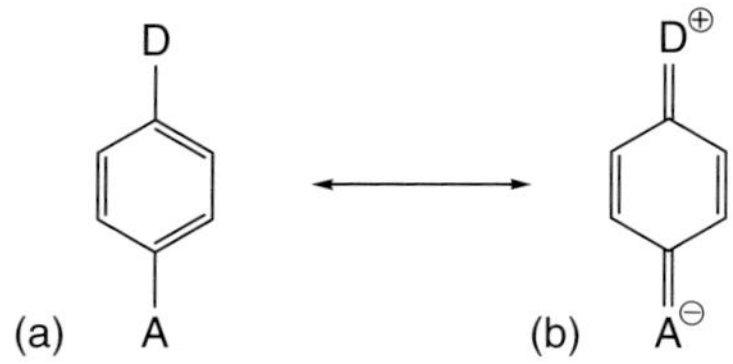

Figure 2.3 The normal (a) and quinoid (b) forms of a D-π-A molecule, where the spacer (π) is a benzene ring.

It is well known that change in the electronic charge distribution can affect the reactivities and acid–base properties of a molecule. If the donor (D) and the acceptor (A) are parts of the same molecule, but barely interact in the ground state, they may give rise to CT in the excited state, provided it is feasible energetically [40–44]. In D-π-A molecules, the excitation of the molecule can be done in one of the two ways that result in CT. In the first case, an excited donor transfers its electron(s) to the acceptor; and in the second case, charge is transferred from a donor to the excited acceptor. So, excitation of either the donor or the acceptor using radiation of proper wavelength will lead to a charge-separated (CS) state (D^+-π-A^-). The electronic structure of this D^+-π-A^- species will correspond to the ground state of the free radical ion pair of opposite charges consisting of a radical cation (D^+) and a radical anion (A^-). We can call this state as pure electron transfer (ET) state, whose zero-order wave function can be described as [1]

$$\psi_{CT} = \psi_{D^+}\psi_{A^-} \tag{2.5}$$

To facilitate an excited-state ICT process, a molecule absorbs radiation of particular energy which is suitable to excite it to its first (S_1) or higher excited states ($S_2, S_3, \ldots$). After the photoexcitation, it rearranges its electron density in the excited state in such a way that a species with lower energy and higher dipole moment originates [1, 37]. In some exceptional cases, the excited state μ of a molecule can be lower than its ground state counterpart. The normal excited species is termed as LE and the new species formed due to rearrangement of its electronic structure in the excited state is called the ICT state. Now, due to the presence of two different species (LE and ICT) in the excited state, the probe can show dual fluorescence. The potential energy surface (PES) of an ICT process in a molecule can be illustrated as shown in Figure 1.2.

The formation of ICT state from the LE state does not happen arbitrarily. It is guided by some rules, which have been expressed in Equations 2.6 and 2.7.

$$E_{LE} - E_{ICT} > 0 \tag{2.6}$$

where E_{LE} and E_{ICT} are energies of the LE and the ICT states.

$$E_{ICT} = IP(D) - EA(A) + C + E_{solv} \tag{2.7}$$

IP(D) and EA(A) are the ionization potential of the donor and electron affinity of the acceptor, respectively. The term C stands for the mutual Coulombic attraction between the linked donor and the acceptor, while E_{solv} is the energy of solvation. So, from Equations 2.6 and 2.7 one can predict that the energy of the LE state must be higher than that of the ICT state and the formation of ICT state from the LE state is not only guided by the properties of the donor and the acceptor but also depends on the properties of the medium. The effects of the properties of the medium on the photophysics of several ICT-based probes are discussed in Chapter 4. Although several ICT-based organic molecules show dual emission in their electronic spectra, some ICT molecules are reported to show emission from either the LE or the ICT species only [40–44].

2.3 A Brief Review of ICT Process in Commonly Studied Organic Molecules

2.3.1 ICT in 4-*N*,*N*-Dimethylaminobenzonitrile and Related Molecules

In the early sixties of previous century, Lippert and coworkers [43, 45] first observed the dual fluorescence in DMABN, a simple donor–acceptor substituted derivative of benzene.

This compound showed two fluorescence bands, out of which one was termed "normal," which was closely related to benzene derivatives. The other redshifted band, termed "abnormal," was of considerably lower energy than the normal emission band. Lippert and coworkers explained [1] the dual emission of DMABN by considering a solvent-induced reversal of the excited state. It was proved later that this explanation is valid only for the molecules having two close-lying excited states. To explain the dual emission from DMABN and related molecules, several other conflicting mechanisms were proposed by different groups [1]. It gradually turned out that those mechanisms are valid only under certain circumstances. Some of the popular explanations and their drawbacks are as follows:

(i) Formation of an exciplex with the solvent was reported to be the main cause of the dual fluorescence. This theory is only valid for solvents having available lone pair(s) with suitable energy to form excited-state complexes with the molecule and thus failed to explain the dual emission of DMABN in solvents that do not have available lone pairs.
(ii) Formation of a dimer either in the ground state or formation of complex between an excited DMABN molecule with its ground-state counterpart. In this case, the dual emission is expected to show concentration dependence as the number of dimer could increase with increase in concentration. Concentration dependence studies on this molecule do not support this hypothesis.
(iii) Specific solvation of DMABN by water molecules was believed to be another reason for the dual emission in DMABN. This requires traces of water in the solvents used. Formation of the ICT state has been observed in highly polar but aprotic solvents and also in the gas phase does not support this claim.

So, it is apparent that these explanations indicate that they are valid only in certain circumstances. Therefore, despite several studies, the geometrical structure of the emitting ICT state as well as the pathway of formation of the ICT state from the LE state of DMABN remained a matter of lively debate till date. Grabowski and coworkers [1] put forward the twisted intramolecular charge-transfer (TICT) model to explain the emission behavior of DMABN that says that in the excited state the $—NMe_2$ group DMABN twists by 90° about the plane perpendicular to the benzene ring so that the dimethylamino group would be electronically decoupled from the benzene π system (Figure 2.4). So, in the TICT structure the dimethylamino group is perpendicular or almost perpendicular relative to the benzene ring. Although this model has some limitations, this is one of the most popular explanations of dual fluorescence in DMABN

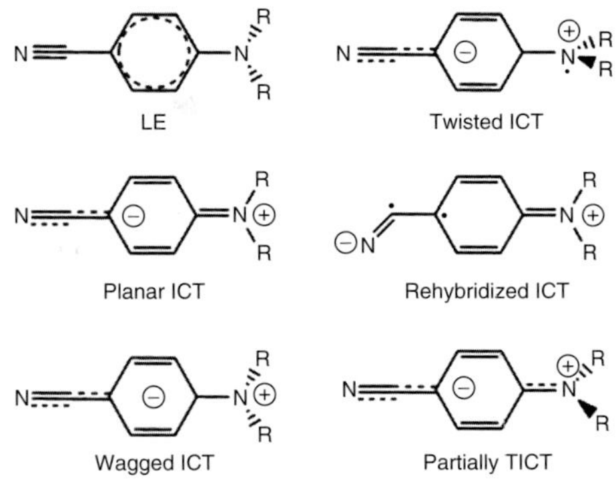

Figure 2.4 The chemical structure of locally excited (LE), TICT, PICT, RICT, WICT and partially twisted ICT (pTICT) states of ABN derivatives. R = —H, ABN and R = —CH_3, DMABN. (Copyright © 2015 American Chemical Society, Adapted with permission from Gómez *et al.* [46].)

and related molecules till date. In the planar intramolecular charge-transfer (PICT) model proposed by Zachariasse [12], the dimethylamino group remains in the same plane of the benzene ring. This mechanism is observed in several ICT molecules and is discussed later on in this chapter. The PICT model, in contrast to the TICT mechanism, does not accept the necessity of the twisting of the dimethylamino group and emphasizes the importance of the S_2 to S_1 energy gap and the vibronic interaction. The wagged intramolecular charge transfer (WICT) model was proposed by Zachariasse and coworkers [47, 48] where rehybridization of the nitrogen atom of the dimethylamino group occurs, resulting in a pyramidal sp^3 structure from a planar sp^2 structure. This model was least supported among all the proposed structures of the ICT state, both theoretically and experimentally. The rehybridized ICT or RICT model [49] involves rehybridization of the carbon atom of the cyano group whose hybridization is changed from sp to sp^2. Only a few studies have supported this mechanism. Recent studies by Lim *et al.* [50] have proposed the formation of a partially twisted intramolecular charge-transfer (pTICT) state based on the experimental and theoretical studies, while some other studies by Lasorne *et al.* have raised doubts over the formation of a stable pTICT state [51] in 4-aminono benzonitrile and related molecules. The TICT mechanism was supported by several experimental studies on CT on some related donor–acceptor substituted benzene derivatives, discussed later. It was seen that molecules in which the donor group is more or less fixed does not show any dual fluorescence and indeed only a single-emission peak in the position of the "normal" band of DMABN was observed. For example, 1-methyl-5-cyanoindoline (MIN), a fixed-donor molecule (Figure 2.5) shows only the "normal" emission from the LE state. On the other hand, the molecules which are already twisted in the plane shows a single fluorescence peak in position of the "abnormal" band of DMABN. Tetramethylamino-benzonitrile (TMABN) is an example of such a molecule,

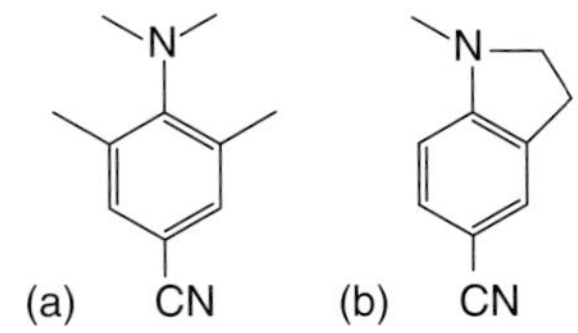

Figure 2.5 Structure of (a) TMABN and (b) MIN.

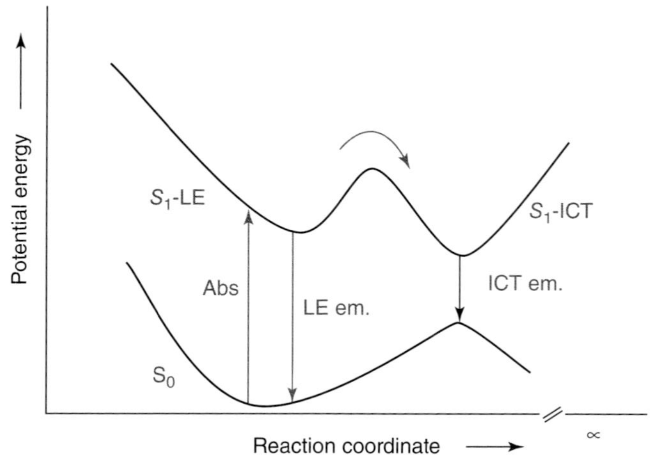

Figure 2.6 The potential energy surfaces (PESs) for intramolecular charge-transfer process. The photophysical processes involved are discussed in the text.

where the $-NMe_2$ group is already twisted in the ground state (Figure 2.5). So, initially, the TICT mechanism appeared to be the most acceptable mechanism and this was supported by other theoretical and experimental studies [1].

Other than the structure of the ICT state the issue that still remains unresolved is the pathway of formation of the ICT state from the LE state [51, 52]. A simple picture for PESs of an ICT molecule only involving the first excited state (S_1) can be as follows: The absorption of radiation leads the molecule to the S_1-LE state, which is connected adiabatically with the S_1-ICT state (Figure 2.6). If the S_1-ICT state has lower energy than the S_1-LE state, excitation to the LE state will populate both the LE and ICT states. So, absorption of radiation will lead to dual emission if both the states are fluorescent.

However, the PESs of DMABN are different from that shown in Figure 2.6. Based on several experimental and theoretical investigations on the ICT process in DMABN and related molecules, the following pathways can be proposed.

Path 1: The absorption of radiation leads the molecule to the higher excited state (S_2). The system then relaxes through a nearby conical intersection (CI) by an internal conversion to the first excited state (S_1) on which the minima of both the LE and ICT states lie. The LE and ICT states are connected adiabatically and therefore the population of the S_1-LE and S_1-ICT states and the dual fluorescence process are dependent on the adiabatic and nonadiabatic paths connecting the minima [53, 54].

Path 2: The population of the ICT state at the Frank–Condon (FC) geometry can be followed by a fast switch that leads to a splitting between a bright (fluorescent) LE and a dark (nonfluorescent) $\pi\sigma^*$ state. The $\pi\sigma^*$ state populates the TICT state from which emission occurs. The lack of dual fluorescence in some ICT chromophores are attributed to the presence of this dark $\pi\sigma^*$ state [55].

Path 3: After population to the ICT state at the FC geometry, the system bifurcates into a bright LE state and a dark $\pi\sigma^*$ state. The LE state in turn populates a fluorescent pTICT state and the dark πσ* state populates a dark TICT state. Therefore, this mechanism claims the existence of two ICT states that are non-communicating. The formation of the pTICT state seems to be controversial [51, 52] and is discussed later on in this chapter.

Fuβ and coworkers [56] have studied the ultrafast relaxation and coherent oscillations in molecules related to 4-aminobenzonitrile (ABN), viz, N-methyl-6-cyano-1,2,3,4-tetrahydroquinoline (NMC6), the corresponding *N-tert*-butyl derivative NTC6 in the gas phase and compared the results with other previously studied benzonitrile derivatives 4-amino benzonitrile (ABN), 4-dimethylamino-benzonitrile (DMABN), 4-piperidino-benzonitrile (PIPBN), 4-pyrrolidino-benzonitrile (PYRBN), and 4-pyrrolyl-benzonitrile (PBN) [57–59]. Using their results, the PES of DMABN that was drawn, has been shown in Figure 2.7. Out of the two emitting states of these types of molecules, one is localized on the benzene ring (called L_b state) and is not affected much due to substitution. The other emitting state has a donor-to-acceptor CT character and is typically higher in energy compared to the L_b state in the gas phase. The energy of this CT state can be modified using the solvent of suitable polarity as this state has a higher dipole moment than does the L_b state. So, in solution, equilibrium between the L_b and CT states can be established and dual emission can be observed if the population is comparable. We should note here that even equal population of the L_b and CT states may not give rise to dual emission as the quantum yield of the CT state is much lower compared to the L_b state. In an earlier report, they have reported the photophysics of DMABN in the gas phase [58]. DMABN was excited to the S_2 (or L_a) state in the gas phase using 270-nm ultrashort pulse obtained from frequency-tripled pulse of a Ti-sapphire laser system and probed by femtosecond time-resolved photoionization at 2 μM. Their results suggest that S_2 relaxes through a CI to the S_1 (or L_b) state, where it

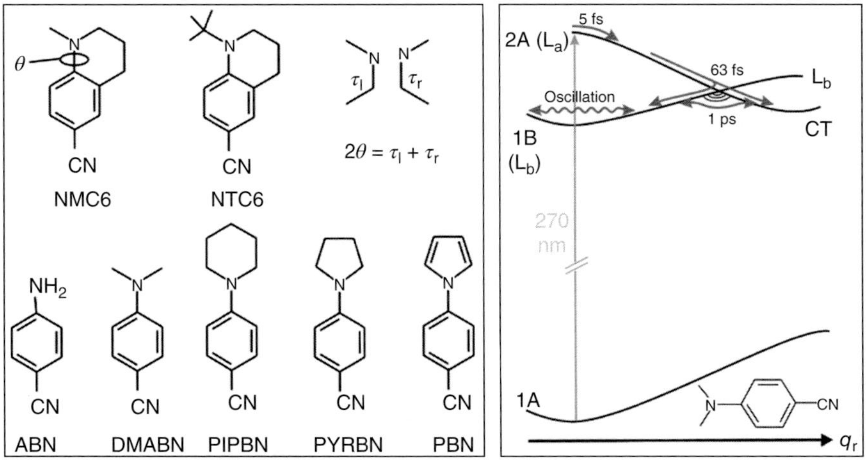

Figure 2.7 Structure of NMC6, NTC6, ABN, DMABN, PIPBN, PYRBN, and BPB (left). The twist angle (θ) is defined as the average of the left (τ_l) and right (τ_r) torsional angles. The potentials and dynamics for DMABN (left). (Fuβ *et al.* 2007 [56]. Reproduced with permission of Royal Society of Chemistry.)

vibrates along the amino group twist and inversion. They also claimed that the CI, and therefore the CT state, is displaced from the L_b minimum in a direction containing the twist and probably the inversion as components also. Although NMC6 does not show dual emission (no ICT emission) even in highly polar solvents, CT emission was reported in NTC6 in which the methyl group of NMC6 is replaced by a *tert*-butyl group [60]. Fuß and coworkers used a weak femtosecond pump pulse at 270 nm to excite the NMC6 and NTC6 molecules (Figure 2.7) to the L_a (S_2) state, followed by an intense infrared (IR) pulse as probe that ionizes these molecules [56]. They measured the ion yield mass selectively as a function of the pump-probe delay using a time-of-flight mass spectrometer. In conjunction with quantum chemical calculations by Robb and coworkers [54], they inferred that the L_b state of DMABN belongs to the $1B_2$ species in c_{2v}, the idealized symmetry group of this molecule. The S_0 to L_a absorption and CT to S_0 emission, in contrast to the S_0 to L_b transition, are parallel polarized where the latter is polarized perpendicular to the long molecular axis. So, the L_a and CT excited states belong to the $2A_1$ species and has the same PESs. Due to the partial CT character of the L_a state, the C—N bond gets partial double-bond character and therefore is less likely to be twisted. After photoexcitation of the S_2 (L_a) state of DMABN and the related molecules, the wave packet passes through a CI. From the CI one part leaches the L_b (S_1) state directly, while the other part populates the CT state temporarily from where it goes around the CI to the L_b well. In DMABN, the population flows from the initially excited L_a state within 5 fs out of the FC region [56]. Within 63 fs it reaches the CI from where one part of it reaches the L_b minimum directly, while the other part of it goes to the CT state. The part that reached the CT state leaves it within 1 ps and reaches the L_b well. The wave packet arriving at the L_b state directly oscillates there along the coordinates involving twisting of amino group and wagging or molecular arching and quinoidal distortion. The authors concluded that the potential and dynamics of all the molecules in Figure 2.7 are the same, except for the timescales, the oscillation frequencies, and the relative energies of the ground and excited states. Their studies detected large anisotropies from the CT state of all the molecules (Figure 2.7), except for NMC6, where it is small. This led them to conclude that there is a substantial twist in the CI of these molecules with large anisotropies. In NMC6, where the anisotropy is small, the value of θ is predicted to be much less than 90° in the CT state. In fact, they argued that the value of θ in this molecule in the CT state could be around 45° that can be accommodated by this molecule, although the amino group is integrated in a six-membered ring that potentially hinders its twisting. Twisting of the amino group by 45° lowers the energy of the CT state of this molecule, compared to the L_a state, although the energy of the CT state will be higher than its unhindered counterparts.

Lim and coworkers [61] studied the ICT process and dual emission in DMABN using femtosecond time-resolved transient absorption (TA) spectroscopy and fluorescence experiments in conjunction with *ab initio* multi-reference perturbation theory CASPT2/complete active space self-consistent field (CASSCF) calculations. Their results suggest that after initial excitation to the L_a state, an ultrafast branching process takes place. This switch leads to population of both the LE and $\pi\sigma^*$ states. Their time-resolved fluorescence spectra show a

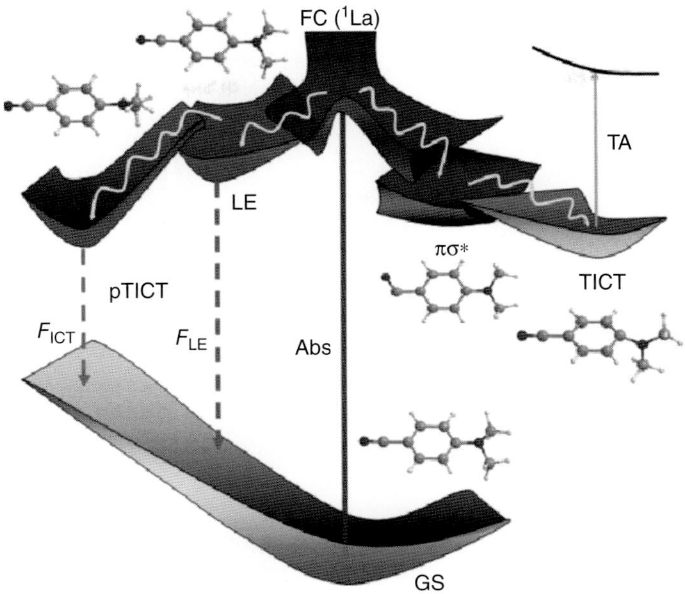

Figure 2.8 A schematic representation of photophysics of DMABN, an ICT molecule, in polar solvent (Coto *et al.* 2011 [61]. Reproduced with permission of Royal Society of Chemistry.)

correlation in decay time of the LE state with the rise time of the fluorescent ICT state. On the other hand, the long-time behavior of the fluorescent ICT state was different from that of the ICT state, leading them to infer that these states are different. Their computational results predict that out of the two TICT states formed in this process, only the pTICT state has significant oscillator strength ($f = 0.230$), comparable with that obtained experimentally. The oscillator strength of the TICT state was found to be negligible ($f = 0.001$). This led them to conclude that the pTICT state is responsible for emission of DMABN, not the TICT state. Figure 2.8 depicts the photophysics of DMABN in polar solvents.

The importance of the formation of the πσ* state was later challenged by Reguero and coworkers, and they suggested that the formation of the pTICT state could be an artifact of the computational method [52, 62]. They used quantum chemical calculations at the *ab initio* CASSCF/CASPT2 level of theory to understand the ICT pathway in several ABN derivatives, namely, ABN, DMABN, and their 2,3,4,5-tetrafluoro derivatives, abbreviated as ABN-4F and DMABN-4F, respectively. They have concluded that although the πσ* state [they refer it as ICT(CN)] could be an intermediate to populate the bright state, involvement of this state in the ICT process is less probable. They further emphasize that the TICT state is the most stable excited-state species of the molecules studied in polar solvents. They have also taken into account the radiation-less deactivation process to explain the fluorescence properties of these molecules.

So, the discussion so far indicates the DMABN and its congeners draw much attention to the scientific community for studies of ICT. This is mainly because

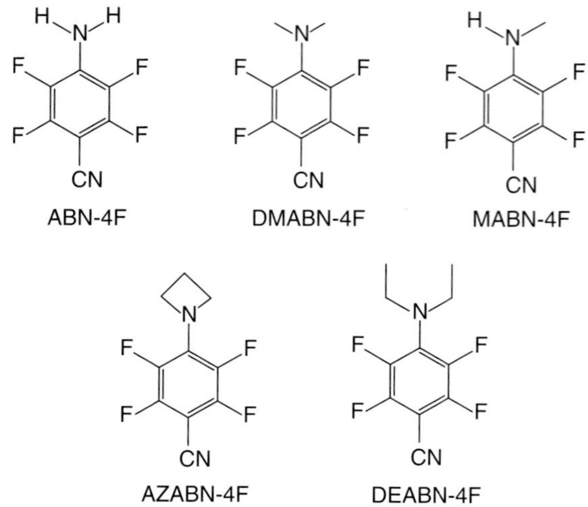

Figure 2.9 Molecular structures of 2,3,5,6-tetrafluoro-4-aminobenzonitriles studied by Zachariasse and coworkers.

studies of ICT processes in these molecules help decipher the nature of the ICT state in other molecules and thus provide unprecedented insight into the formation of the ICT state. ABN (Figure 2.7), where the methyl groups of the dimethylamino group of DMABN have been replaced with hydrogen atoms, has been under the scanner for ICT studies [52]. Although DMABN and ABN show fluorescence from the LE state in nonpolar solvents, unlike DMABN, ABN does not show dual emission even in polar solvents. Reguero *et al.* [52, 62] studied the excited-state photophysics of ABN and its 2,3,4,5-tetrafluoro counterpart, acronymed ABN-4F using *ab initio* calculations at the CASSCF/CASPT2 level of theory.

Zachariasse *et al.* [63] studied the effect of fluorine atom substitutions on the photophysics of ABN and related molecules (XABN-4F). The molecules are named according to their donor groups – ABN-4F for amino, DMABN-4F for dimethylamino, DEABN-4F for diethylamino, AZABN-4F for azetidinyl, and MABN-4F for methylamino (Figure 2.9). The authors have found that all the XABN-4F molecules undergo the ICT process at room temperature in acetonitrile, a polar solvent as well as in *n*-hexane, a nonpolar solvent. The ICT process takes place in their non-fluorinated counterparts DMANB, DEABN, and AZABN, but not in ABN and MABN.

The authors estimated the dipole moments of the ICT state of XABN-4F molecules to be about 14 D, which is less than that of DMABN (17 D) and they attributed this to electron withdrawal from the cyano group to the phenyl ring by the fluorine substituents. The XABN-4F molecules form an LE state upon photoexcitation from which the ICT state is formed in the excited state. The authors reported that in *n*-hexane the LE to ICT reaction takes place in 0.35 ps in DMABN-4F, while it takes 0.29 ps and 0.13 ps in DEABN-4F and AZABN-4F, respectively. The LE to ICT process occurs in about 90 fs in acetonitrile in all the XABN-4F molecules studied in this report. The solvent relaxation time of acetonitrile is about 90 fs; therefore, the authors conclude that the ICT reaction time in this solvent is limited by the solvent relaxation time. The authors argue that during this ultrashort period a large-amplitude motion such as full 90° twist of the amino group is unlikely to occur during the ICT process of XABN-4F molecules. The ICT lifetimes of the XABN-4F molecules are quite shorter (4–535 ps) than their corresponding non-fluorinated counterparts (3–4 ns for DMABN) in acetonitrile. Therefore, they postulate that the internal conversion process has been increased due to fluorine substitution, which possibly enhances the accessibility of a CI leading to the electronic ground state.

Reguero *et al.* [52, 62] theoretically studied the emission properties of ABN, DMABN, and their tetrafluoro derivatives, namely, ABN-4F and DMABN-4F (Figure 2.9). The authors also considered the possibility of the existence of a $\pi\sigma^*$ state, proposed by other authors, which they call ICT(CN) state as an intermediate responsible for ICT emission of DMABN. They also considered the possibility of the formation of a pTICT state in these molecules. They claimed that not only the relative energies of the excited state species but also understanding the pathway connecting them are necessary to rationalize the emission behavior of these molecules. As the radiation-less deactivation competes with the radiative process (fluorescence) in these molecules, the authors considered these non-radiative pathways while explaining the experimentally obtained emission behaviors of these molecules. It has been reported that DMABN shows emission from mainly the LE state in nonpolar solvents with a redshifted band of very low intensity originating from the ICT species. As the polarity of the solvent increases, the ICT band becomes more prominent and dual fluorescence of DMABN are observed in polar solvents. The ABN, however, shows only a single emission peak from the LE state in both polar and nonpolar solvents. On the contrary, the tetrafluoro derivatives of ABN and DMABN show emission from ICT only irrespective of solvent polarity and no sign of LE band is detected. These results indicate that the CT states of these fluoro-substituted derivatives can be populated easily compared to their unsubstituted counterparts. The authors explained these results using their quantum chemical studies summarized later (see Figure 2.10 for pictorial representation). The authors first considered the possible low-lying states involved in the CT process. Their results show that the LE state results from a $\pi\pi^*$ excitation of the orbitals located on the ring. There are possibilities of the formation of two excited states due to excitation of an electron from the lone pair of the amino group to a quinoid and an anti-quinoid antibonding of the ring that they named ICT(Q) and ICT (AQ), respectively. Relaxation on the ICT(Q) and ICT (AQ) states lead to high-energy minima where the molecules under investigation get a PICT structure. This relaxation process can lead the molecule to a low-energy minima where the amino group is twisted, finally giving TICT(Q) and TICT(AQ) species. The authors found that the ICT(CN) state is formed due to excitation of an electron from the lone pair of the nitrogen atom of the amino group to an anti-bonding π-orbital of the cyano (—CN) group. This process leads to rehybridization of the nitrogen atom and also bending of the cyano group and this state is called rehybridized intramolecular charge-transfer (RICT) state. In ABN, excitation populates the molecule to the S_2-ICT(Q) state, although absorption to the first excited state with LE character is also possible. The authors found that the LE, TICT(Q), and RICT states correspond to the minima located at the PESs of the

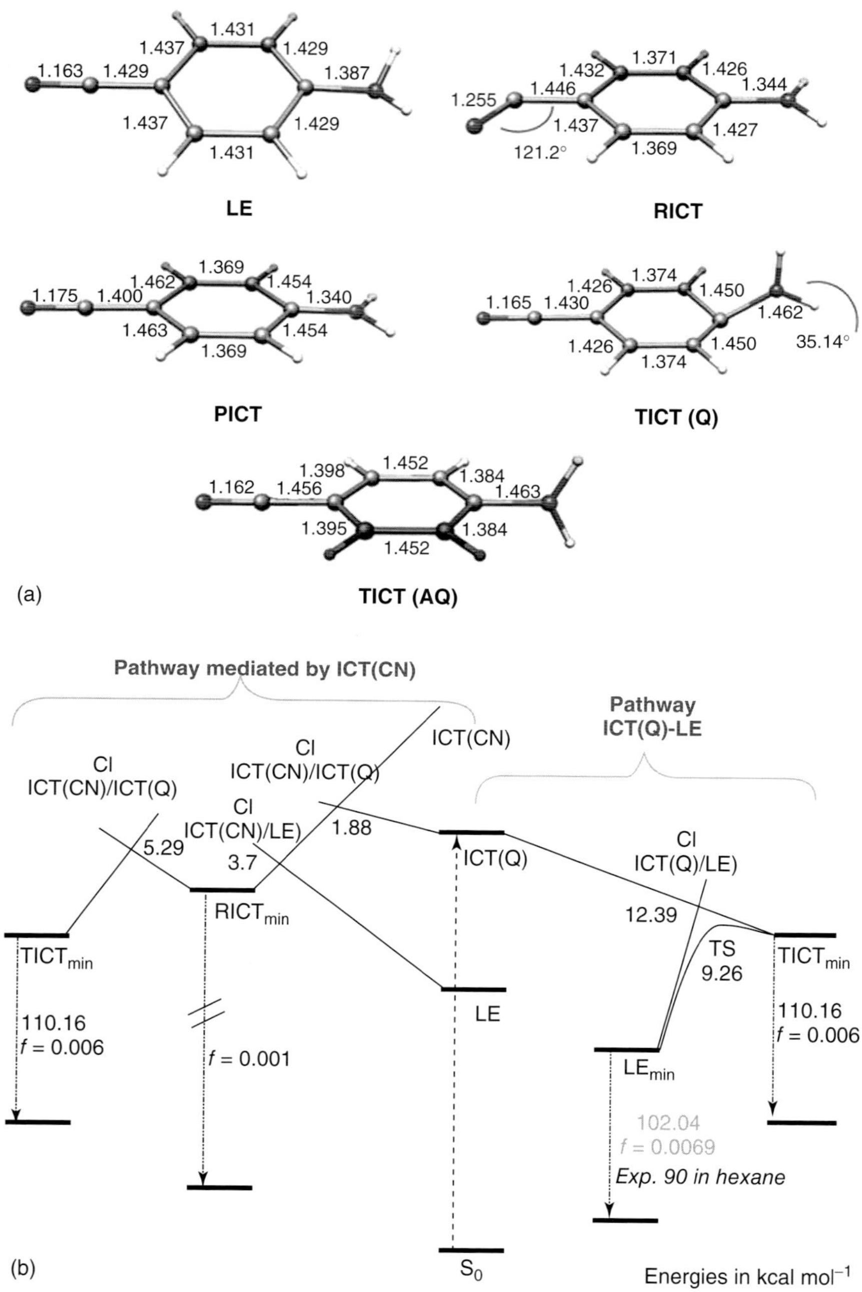

Figure 2.10 (a) Optimized structures of excited-state minima of ABN. Some important bond lengths are shown in angstroms. (b) Schematic representation of excited-state reactions of ABN (relative energies are not in scale). (Segado *et al.* 2016 [52]. Reproduced with permission of Royal Society of Chemistry.)

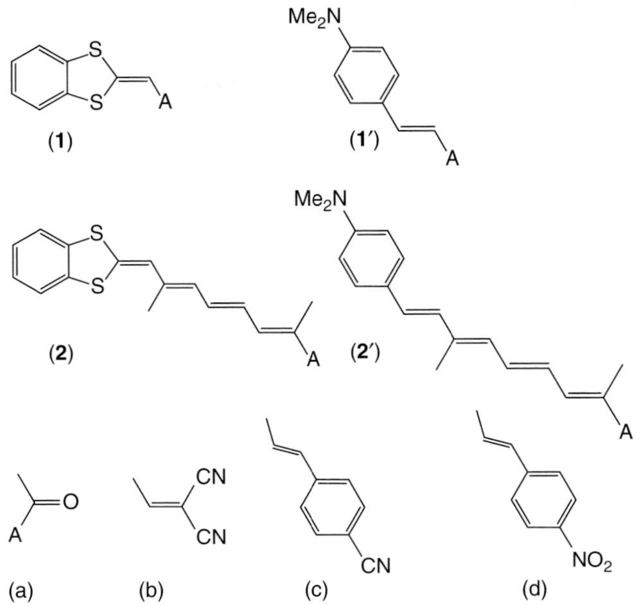

Figure 2.12 Some of the donors (1–2 and 1′–2′) and acceptor (a–d) units of the polyenes and carotenoids studied by Lehn and coworkers. (Slama-Schwok *et al.* 1990 [69]. Reproduced with permission of American Chemical Society; Redrawn for clarity.)

signal processing and transfer, fluorescence sensor, organic semiconductor, NLO properties, among many others (see Chapter 6 for details). Lehn and coworkers [69] studied the ICT process in several "push–pull" polyenes and carotenoids using absorption and emission spectral studies (Figure 2.12). The fast ICT process in molecules obtained by anchoring donor and acceptor units at two ends of polyenic chains are important in designing molecular devises for long-distance signal processing and transfer. The ICT process in these molecules can be tuned by altering the donor and acceptor strengths and the properties of the medium and can push the ICT emission up to the near-IR region [69].

The ICT process of Rhodamine derivatives was studied by Gofman [70] and later by Sun and coworkers [71]. Out of several molecules of this class, Rhodamine B (Figure 2.13) and Rhodamine 6G are very popular among the scientific community for studies of CT in these molecules. Using steady-state

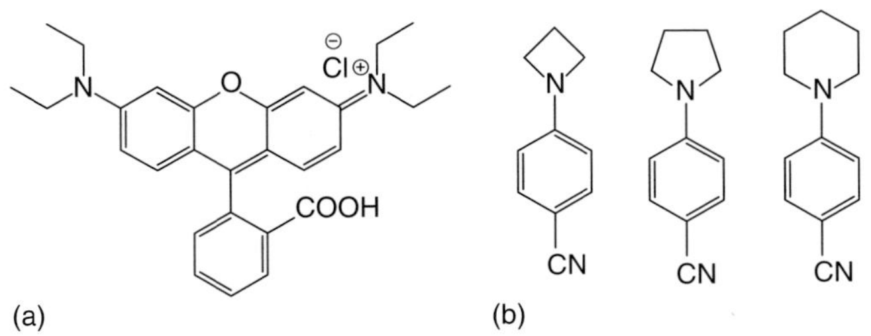

Figure 2.13 (a) Chemical structure of Rhodamine B and (b) structure of P4C, P5C, and P6C.

spectroscopic techniques, Maroncelli and coworkers [72] studied the ICT process in some (alkylamino)-benzonitriles, which are structurally quite similar to ABN. These molecules are 4-(1-azetidinyl)-benzonitrile [P4C], 4-(1-pyrrolidinyl)-benzonitrile [P5C], and 4-(1-piperidinyl)-benzonitrile [P6C]. Although these compounds are structurally quite similar to ABN, they show dual emission in moderately polar solvents, while ABN shows emission from the LE state only even in polar solvents [73]. The structures of P4C, P5C, and P6C are shown in Figure 2.13.

1,2-Benzopyrone dyes, also known as Coumarin dyes [74–77], have been reported to be another class of organic molecules for studies of the ICT process. Several examples can be found in literature on the studies of medium implications on the ICT processes in these molecules. The effects of solvent polarity and/or hydrogen bonding on the ICT processes of some of these molecules are discussed in Chapter 4. Out of the many molecules of this class, the C152, C153, and C481 are studied very widely. Pal and coworkers have reported the effect of the medium (temperature and solvent) on C152 and C481 dyes [76]. The structures of C152, C153, and C481 are shown in Figure 2.14.

Recently, molecules of the oxazine molecular [7, 78] family have got much attention as ICT probes. Han and coworkers [7] studied the effect of hydrogen bonding and polarity of the medium on the photophysics of oxazine 750 (Figure 2.15) using spectroscopic methods and theoretical tools.

The thermally activated delayed fluorescence (TADF) process has caught the attention of the scientific community due to its potential for application in producing white light at low cost. For designing efficient red TADF emitters for OLEDs, the molecule must have a large fluorescence rate and the energy difference between the lowest singlet and lowest triplet excited states should be low [79]. Adachi and coworkers [80] reported dual emission from a material containing phenothiazine (PTZ) as electron-donor unit and 2,4,6-triphenyl-1,3,5-triazine (TRZ) as electron-acceptor unit. Acronymed PTZ-TRZ (Figure 2.16), this novel ICT molecule exists in two forms in the ground

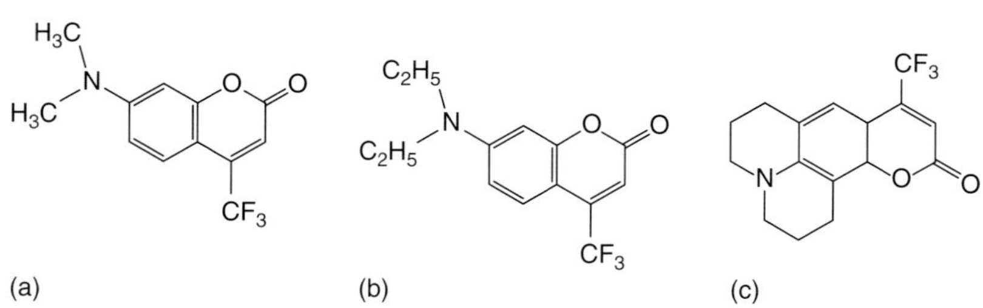

Figure 2.14 Structure of C152, C153, and C481.

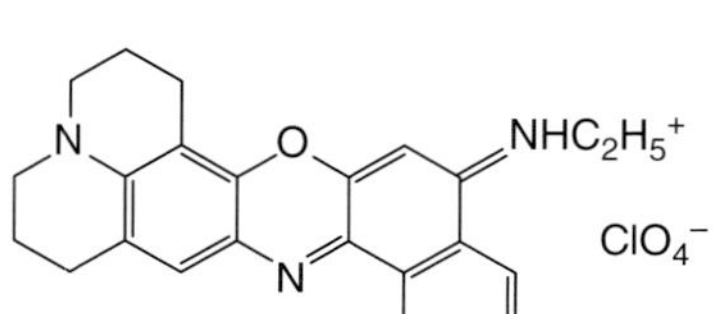

Figure 2.15 Structure of Oxazine 750.

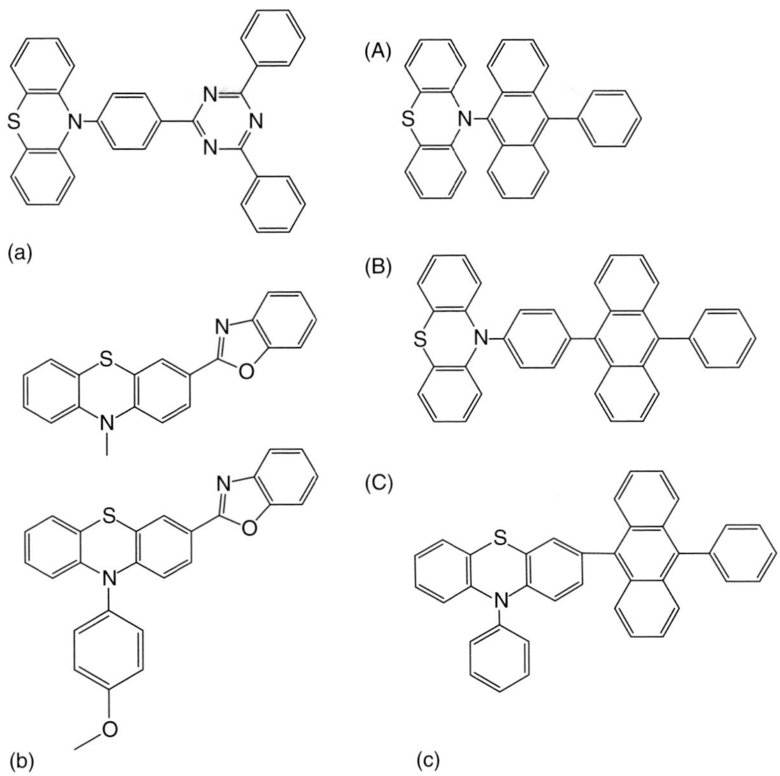

Figure 2.16 Structure of (a) PTZ-TRZ; (b) phenothiazine core electron-donor and oxazole-acceptor molecules studied by Ghosh and coworkers and (c) PTZ-10-AnP (A), PTZ-10P-AnP (B), and PTZ-3-Anp (C).

state that have different energy gap between lowest singlet and lowest triplet excited states due to distortion of the TPZ unit. ICT process in a donor–acceptor type of molecule that has PTZ core as electron donor and oxazole as an electron acceptor (Figure 2.16) was studied by Ghosh and coworkers [81].

The electron-donor strengths were altered using methyl and anisole donor groups. Their aim was to diminish the dark TICT emission using a zero-twist D-A system. Yang and coworkers [82] studied the CT process in three molecules having PTZ as donor and Anp as acceptor units, linked through different molecular architectures. Anthracene was introduced into the 10-position (PTZ-10-AnP) and 3-position (PTZ-3-AnP) of PTZ, respectively (Figure 2.16). Phenylene group was inserted between the 10-position of PTZ and anthracene into PTZ-10-AnP to obtain the PTZ-10P-AnP. They reported that the distance between the D-A units and the torsion angles dictate the energy of the CT state of these molecules. Using combined experimental and theoretical studies, Zhu *et al.* [83] reported the ICT process in several tetrahydro [5] helicene-based imide (THHBI) dyes (Figure 2.17). A large increase in dipole moment in the excited state from the ground state in these molecules is attributed to the ICT process from the electron-donating groups to the helicine-imide group. Their time-resolved TA

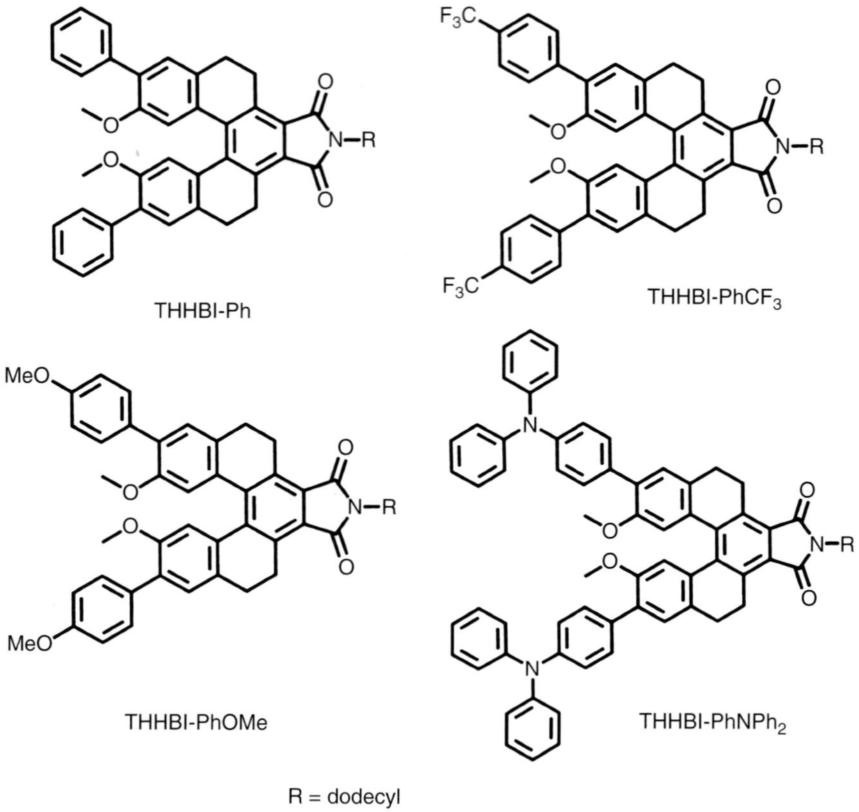

Figure 2.17 Structure of TTHBI-Ph, TTHBI-PhCF$_3$, TTHBI-PhOMe, and TTHBI-PhNPh$_2$. (Zhu *et al.* 2016 [83]. Reproduced with permission of Nature Publishing Group.

spectral studies reveal that the dye with stronger donor (THHBI-PhNPh$_2$) forms solvent-stabilized ICT state through a fast excited-state ICT relaxation in polar solvents. The dyes with relatively weaker donor groups, namely, THHBI-Ph, THHBI-PhCF$_3$, and THHBI-PhOMe show much less dependence with solvent polarity, and ISC was observed from the initial excited state to the triplet state.

Wolf and coworkers [84] studied the CT process in D-A and D-A-D types of molecules containing anthracene as donor unit and sulfur-containing acceptor group (Figure 2.18). Their results show that the oxidation state of the sulfur atom affects the CT process in these molecules.

Saha and coworkers [85] studied the ICT process in *trans*-4-[4′-(*N*,*N*′-dimethylamino)styryl] pyridine molecules (Figure 2.18). Their theoretical calculations predict twisting of the donor group with respect to the acceptor group, leading to the formation of TICT states. A large change in dipole moment in the excited state from the ground state was also predicted through quantum chemical calculations, which was further supported through experimental results. Jana *et al.* [86] reported the ICT process in 5-(4-dimethylamino-phenyl)-penta-2,4-dienoic acid ethyl ester (DMAPPDE). The dual emission of these molecules (Figure 2.18) was observed in polar solvents that were attributed

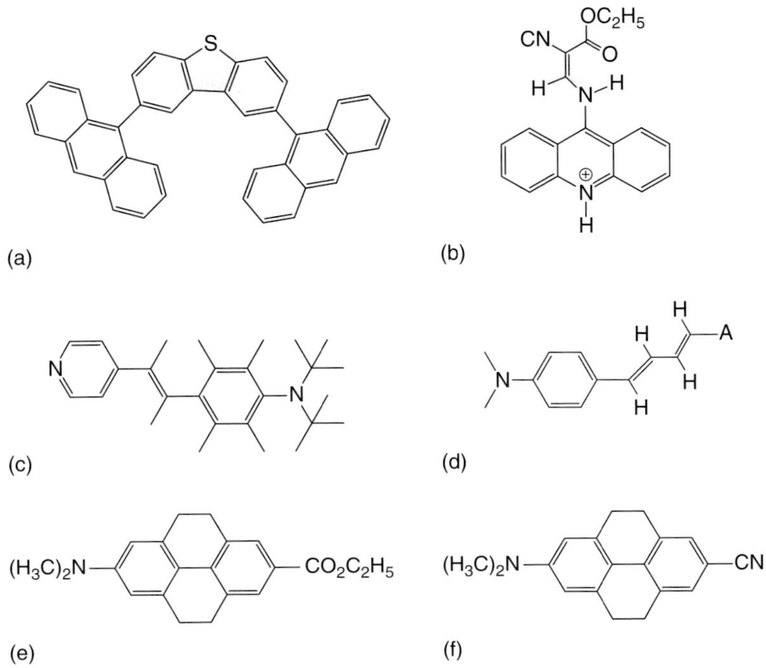

Figure 2.18 Structure of the molecules studied by (a) Wolf, (b) Pereira, (c) Saha and coworkers. (d) DMAPPDE molecules studied by Jana *et al.* (A = —CN, —CO_2Et and —CO_2H) and (e and f) tetrahydropyrene derivatives were studied by Gopidas and coworkers.

to an LE state and an ICT state formed in the excited state. Variation of acceptor strength affects the absorption and emission spectral profile of these molecules. The ICT process in tetrahydropyrene derivatives (Figure 2.18) were studied by Gopidas and coworkers [87]. Their studies reveal the planar nature of the CT species. The absorption and emission properties of those molecules are pH-dependent, making them potential candidates for pH-sensing applications. The same group has experimentally studied the ICT process in donor–acceptor substituted vinyltetrahydropyrene derivatives [88]. The structures of the excited states of these molecules are dependent on the nature of the medium. Using steady-state and time-resolved spectroscopic studies in conjunction with theoretical calculations, Kim and coworkers [89] studied the effect of bulky meso-substituents on the ICT process of *meso*-diarylamino subporphyrins. Carlotti *et al.* [90] reported the ICT process in three cationic D-π-A type of molecules, where dimethyl amino group is used as donor and methyl pyridinium or quinolinium group as acceptor (Figure 2.19). They used either ethylene or butadiene group as spacers. Time-resolved TA spectral studies reveal the formation of the ICT state in these molecules.

Pereira *et al.* [91] studied the ICT process in 9-aminoacridine derivative that was obtained by a reaction between 9-aminoacridine and ethyl-2-cyano-3-ethoxyacrylate (Figure 2.20) using steady-state and time-resolved emission spectral studies. This molecule appears in both enol and keto forms reversibly

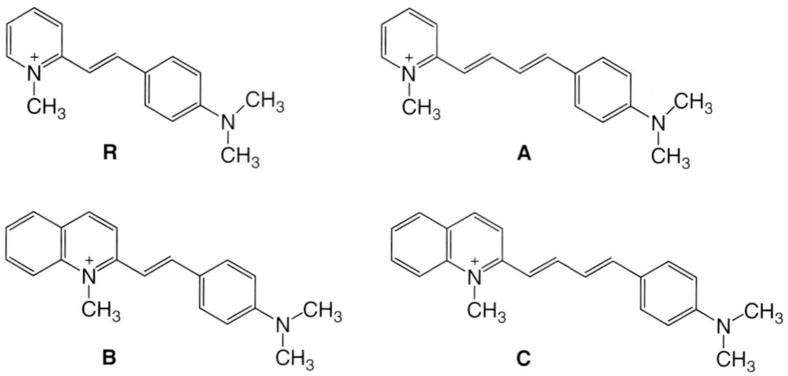

Figure 2.19 Structure of the molecules (A–C) studied by Carlotti et al., along with the reference molecule (R). (Carlotti *et al.* 2015 [90]. Reproduced with permission of John Wiley & Sons.)

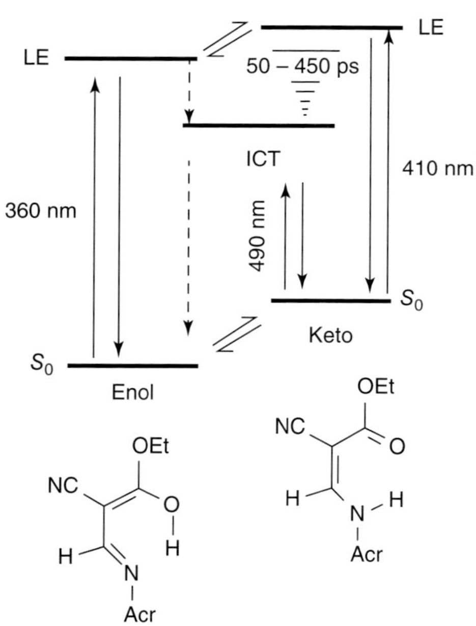

Figure 2.20 A typical electronic energy diagram of 9AAECA; keto-enol equilibrium and excited-state processes. The absorption wavelengths are reported in nanometer. Acr stands for acridine. (Pereira *et al.* 2005 [91]. Reproduced with permission of American Chemical Society.)

in the ground and excited states. The emission of this molecule is complex and is assigned to an LE state of acridine chromophore as well as to an ICT species formed upon excitation.

ICT in donor–acceptor–donor (D-A-D) chromophores, consisting of two donor units and an acceptor unit has been popular very recently. Lu *et al.* [92] studied the ICT process in a series of symmetrical D-A-D chromophores having triphenylamine (TPA) as electron donor and several electron-withdrawing groups, such as quinoxaline, benzo[g]quinoxaline, phenazine, benzo[*b*]phenazine, thieno[3,4-*b*]pyrazine, and thieno[3,4-*b*]quinoxaline as electron acceptors. Toyota and coworkers [93] reported the ICT process in a D-A-D molecule having two TPA groups connected to 9- and 10-positions of

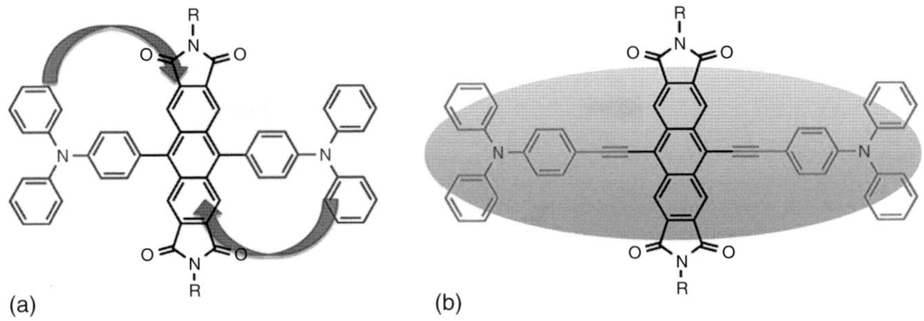

Figure 2.21 (a) The ICT process in D-A-D molecule studied by Toyota and coworkers. (b) The expanded π-conjugated system in this molecule. (Copyright © 2005 American Chemical Society, Adapted with permission from Iwanaga *et al.* [93].)

anthracene bisimide (ABI) group (Figure 2.21). The π-conjugation network was further extended using ethynylene linkers between the donor and acceptor units.

Bai *et al.* [94] studied the ICT processes in two molecules – one with asymmetrical DA structure and another with symmetrical DAD structure (Figure 2.22). Their results indicate that the increase in dipole moment due to excitation was higher in the DA molecule than that in the DAD molecule. The authors used the Lippert–Mataga analysis (see Chapter 4) to estimate the change in dipole moment between the ground and excited states. Their results show that the shift in absorption maxima with increase in solvent polarity was small, while the change in emission maxima was large. The combination of these two results led them to conclude that the changes in dipole moments of these molecules are much larger in the excited state than those in ground state, an indication of the ICT process. The authors carried out femtosecond fluorescence depletion

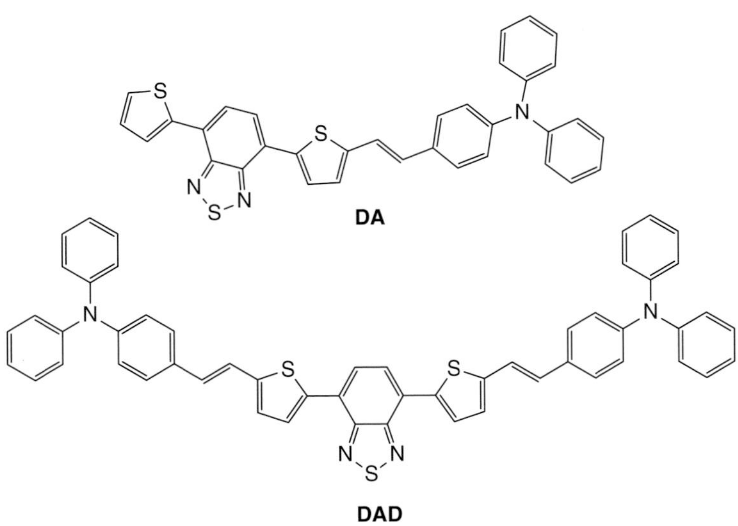

Figure 2.22 Structures of DA and DAD molecules studied by Bai and coworkers. (Gong *et al.* 2007 [94]. Reproduced with permission of American Chemical Society.)

measurements to understand the solvation process and dynamics of these ICT molecules. Their results show that the DA molecule had faster solvation time than the DAD molecule, which corresponds to a relaxed ICT state. Therefore, they concluded that the DA molecule has larger charge distribution in the excited state than its DAD counterpart. As performance of many photovoltaic devices depends on the push–pull ability of the chromophore, the asymmetric DA chromophore is expected to function better than the symmetric DAD chromophore in those devices.

The ICT process has been mostly studied in the condensed phase. Several theoretical studies on CT in the gas phase are reported in literature, but experimental studies in this direction are scarce [56–59, 95, 96]. Scalmani and coworkers [97] have used time-dependent density functional theory (TDDFT) calculations to study the excited-state behavior of two ICT molecules, namely, PNA and DMABN. They have also compared their gas-phase results with those obtained in the solution phase. Zachariasse and coworkers [95] have studied gas-phase CT in jet-cooled 4-(diisopropylamino)benzonitrile [DIABN] using fluorescence excitation spectroscopy. Weinkauf *et al.* reported the gas-phase CT in peptide cation [96]. Using resonant UV two-photon ionization spectroscopy, they showed that the CT in this system occurs by a through-bond process. Some studies concerning ICT process in the solid state are summarized in Chapter 4.

2.4 Structure of the ICT State: Twisted or Not?

From the discussion in the preceding section it is clear that the debate regarding the structure of the ICT emitting state is still active. The TICT model, proposed by Grabowski and coworkers, seems to be the most plausible structure in the excited state [1, 43], although its validity was challenged by several studies [12], introducing a PICT model. Domcke and coworkers proposed a different mechanism, known as RICT mechanism for the dual emission of DMABN. Using their theoretical calculation, they claimed that rather than the twisting of the —NMe_2 group of DMABN, the —CN group bends to get the energy equal to that of the so-called TICT state. Experimental studies [98–100] on DMABN could not confirm the formation of the RICT state and not many studies are done to confirm the formation of the RICT state. As claimed by the PICT model, several theoretical and experimental studies refuted the necessity of formation of a twisted excited state in the CT mechanism. For example, Yoshihara *et al.* [101] experimentally studied the ICT process in N-phenylpyrrole (PP) and its planar analog, fluorazene phenylpyrrole (FPP). They showed that in FPP the CT occurs in a few picoseconds as in its flexible counterpart, PP (Figure 2.23). This led them to refute the idea of the TICT state and a PICT state of the concerned molecules was proposed.

There are some other theoretical studies in literature that claim that the CT state of a molecule may not necessarily be twisted [54, 102, 103]. For example, using TDDFT theory, Adamo and coworkers [103] studied the origin of dual fluorescence in Nile red (Figure 2.23), an ICT molecule considering

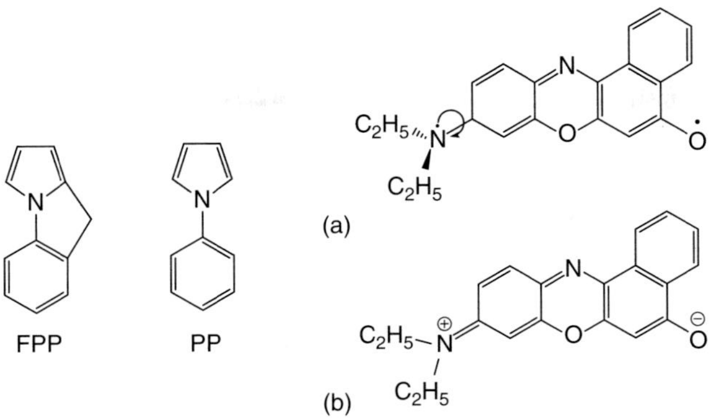

Figure 2.23 Left: Structures of FPP and PP. Right panel shows the (a) TICT and (b) PICT forms of Nile red.

the solvent effect. Their studies using long-range CAM-B3LYP functional predicted that the dual fluorescence in Nile red arises due to strongly active vibronic coupling, supporting the PICT mechanism. Use of B3LYP functional supported the TICT mechanics, which they found to be due to computational artifact. Robb and coworkers [54] used CASSCF level of theory to understand the ICT mechanism of ABN and DMABN. Their studies have concluded that the charge-transferred state of these molecules need not be twisted. Haas and coworkers [102] studied the ICT process of DMABN and some other benzene derivatives and concluded that the ICT state of a molecule always need not be twisted, as the planar ICT state of a molecule can have lower energy than the LE state, facilitating CT in that molecule. The energy of an intermolecular complex $E(A^-D^+)$ that has weakly interacting donor and acceptor chromophores is calculated using the semiempirical Weller equation, shown in Equation 2.8 [104]. Although this equation is valid in *n*-hexane and for electronically decoupled A and D subgroups, it can be used in the context of TICT, as in the TICT model the donor and acceptor subgroups lie in perpendicular conformation (e.g., in DMABN), leading to electronic decoupling of the donor and the acceptor groups [105].

$$E(A^-D^+) = E(D/D^+) - E(A^-/A) + 0.15 \pm 0.10\ \text{eV} \tag{2.8}$$

where $E(D/D^+)$ and $E(A^-/A)$ are the oxidation and reduction potentials, respectively. The factor 0.15 was determined by comparing the experimental value of $E(A^-D^+)$ with the difference in redox potentials $E(D/D^+)$ and $E(A^-/A)$. Zachariasse *et al.* [105] determined the values of $E(A^-D^+)$ from the experimental enthalpy difference for the ET (ΔH) reaction whether from the excited donor ($^1D^*$) or excited acceptor ($^1A^*$) to (A^-D+) in *n*-hexane at room temperature, so that

$$E(A^-D^+) = E(S_1) + \Delta H \tag{2.9}$$

$E(S_1)$ is the energy of $^1A^*$ or $^1D^*$ in first singlet excited state.

Initial studies on ICT molecules suggested that the formation of the TICT state was driven by dipolar solute–solvent interaction as the TICT state gets more stabilized in polar solvents, owing to high dipole moment than the nonpolar solvents. So, if the statement is true, a molecule that has a large difference in dipole moment between TICT and PICT states will favorably be twisted in the excited state in polar solvents. This also indicates large extent of CT and higher distance between the donor and acceptor in the TICT state than in its PICT counterpart. This theory is not universal as pentacyano-*N*,*N*-dimethylaniline (PCDMA) does not show any sign of ICT even in polar solvents (acetonitrile) even if it has a stronger acceptor than that of DMABN [105]. The pentacyano moiety of PCDMA has much higher electron affinity than that of benzonitrile group of DMABN. So, according to the TICT model (Equation 2.8), the energy of the ICT state of PCDMA should be lower than that of DMABN. Absence of ICT in PCDMA, therefore, raises doubts over the formation of the TICT state. ICT studies on several other molecules, like, 6-propionyl-2-(dimethylamino)naphthalene (PRODAN) [106], planar quinoxaline derivatives [107], 4-*N*,*N*-dimethylamino-4′-cyanobiphenyls (DMABC) [95], and so on, have proved that elongating the acceptor can prohibit the TICT formation, leading to the conclusion that TICT formation is favored in molecules with shorter conjugation pathway, for example, one aromatic ring. A recent report by Zhong [108] proposed that the formation of TICT or PICT state can be explained in terms of the following three parameters: energy gap, hole–electron interaction, and excited state relaxation. The author proposed that the energy of the relaxed first excited state (E_{S_1}) can be expressed as follows:

$$E_{S_1} = E_{\mathrm{G}} + E_{\mathrm{Gap}} + E_{\mathrm{he}} + E_{\mathrm{R}} \tag{2.10}$$

In this expression, E_{G} is the energy of the molecule in the ground state, E_{Gap} stands for highest occupied molecular orbital–lowest unoccupied molecular orbital (HOMO–LUMO) gap calculated at the ground-state geometry, E_{he} is the energy for hole–electron interaction, and E_{R} is relaxation energy of the excited state. Using Equation (2.10), the difference in energy between TICT and PICT states ($\Delta E^{\text{T-P}}$) can be expressed as follows:

$$\Delta E^{\text{T-P}} = \Delta E_{\mathrm{G}}^{\text{T-P}} + \Delta E_{\mathrm{Gap}}^{\text{T-P}} + \Delta E_{\mathrm{he}}^{\text{T-P}} + \Delta E_{\mathrm{R}}^{\text{T-P}} \tag{2.11}$$

A negative $\Delta E^{\text{T-P}}$ value predicts formation of the TICT state, while a positive $\Delta E^{\text{T-P}}$ favors the PICT state. When the energy gap between the HOMO of the donor with the LUMO of the acceptor is small, we can predict a strong interaction between them that in turn increases the energy gap in planar conformation, facilitating the TICT conformation (Figure 2.24). There are several approximations involved in this model and one needs to be careful while predicting the excited-state geometry using Equation (2.11) and comparative studies among several known molecules was advised. Zhong [108] also concluded that while predicting the excited-state geometry one needs to consider the effect of basis set in the calculation, as use of wrong basis set may lead to erroneous results. For example, use of B3LYP and BLYP functionals generally tend to predict TICT state. In general, functionals with less Hartree–Fock (HF) exchange favor the TICT state over the PICT state.

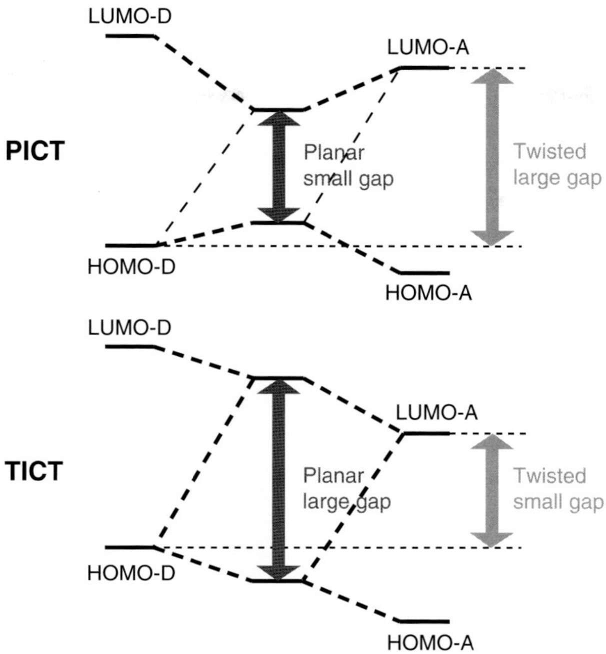

Figure 2.24 A conceptual representation on how frontier orbital interactions determine the energy gap of the twisted or planar state. (Zhong 2015 [108]. Reproduced with permission of Royal Society of Chemistry.)

2.5 Through-Space Charge Transfer

Through-space CT can be defined as CT from a donor to an acceptor part in a same molecule, not in conjugation with each other. Although several studies are reported on intermolecular through-space CT process, reports on intramolecular through-space CT are scarce [109, 110]. The intermolecular through-space CT dictates the properties of many π-stacked molecular systems [109]. Of late, Zhu *et al.* [110] reported intramolecular through-space CT and tunable dual emission in several terfluorene derivatives (Figure 2.25).

2.6 Charge Transfer in Inorganic Complexes

Light-induced ET process in metal complexes has been studied for the past few decades by the scientific community [111]. The major driving force behind studies of CT process in these complexes was to develop an artificial photosynthetic system for conversion and chemical storage of solar energy. The CT process in these molecules can be facilitated by either a direct optical transition or by an excited-state ET that may occur through electronic excitation of the oxidant or the reductant. Natural photosynthesis occurs via a light-induced ET

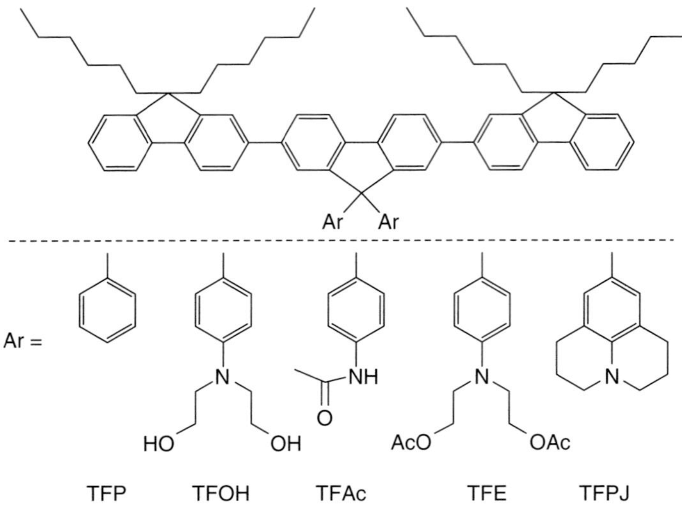

Figure 2.25 The structures of tetrafluorene molecules. (Zhu *et al.* 2010 [110]. Reproduced with permission of Royal Society of Chemistry.)

reaction in which the chlorophyll acts as a key compound. Therefore, to prepare an artificial photosynthetic system, one needs to understand the ET process. Studies of CT in metal complexes provide an opportunity toward this goal. Like the organic molecules, the ET process in the metal complexes can be classified into intramolecular and intermolecular processes. The electronic coupling between the electron donor (D) and electron acceptor (A) is important, as a weak electronic coupling generally leads to transfer of the whole electron from the donor to the acceptor while a strong coupling between the D and A units leads to partial electron delocalization from the donor to the acceptor. In some cases, the primary light-induced ET is followed by a back ET to get back to the initial state; while in others secondary reaction may occur, leading to formation of a stable photoproduct. The photophysics of these reactions have been studied using several steady-state as well as time-resolved spectroscopic techniques.

Direct optical CT occurs due to electronic interaction between metal centers and ligands that are called the redox sites [111]. This process can be classified into several categories. The metal-to-ligand charge transfer (MLCT) and ligand-to-metal charge transfer (LMCT) are the most studied among these, while several studies are devoted to explore the metal-to-metal charge transfer (MMCT), ligand-to-ligand charge transfer (LLCT), and intraligand charge transfer (ILCT) processes. The colors of oxametallets like MnO_4^- and CrO_4^- arise due to LMCT process; otherwise, one may expect their compounds to be colorless due to the d^0 electronic configuration of the metal centers. In several cases, where the metal is reducing and the ligand has empty orbitals, MLCT can take place. In Fe^{II} complexes like $Fe^{II}(CN)_6^{4-}$ MLCT occurs, while in most of the Fe^{III} complexes the LMCT is expected. Apart from direct optical CT, the photoredox process can occur in the excited state. This excited-state ET occurs when a chromophore in the complex gets excited, subsequently changing its reducing or oxidizing strength. In $[Co^{III}(NH_3)_5O_2CR]^{2+}$, for example, where R can be 1,2-napthalene, 4-stilbene, 4-biphenyl, and so on, the electronic coupling between the donor and acceptor is weak. In these compounds, excited-state ET occurs from the singlet (ππ*) state of the R group to the Co^{III} centers [111]. This process facilitates the fluorescence quenching, an indication of the ET process. The excited-state CT in these compounds occurs as excited aromatic groups are strong reductants, while Co^{III} amines are weakly oxidizing.

To counter global warming and climate change due to use of fossil fuel led the scientific community to find some alternative energy sources that are clean, affordable, and abundant. Inspired by nature, researchers are trying to create artificial photosynthetic systems that can be efficiently used to convert solar energy to chemical energy [112]. A typical artificial photosynthetic system includes an antenna system made of several chromophores that converts light energy to electronic energy, a reaction center, and multi-ET catalysts [112]. The role of the antenna is to collect the light energy and funnel it to a specific subunit, called an energy trap through a series of energy-transfer processes. The reaction center consists of the energy traps as well as some electron-donor and acceptor (EDA) subunits. The EDA subunits are coupled to the energy trap so that a sequence of photoinduced electron transfer (PET) can take place at the reaction center that produces a CS state. Therefore, the reaction center converts electronic energy to redox energy. It has been reported that one of the major challenges of designing artificial photosynthetic systems is to build molecular assemblies that have a photoinduced long-lived CS state [113]. The excited CS state corresponds to transient conversion of light into an electrochemical potential that can be used to store energy or electricity production. The catalysts collect the holes and electrons that drive multi-ET process, producing chemical energy. So, using these multistep processes it is possible to prepare high-energy-content chemicals (called fuel) from low-energy-content materials. Production of molecular hydrogen and oxygen through the water splitting process is an important example of this kind of process. There are several reports studying the PET process in inorganic complexes [112–115]. Some of the representative examples are summarized in this chapter, while some other examples can be found in Chapter 3.

Campagna and coworkers studied the PET process in several linearly arranged donor–bridge–acceptor (D-B-A) dyads [114]. They reported the synthesis and PET in D-B-A dyads where Ru(II)- or OS(II) bis(perpyridine) acts as photoactivable electron donor, a biphenylene or phenylene-xynylene fragment as the spacer (or bridge), and expanded (bi)pyridinium subunits as electron acceptors. They reported that replacing the biphenylene bridges with monophenylene is likely to favor the formation of the CS state. They conclude that the less-pronounced electron-donating nature of the monophenylene spacers compared to the biphenylene counterparts allows circumventing of the bridge-assisted mechanism of charge-recombination (CR) process in cases where the photoinduced process leading to formation of CS states is an oxidative ET process.

The same group [115] reported the ET process in several molecular dyads containing ruthenium(II) or osmium(II)-bis-(terpyridine) subunits as the chromophore and expanded pyridinium subunits as electron acceptor (Figure 2.26).

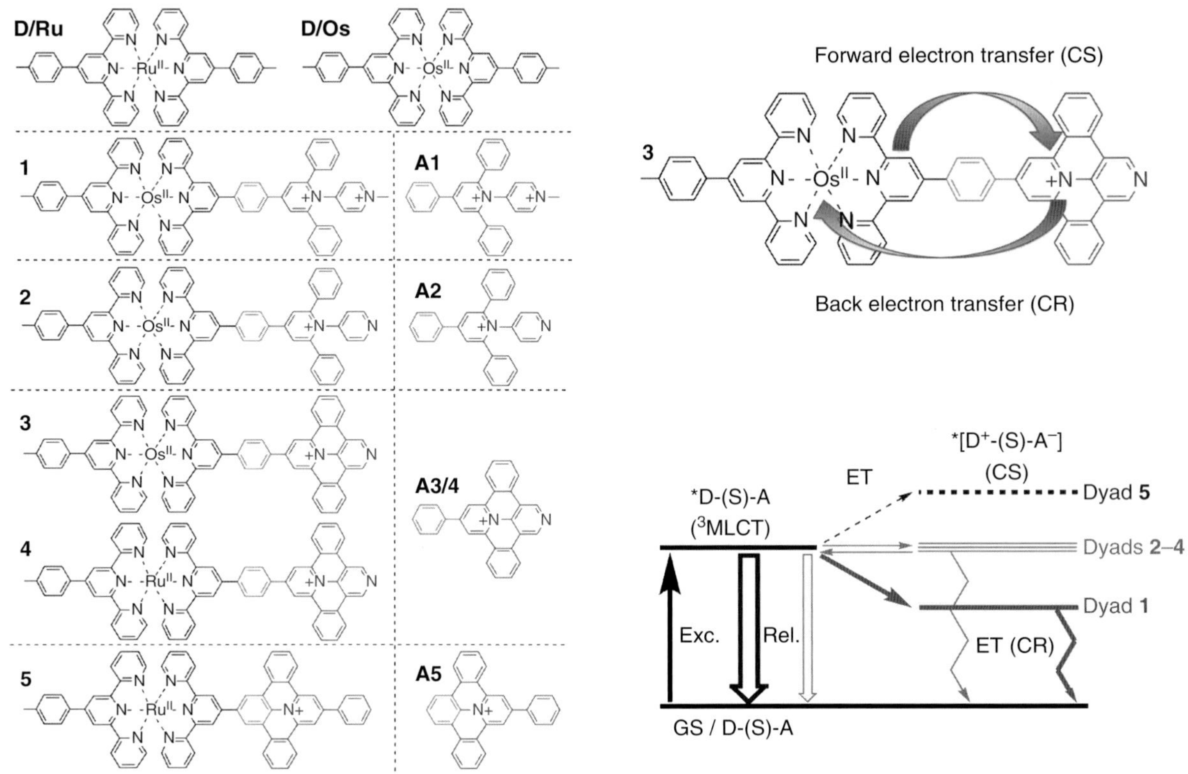

Figure 2.26 The structures of the molecular dyads (1–5) studied by Fortage *et al.* along with their model compounds. The donor (D) and acceptor (A) units are also shown (left). The forward electron transfer (charge separation) and back electron transfer (charge recombination) processes are shown in the right upper panel. The right lower panel shows the schematic presentation of energy levels in dyads (1–5), where GS is the ground state, Exc. and Rel. are excitation, relaxation processes, respectively, and ET is electron transfer process. (Fortage *et al.* 2013 [115]. Reproduced with permission of American Chemical Society.)

In these dyads, a single phenylene moiety is used as spacer. They have designed these dyads such that the driving force of PET is almost zero. That is, the driving force between the initially prepared MLCT excited state to produce [D^+-S-A^-] CT state is close to zero. They have reported that in dyad 1 (Figure 2.26), forward PET and the subsequent CR process occur with time constants about 7 and 5 ps, respectively. The dyads 2–4 have their triplet MLCT and CS states close in energy, and excited state equilibration between these two states are expected to happen. In dyad 5, the PET is thermodynamically forbidden, the triplet MLCT is the lowest excited state for this dyad. Therefore, once the MLCT is formed in dyad 5 after photoexcitation, it directly decays to the ground state. While studying the redox behavior of the aforesaid dyads, the authors found that all these dyads undergo a reversible one-electron oxidation process and several reversible reduction processes that are bielectric in some cases. They ascribed the single-electron oxidation to a metal-centered process and the reduction to ligand-centered processes. They studied the luminescence of dyads 2–5 (Figure 2.26) in acetonitrile at room temperature as well as in butyronitrile rigid matrix formed at 77 K. For Os(II)-based dyads they found that the emission maxima of dyads 2–5 are slightly redshifted vis-à-vis the model species (D/Os) both at room temperature solution and 77 K. The redshift in the Ru(II)-based dyads are more than that in Os(II)-based dyads. Room temperature emission is absent in dyad 1. They assigned the emission of these dyads to the triplet MLCT states of the corresponding dyads. At 77 K, the luminescent lifetimes of the osmium-containing species are in the range of 2–3 μs, while that of ruthenium-containing species are in the range of 10–15 μs. These values are close to their corresponding model species, indicating that no new decay process occurs in these dyads at 77 K with respect to that already present in the model dyads. At room temperature, the emission quantum yield and lifetime of dyads 2 and 3 decreased notably with respect to their model dyads, while these quantities increase for ruthenium-containing dyads. The small bathochromic shift in the emission spectra of dyads 1–3 compared to their model compound (D/Os) was assigned to the perturbation of the MLCT state due to the presence of expanded pyridinium subunits.

The authors calculated the driving force of the photoinduced CT (ΔG^0_{ET}) in these molecules using the equations:

$$\Delta G^0_{ET} = (*\,E_{ox} - E_{red}) + W \quad (2.12)$$

$$*\,E_{ox} = E_{ox} - E^{00} \quad (2.13)$$

where $^*E_{ox}$ is the oxidation potential of the chromophoric electron donor in the excited state and E_{ox} is the oxidation potential of the donor ground states, expressed as one-electron energies in eV. E^{00} is the excited-state energy (MLCT) and the E_{red} stands for the reduction potential of the acceptor. The term W (called, work term) in Equation 2.12 is the difference between Coulombic stabilization energies of reactants and products. The work term functions to reduce the driving force slightly and is often neglected.

The authors used pump-probe TA spectroscopy to study the PET processes in these molecules. They attributed the growing TA of dyad 1–4 at the 500- to

550-nm ranges with a concomitant decrease in the absorption related to the reduced terpyridine-based fragment in the 550- to 700-nm ranges to formation of the CS state. This transient spectrum directly decays to the ground state through a CR process. Their results also suggest that the CS state is formed from the initially prepared MLCT states in these dyads. In dyad 5, the ET cannot take place as this process is endoergic in this dyad. This was confirmed from the TA measurements that do not show any time evolution, unlike dyads 1–4. In fact, the relaxed MLCT directly decays to the ground state on a timescale longer than 3.2 ns. The incompletely quenched or enhanced donor-centered emission in Os(II)- and Ru(II)-based dyads 2–4, with respect to their model compounds, respectively, indicated the CS and CR processes in these dyads were somewhat different than that in dyad 1. The authors argued that the energy levels of MLCT and CS states in dyads 2–4 are so close in energy that thermal equilibration between these states are quite likely to occur. This equilibration needs a back ET from the CS to the MLCT state faster than that from the CR state to the ground state. They also found that the observed rate constants of formation of CS states in these dyads are not the rate constant for the forward ET process alone but summation of forward and backward ET processes. So, it is apparent that the quantum yields and lifetimes of MLCT states of dyads 2–4 are dependent on the equilibration ratios between the equilibrated states.

Adamo and coworkers [113] studied the PET process in several ruthenium and osmium polyads (Figure 2.27) by theoretical means to obtain insight that will help in designing efficient dyads for long-lived CS state. The specially developed photochemical molecular devices (called polyads) are typically consisting of electron-donating (D) and/or electron-accepting (A) units connected to a photosensitizing unit (P). These systems are called "multicomponent systems" in the framework of supramolecular chemistry, where the D, A, and P units are expected to preserve their functional character (electronic properties) even when they are linked together. These units are generally arranged in a linear manner according to D-P-A sequence and are generally held together through covalent yet saturated links, hydrogen bonding, and so on. Through excitation of the P unit, called primary donor, a cascade of ICT can take place, resulting in a *[D^+-P-A^-], state, called the CS state. As one of the major goals is to design efficient dyads to get long-lived CS states, one needs to design a strategy to delay the CR process. In the D-P-A dyads, the metal cation (M) of P allows fine-tuning of the electronic as well as electrochemical properties of the polyads. The Ru(II) and Os(II) complexes of [4′-(*p*-phenyl)]terpyridyl ligand (ptpy) derivatized with an electron acceptor of the triphenylpyridinium type ($H3TP^+$), used by the authors satisfied two major criteria for obtaining long-lived CS state – a linear (rodlike) shape and controlled overall architecture. The authors studied the ground-state properties of the dyad systems (P1A/M) and their model compounds (A and P1/M) in their "native" as well as in the reduced forms. Their results indicate that on one-electron reduction, the conjugation between the P1 and A units is increased. Their results also suggest that in P1A/Os, the photoinduced intramolecular and/or direct optical ET leads to the CS state. They have also considered formation of a hypothetical CS state that can be reached through classical stepwise PET. To generate long-lived CS states and also to

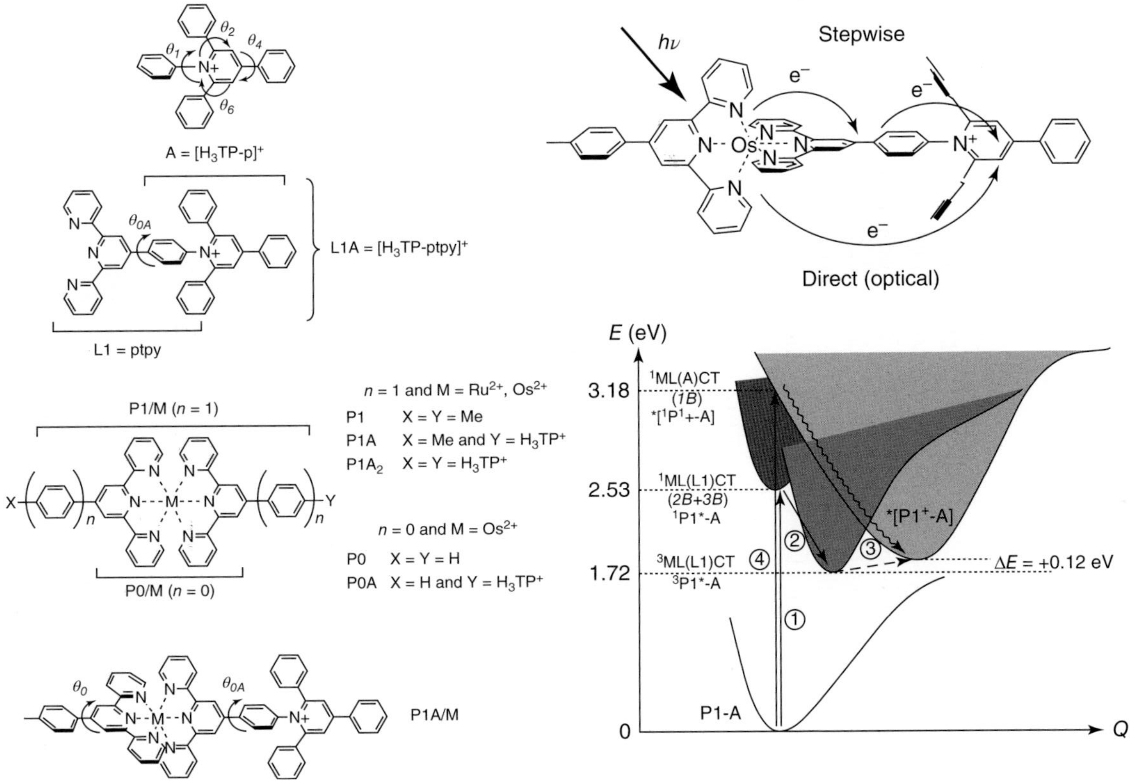

Figure 2.27 The structures of the organic and coordination compounds studied by Adamo and coworkers (left). In the right upper panel, the mechanism of direct and stepwise electron transfer processes are shown, while the right lower panel shows the potential energy diagram of these two pathways. (Ciofini *et al*. 2004 [113]. Reproduced with permission of American Chemical Society.)

overcome the shortcomings caused by residual intercomponent conjugation, they have envisioned several synthetic approaches, including increasing steric hindrance around the intercomponent linkage that may lead to geometrical decoupling, increasing electron-withdrawing strength of the acceptor, and designing redox cascades that are capable of delaying the CR (backward ET) through long-range electron hopping.

2.7 Electron Transfer in Biomolecules

The ET process plays a central role in many biological processes, including energy conversion processes like photosynthesis and respiration [2, 3]. Recent studies by Zigmantas and coworkers [116] suggested that carotenoid to chlorophyll energy transfer in the peridinin-chlorophyll-a-protein (PCP) complex involves an ICT state. The authors employed the femtosecond TA spectroscopic technique to explore the energy transfer pathways in the PCP complex containing the highly substituted carotenoid peridinin. In photosynthetic systems, carotenoids function as light-harvesting antennae, along with their regulatory role in the flow of absorbed energy. Carotenoids transfer the absorbed photon energy to chlorophyll or bacteriochlorophyll molecules that funnel energy to the reaction center where the charge separation takes place [117]. In general, the absorption of carotenoids in the blue-green region of visible spectrum is caused by a strongly allowed S_0 ($1A_g^-$) state to S_2 ($1B_u^+$) state transition [116] and the emission from the S_2 and S_1 states of carotenoids has good overlap with the Q_y and Q_x bands of chlorophyll or bacteriochlorophyll (Figure 2.28). The authors mentioned earlier studies [118, 119] that showed that the lifetime of the lowest excited state of peridinin to be strongly dependent on solvent polarity – the lifetime being decreased with increase in solvent polarity. They attributed this behavior to the ICT of peridinin in the excited state.

Zigmantas *et al.* [116] found that the energy-transfer efficiency in the PCP complex is enhanced by the lowest excited ICT state of peridinin. Therefore, the authors also noted that the energy-transfer process in this complex could be tuned using the polarity of the medium. They proposed that a very high or very low polarity of the medium could be detrimental to the energy-transfer process and a medium of balanced polarity could lead to the maximum energy-transfer efficiency. ET reaction is also known to mediate several enzyme-catalyzed chemical transformations. ET proteins that use oxidation-reduction chemistry to transfer electrons from a donor site to an acceptor site serve an important role in the transport and utilization of cellular energy. For example, the blue copper protein, Azurin, a single copper protein that functions as electron carrier in the bacterial energy conversion systems of many bacteria [120]. The intramolecular ET reactions in Azurin have been studied extensively in solution employing a variety of methods [121]. For example, the ET process between the disulfide radical ($RSSR^-$) and the Cu^{II} center was studied by pulse radiolysis technique and the ET between one of the histidines that was Ru-labeled and the Cu^{II} center was studied with the help of flash-quench technique. The active centers of the ET proteins, like azurin, generally contain transition metal ions. The presence of cofactors like nicotinamide, flavins, and adenine dinucleotide (NAD) also

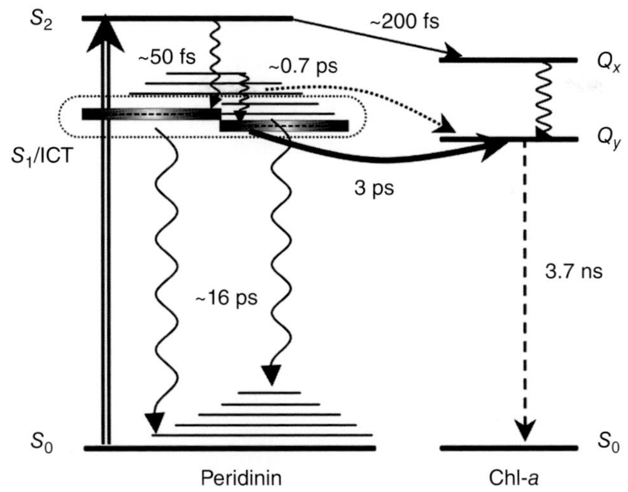

Figure 2.28 Schematic representation of the energy levels and energy-transfer pathways between the peridinin and chlorophyll-a in the PCP complex. Excitation at 535 nm is shown as a double arrow. The wavy arrows and the dashed arrow represent intramolecular relaxation processes and the long-lived chlorophyll-a fluorescence, respectively. The energy-transfer channels are shown as solid arrows (S_2 to Q_x channel, 25%; S_1/ICT to Q_y channel, 63%), while the possible minor energy-transfer channel involving higher vibrational levels of the S_1/ICT state is shown using a dotted line. The corresponding time constant of the processes mentioned are also shown. (Zigmantas *et al.* 2001 [116]. Reproduced with permission of National Academy of Sciences.)

facilitate the ET process in the proteins. A review of ET processes in copper- and iron-containing proteins using pulse electrolysis can be found in Reference [122]. Recently, Ichiye and coworkers studied the intramolecular ET process in [4Fe-4S] ferredoxin, a class of iron-sulfur ET protein, employing molecular dynamics simulations [123]. To minimize the role of solvents in the ET of protein, the electron transport properties are studied in the solid state. Cahen and coworkers [124] summarized the electron transport process in several proteins and compared them with organic molecules. In general, proteins are better conductors than saturated organic molecules, while they are poorer conductors with respect to conjugated molecules. The presence of cofactors enhances the conductance of proteins. The authors mentioned that at high temperature the conduction occurs through hopping and through tunneling at low temperatures, below about 150–200 K. Interested readers may go through a detailed review on the intramolecular ET and electron transport processes in protein reported by Cahen and coworkers [125].

References

1 Grabowski, Z.R., Rotkiewicz, K., and Rettig, W. (2003) *Chem. Rev.*, **103**, 3899.

2 Closs, G.L. and Miller, J.R. (1988) *Science*, **240**, 440.

3 Barbara, P.F., Walker, G.C., and Smith, T.P. (1992) *Science*, **256**, 975.

4 Sedghi, G., Sawada, K., Esdaile, L.J., Hoffmann, M., Anderson, H.L., Bethell, D., Haiss, W., Higgins, S.J., and Nichols, R.J. (2008) *J. Am. Chem. Soc.*, **130**, 8582.
5 Kobori, Y., Yamauchi, S., Akiyama, K., Tero-Kubota, S., Imahori, H., Fukuzumi, S., and Norris, J.R. (2005) *Proc. Natl. Acad. Sci. U.S.A.*, **102**, 10017.
6 Li, G., Josowicz, M., Janata, J., and Semancik, S. (2004) *Appl. Phys. Lett.*, **85**, 1187.
7 Zhao, G.-J., Liu, J.Y., Zhou, L.-C., and Han, K.-L. (2007) *J. Phys. Chem. B*, **111**, 8940.
8 Bulheller, B.M., Miles, A.J., Wallace, B.A., and Hirst, J.D. (2008) *J. Phys. Chem. B*, **112**, 1866.
9 Arzhantsev, S., Zachariasse, K.A., and Maroncelli, M. (2006) *J. Phys. Chem. A*, **110**, 3454.
10 Cao, X., Tolbert, R.W., McHale, J.L., and Edwards, W.D. (1998) *J. Phys. Chem. A*, **102**, 2739.
11 Thar, J., Zahn, S., and Kirchner, B. (2008) *J. Phys. Chem. B*, **112**, 1456.
12 Zachariasse, K.A. (2000) *Chem. Phys. Lett.*, **320**, 8.
13 Rettig, W., Bliss, B., and Dirnberger, K. (1999) *Chem. Phys. Lett.*, **305**, 8.
14 Akemann, W., Laage, D., Plaza, P., Martin, M.M., and Blanchard-Desce, M. (2008) *J. Phys. Chem. B*, **112**, 358.
15 Zilberg, S. and Haas, Y. (2002) *J. Phys. Chem. A*, **106**, 1.
16 Sasaki, S., Drummen, G.P.C., and Konishi, G. (2016) *J. Mater. Chem. C*, **4**, 2731.
17 Edvinsson, T., Li, C., Pschirer, N., Schoneboom, J., Eickenmeyer, F., Sens, R., Boschloo, G., Herrmann, A., Mullen, K., and Hagfeldt, A. (2007) *J. Phys. Chem. A*, **111**, 15137.
18 Zyss, J., Ledoux, I., Volkov, S., Chernyak, V., Mukamel, S., Bartholomew, G.P., and Bazan, G.C. (2000) *J. Am. Chem. Soc.*, **122**, 11956.
19 Xu, J., Liu, X., Lu, J., Zhu, M., Huang, C., Zhou, W., Yin, X., Liu, H., Li, Y., and Ye, J. (2008) *Langmuir*, **24**, 4231.
20 Lehn, J.M. (1999) *Angew. Chem. Int. Ed.*, **29**, 1304.
21 Zyss, J., Ledoux, I., Volkov, S., Chernyak, V., Mukamel, S., Bartholomew, G.P., and Bazan, G.C. (2000) *J. Am. Chem. Soc.*, **122**, 61195.
22 Tsiperman, E., Regev, T., Becker, J.Y., Bernstein, J., Ellern, A., KhodorKovsky, V., Shames, A., and Shapiro, L. (1999) *Chem. Commun.*, 1125.
23 Hill, W.T. and Lee, C.H. (2008) *Light-Matter Interaction: Atoms and Molecules in External Fields and Nonlinear Optics*, John Wiley & Sons, Inc.
24 Turro, N.J., Ramamurthy, V., and Scaiano, J.C. (2010) *Modern Molecular Photochemistry of Organic Molecules*, University Science Books.
25 Birks, J.B. (1970) *Photophysics of Aromatic Molecules*, Wiley Interscience.
26 Birks, J.B. (1975) *Organic Molecular Photophysics*, vol. **2**, John Wiley & Sons, Ltd.
27 Turro, N.J. (1991) *Modern Molecular Photochemistry*, University Science Books.

28 Lakowicz, J.R. (2006) *Principles of Fluorescence Spectroscopy*, 3rd edn, Springer, New York.

29 Chang, C.-W. and Mycek, M.A. (2012) Quantitative molecular imaging in living cell via FLIM, in *Reviews in Fluorescence 2010* (ed. C.D. Geddes), Springer.

30 Geddes, C.D. (2001) *Meas. Sci. Technol.*, **12**, R53.

31 Lin, C., Wu, K., Sa, R., Mang, C., Liu, P., and Zhuang, B. (2002) *Chem. Phys. Lett.*, **363**, 343.

32 Poisson, L., Roubin, P., Coussan, S., Soep, B., and Mestdagh, J.-A. (2008) *J. Am. Chem. Soc.*, **130**, 2974.

33 Mitra, S., Das, R., Bhattacharyya, S.P., and Mukherjee, S. (1997) *J. Phys. Chem. A*, **101**, 293.

34 Misra, R., Mandal, A., and Bhattacharyya, S.P. (2011) *J. Phys. Chem. A*, **115**, 11840.

35 Druzhinin, S.I., Mayer, P., Stalke, D., von Bulow, R., Noltemeyer, M., and Zachariasse, K.A. (2010) *J. Am. Chem. Soc.*, **132**, 7730.

36 Pal, B.K., Samanta, A., and Guchhait, N. (2010) *J. Phys. Chem. B*, **114**, 6183.

37 Cornelissen-Gude, C. and Rettig, W. (1998) *J. Phys. Chem. A*, **102**, 7754.

38 Forster, T. (1939) *Z. Elektrochem. Angew. Phys. Chem.*, **45**, 571.

39 Wagner, N.L., Greco, J.A., Enriquez, M.M., Frank, H.A., and Birge, R.R. (2013) *Biophys. J.*, **104**, 1314.

40 Saha, S.K., Purkayastha, P., and Das, A.B. (2007) *J. Photochem. Photobiol. C*, **8**, 109.

41 Stsiapura, V.I., Maskevich, A.A., Kuzmitsky, V.A., Turoverov, K.K., and Kunetsova, I.M. (2007) *J. Phys. Chem. A*, **111**, 4829.

42 Soujanya, T., Fesseanden, R.W., and Samanta, A. (1996) *J. Phys. Chem.*, **100**, 3507.

43 Rettig, W. (1986) *Angew. Chem. Int. Ed. Engl.*, **25**, 971.

44 Haidekker, M.A., Brady, T.P., Lichlyter, D., and Theodorakis, E.A. (2005) *Bioorg. Chem.*, **33**, 415.

45 Kwok, W.M., Ma, C., George, M.W., Grills, D.C., Matousek, P., Parker, A.W., Phillips, D., Toner, W.T., and Towrie, M. (2007) *Photochem. Photobiol. Sci.*, **6**, 987.

46 Gómez, I. *et al.* (2015) *J. Phys. Chem. A*, **119**, 1983–1995.

47 Schuddeboom, W., Jonker, S.A., Warman, J.H., Leinhos, U., Kunhle, W., and Zachariasse, K.A. (1992) *J. Phys. Chem.*, **96**, 10809.

48 Zachariasse, K.A., Von Der Haar, T., Hebecker, A., Leinhos, U., and Kunhle, W. (1993) *Pure Appl. Chem.*, **65**, 1745.

49 Sobolewski, A.L. and Domcke, W. (1996) *Chem. Phys. Lett.*, **259**, 119.

50 Gustavsson, T., Coto, P.B., Serrano-Andres, L., Fujiwara, T., and Lim, E.C. (2009) *J. Chem. Phys.*, **131**, 031101.

51 Perveaux, A., Castro, P.J., Lauvergnat, D., Reguero, M., and Lasorne, B. (2015) *J. Phys. Chem. Lett.*, **6**, 1316.

52 Segado, M., Gomez, I., and Reguero, M. (2016) *Phys. Chem. Chem. Phys.*, **18**, 6861.

53 Druzhinin, S.I., Ernsting, N.P., Kovalenko, S.A., Perez Lustres, L., Senyushkina, T.A., and Zachariasse, K.A. (2006) *J. Phys. Chem. A*, **110**, 2955.

54 Gomez, I., Reguero, M., Boggio-Pasqua, M., and Robb, M.A. (2005) *J. Am. Chem. Soc.*, **127**, 7119.

55 Ramos, R., Fujiwars, T., Zgierski, M.Z., and Lim, E.C. (2005) *J. Phys. Chem. A*, **109**, 7121.

56 Fuβ, W., Schmid, W.E., Pushpa, K.K., Trushin, S.A., and Yatsuhashi, T. (2007) *Phys. Chem. Chem. Phys.*, **9**, 1151.

57 Fuβ, W., Pushpa, K.K., Rettig, W., Schmid, W.E., and Trushin, S.A. (2002) *Photochem. Photobiol. Sci.*, **1**, 255.

58 Trushin, S.A., Yatsuhashi, T., Fuβ, W., and Schmid, W.E. (2003) *Chem. Phys. Lett.*, **376**, 282.

59 Yatsuhashi, T., Trushin, S.A., Fuβ, W., Rettig, W., Schmid, W.E., and Zilberg, S. (2004) *Chem. Phys.*, **296**, 1.

60 Zachariasse, K.A., Druzhinin, S.I., Bosch, W., and Machinek, R. (2004) *J. Am. Chem. Soc.*, **126**, 1705.

61 Coto, P.B., Serrano-Andrés, L., Gustavsson, T., Fujiwara, T., and Lim, E.C. (2011) *Phys. Chem. Chem. Phys.*, **13**, 15182.

62 Segado, M., Mercier, Y., Gomez, I., and Reguero, M. (2016) *Phys. Chem. Chem. Phys.*, **18**, 6875.

63 Galievsky, V.A., Druzhinin, S.I., Demeter, A., Jiang, Y.B., Kovalenko, S.A., Lustres, L.P., Venugopal, K., Ernsting, N.P., Allonas, X., Noltemeyer, M., Machinek, R., and Zachariasse, K.A. (2005) *Chem. Phys. Chem.*, **6**, 2307.

64 Il'ichev, Y.V., Kühnle, W., and Zachariasse, K.A. (1998) *J. Phys. Chem. A*, **102**, 5670.

65 Demeter, A., Druzhinin, S., George, M., Haselbach, E., Roulin, J.-L., and Zachariasse, K.A. (2000) *Chem. Phys. Lett.*, **323**, 351.

66 Parusel, A.B.A., Kohler, G., and Grimme, S. (1998) *J. Phys. Chem. A*, **102**, 6297.

67 Mishra, A., Sahu, S., Tripathi, S., and Krishnamoorthy, G. (2014) *Photochem. Photobiol. Sci.*, **13**, 1476.

68 Nalwa, H.S. and Miyata, S. (1996) *Nonlinear Optics of Organic Molecules and Polymers*, CRC Press.

69 Slama-Schwok, A., Blanchard-Desce, M., and Lehn, J.-M. (1990) *J. Phys. Chem.*, **94**, 3894.

70 Gofman, I.A. (1970) *J. Appl. Spec.*, **12**, 798.

71 Liu, S., Wan, S., Chen, M., and Sun, M. (2008) *J. Raman Spectrosc.*, **39**, 1170.

72 Dahl, K., Biswas, R., Ito, N., and Maroncelli, M. (2005) *J. Phys. Chem. B*, **109**, 1563.

73 Zachariasse, K.A., Grobys, M., von der Haar, T., Hebecker, A., II'ichev, Y.V., Jiang, Y.-B., Morawski, O., and Kuhnle, W. (1996) *J. Photochem. Photobiol. A*, **102**, 59.

74 Jones, G. II, Jackson, W.R., and Halpern, A.M. (1980) *Chem. Phys. Lett.*, **72**, 391.

75 Jones, G. II, Jackson, W.R., Choi, C.Y., and Bergmark, W.R. (1985) *J. Phys. Chem.*, **89**, 294.

76 Nad, S., Kumbhakar, M., and Pal, H. (2003) *J. Phys. Chem. A*, **107**, 4808.

77 Nath, S., Kumbhakar, M., and Pal, H. (2010) Effect of H-bonding on the photophysical behaviour of coumarin dyes, in *Hydrogen Bonding and Transfer in the Excited State*, vol. **I & II** (eds K.-L. Han and G.-J. Zhao), John Wiley & Sons, Ltd.

78 Vogel, M., Rettig, W., Fiedeldei, U., and Baumgartel, H. (1988) *Chem. Phys. Lett.*, **148**, 347.

79 Zhang, Q., Kuwabara, H., Potscavage, W.J., Huang, S., Hatae, Y., Shibata, T., and Adachi, C. (2004) *J. Am. Chem. Soc.*, **136**, 18070.

80 Tanaka, H., Shizu, K., Nakanotani, H., and Adachi, C. (2014) *J. Phys. Chem. C*, **118**, 15985.

81 Kumar, S., Singh, P., Kumar, P., Srivastava, R., Pal, S.K., and Ghosh, S. (2016) *J. Phys. Chem. C*, **120**, 12723.

82 Yao, L., Pan, Y., Tang, X., Bai, Q., Shen, F., Li, F., Lu, P., Yang, B., and Ma, Y. (2015) *J. Phys. Chem. C*, **119**, 17800.

83 Zhu, H., Li, M., Hu, J., Wang, X., Jie, J., Guo, Q., Chen, C., and Xia, A. (2016) *Sci. Rep.*, **4**, 24213.

84 Pahlavanlu, P., Christensen, P.R., Therrien, J.A., and Wolf, M.O. (2016) *J. Phys. Chem. C*, **120**, 70.

85 Sowmia, M., Tiwari, A.K., and Saha, S.K. (2011) *J. Photochem. Photobiol. A*, **218**, 76.

86 Jana, S., Dalapati, S., Ghosh, S., and Guchhait, N. (2013) *J. Photochem. Photobiol. A*, **261**, 31.

87 Sumalekshmy, S. and Gopidas, K.R. (2004) *J. Phys. Chem. B*, **108**, 3705.

88 Sumalekshmy, S. and Gopidas, K.R. (2005) *Photochem. Photobiol. Sci.*, **4**, 539.

89 Lee, S.-K., Kim, J.O., Shimizu, D., Osuka, A., and Kim, D. (2016) *J. Porphyrins Phthalocyanines*, **20**, 663.

90 Carlotti, B., Benassi, E., Barone, V., Consiglio, G., Elisei, F., Mazzoli, A., and Spalletti, A. (2015) *Chem. Phys. Chem.*, **16**, 1440.

91 Pereira, R.V., Ferreira, A.P.G., and Gehlen, M.H. (2005) *J. Phys. Chem. A*, **109**, 5978.

92 Lu, X., Fan, S., Wu, J., Jia, X., Wang, Z.S., and Zhou, G. (2014) *J. Org. Chem.*, **79**, 6480.

93 Iwanaga, T., Ogawa, M., Yamauchi, T., and Toyota, S. (2016) *J. Org. Chem.*, **81**, 4076.

94 Gong, Y., Guo, X., Wang, S., Su, H., Xia, A., He, Q., and Bai, F. (2007) *J. Phys. Chem. A*, **111**, 5806.

95 Daum, R., Druzhinin, S., Ernst, D., Rupp, L., Schroeder, J., and Zachariasse, K.A. (2001) *Chem. Phys. Lett.*, **341**, 272.

96 Weinkauf, R., Schanen, P., Metsala, A., Schlag, E.W., Bugrle, M., and Kessler, H. (1996) *J. Phys. Chem.*, **100**, 18567.

97 Scalmani, G., Frisch, M.J., Mennucci, B., Tomasi, J., Cammi, R., and Barone, V. (2006) *J. Chem. Phys.*, **124**, 094107.

98 Maus, M., Rettig, W., Bonafaux, D., and Lapouyade, R. (1999) *J. Phys. Chem.*, **103**, 3388.

99 Zachariasse, K.A., Grobys, M., and Tauer, E. (1997) *Chem. Phys. Lett.*, **274**, 372.

100 Chattopadhyay, N., Serpa, C., Perira, M.M., de Melo, J.S., Arnaut, L.G., and Formosinho, S.J. (2001) *J. Phys. Chem. A*, **105**, 10025.
101 Yoshihara, T., Druzhinin, S.I., and Zachariasse, K.A. (2004) *J. Am. Chem. Soc.*, **126**, 8535.
102 Cogan, S., Zilberg, S., and Haas, Y. (2006) *J. Am. Chem. Soc.*, **128**, 3335.
103 Guido, C.A., Mennucci, B., Jacquemin, D., and Adamo, C. (2010) *Phys. Chem. Chem. Phys.*, **12**, 8016.
104 Weller, A.Z. (1982) *Phys. Chem.*, **133**, 93.
105 Zachariasse, K.A., Druzhinin, S.I., Galievsky, V.A., Demeter, A., Allonas, X., Kovalenko, S.A., and Senyushkina, T.A. (2010) *J. Phys. Chem. A*, **114**, 13031.
106 Everett, R.K., Nguyen, A.A., and Abelt, C.J. (2010) *J. Phys. Chem. A*, **114**, 4946.
107 Czerwieniec, R., Herbich, J., Kapturkiewicz, A., and Nowacki, J. (2000) *Chem. Phys. Lett.*, **325**, 589.
108 Zhong, C. (2015) *Phys. Chem. Chem. Phys.*, **17**, 9248.
109 Figueira-Duarte, T.M., Lloveras, V., Vidal-Gancedo, J., Gegout, A., Delavaux-Nicot, B., Welter, R., Veciana, J., Rovira, C., and Nierengarten, J.-F. (2007) *Chem. Commun.*, 4345.
110 Zhu, L., Zhong, C., Liu, Z., Yang, C., and Qin, J. (2010) *Chem. Commun.*, 6666.
111 Vogler, A. and Kunkely, H. (1992) Light induced electron transfer of metal complexes, in *Electron Transfer in Chemistry and Biology* (eds A. Muller *et al*), Elsevier.
112 Arrigo, A., Santoro, A., Puntoriero, F., Laine, P.P., and Campagna, S. (2015) *Chem. Soc. Rev.*, **304**, 109.
113 Ciofini, I., Laine, P.P., Bedioui, F., and Adamo, C. (2004) *J. Am. Chem. Soc.*, **126**, 10763.
114 Fortage, J., Puntoriero, F., Tuyeras, F., Dupeyre, G., Arrigo, A., Ciofini, I., Laine, P.P., and Campagna, S. (2012) *Inorg. Chem.*, **51**, 5342.
115 Fortage, J., Dupeyre, G., Tuyeras, F., Marvaud, V., Ochsenbein, P., Ciofini, I., Hromadova, M., Pospisil, L., Arrigo, A., Trovato, E., Puntoriero, F., Laine, P.P., and Campagna, S. (2013) *Inorg. Chem.*, **52**, 11944.
116 Zigmantas, D., Hiller, R.G., Sundstrom, V., and Polivka, T. (2002) *Proc. Natl. Acad. Sci. U.S.A.*, **99**, 16760.
117 Frank, H.A. and Cogdell, R.J. (1996) *Photochem. Photobiol.*, **63**, 257.
118 Bautista, J.A., Connors, R.E., Raju, B.B., Hiller, R.G., Sharpless, F.P., Gosztola, D., Wasuelewski, M.R., and Frank, H.A. (1999) *J. Phys. Chem. B*, **103**, 8751.
119 Zigmantas, D., Polivka, T., Yartsev, A., Hiller, R.G., and Sundstrom, V. (2001) *J. Phys. Chem. A*, **105**, 10296.
120 Amdursky, N., Precht, I., Sheves, M., and Cahen, D. (2013) *Proc. Natl. Acad. Sci. U.S.A.*, **110**, 507.
121 Zhao, J.W., Davis, J.J., Sansom, M.S.P., and Hung, A. (2004) *J. Am. Chem. Soc.*, **126**, 5601.

122 Farver, O. and Pecht, I. (2007) *Prog. Inorg. Chem.*, **55**, 1.

123 Tan, M.-L., Dolan, E.A., and Ichiye, T. (2004) *J. Phys. Chem. B*, **108**, 20435.

124 Amdursky, N., Marchak, D., Sepunaru, L., Precht, I., Sheves, M., and Cahen, D. (2014) *Adv. Mater.*, **26**, 7142.

125 Bostick, C.D., Mukhopadhyay, S., Precht, I., Sheves, M., Cahen, D., and Lederman, D. (2017) *ArXiv*, arXiv:1702.05028.

3

New Theoretical and Experimental Techniques for Studying the ICT Process

3.1 Introduction

Intramolecular charge transfer (ICT) is one of the primary modes of relaxation of energy and chemical transformation in an electronically excited state of an organic molecule following excitation by light (photoexcitation). Before photoexcitation, the molecule is in the minimum energy configuration or the nuclear arrangement of the ground electronic state in which net forces acting on each atom is zero and the entire system of the electrons and nuclei are held in equilibrium. In an ICT molecule, the photoexcitation is often marked by extensive redistribution of electron density within the candidate molecule, which often produces a large dipole moment in the excited state (much larger than the ground-state dipole moment). The large-scale redistribution of electronic charge density alters forces acting on different atoms, forcing the molecule to assume a new configuration at which the net force acting on each atom again becomes zero. The relaxed excited-state geometry may therefore be quite different from the ground-state molecular geometry. Let us note that the electronic transition takes place quite rapidly, while the much more massive nuclei move slowly and assume the new configuration after a while (in picoseconds). If an energy barrier separates the two configurations of the molecule, namely, the initially excited one (the Franck–Condon (FC) state) and the energetically relaxed one, the passage from one form to the other can be visualized as a rate process or a chemical reaction the kinetics of which is shaped, among other things, by the height and the location of the barrier. If the ICT molecule is not in the gas phase and in isolation but immersed in a polar medium (a solvent), the large dipole moment of the molecule in the charge-separated state may cause further stabilization of the relaxed ICT state by macroscopic solvent polarization. Thus, depending on the barrier height and location, temperature, and the strength of stabilizing interactions with the solvent, an ICT molecule may be expected to display, at least theoretically, two emissions – one from the unrelaxed excited state (normal or the short-wavelength emission) and another from the relaxed ICT state (longer wavelength ICT emission). Of course, the rate of relaxation from the unrelaxed to relaxed excited state must be fast enough to compete with the normal mode of deactivation of the initially excited state of the ICT molecule by emitting radiation. A class of molecules in which we may expect to

Intramolecular Charge Transfer: Theory and Applications,
First Edition. Ramprasad Misra and S. P. Bhattacharyya.

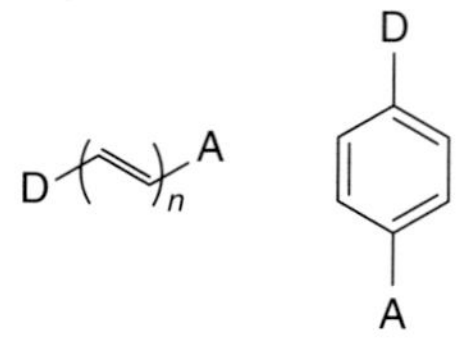

I II

Figure 3.1 Two representative ICT molecules: (I) a linear polyene and (II) an aromatic donor–acceptor molecule.

see "the dual emission" comprises of molecules in which a donor group (D) and an acceptor group (A) are linked by a mobile π-electron network at its opposite ends (see I and II in Figure 3.1 as examples of possible candidate molecules.

Indeed, the phenomenon of dual emission was first observed by Lippert *et al.* [1–3] in *para-N,N* dimethylaminobenzonitrile (DMABN) – a simple example of ICT molecule of type II displayed in Figure 3.1 with D = $—N(CH_3)_2$ and A = —CN. DMABN at room temperature shows two fluorescence bands – a normal fluorescence band that is expected of benzene derivatives and an abnormal ICT band of much lower energy. The origin of the abnormal or the so-called ICT emission was debated for a long while and the most acceptable (apparently) and plausible explanation of its occurrence was advanced by Grabowski *et al.* [4]. Their proposition that in the excited electronic state, called the locally excited (LE) state, the $—N(CH_3)_2$ group twists out of the plane of the ring by 90° stabilizing the charge-separated state (the ICT state) by decoupling the two interacting π-systems, caught the fancy of the workers in the field and became known as twisted intramolecular charge-transfer (TICT) theory of dual emission in DMABN and related molecules. The basic idea that there are two excited states, the LE and the ICT states, the later possessing a high dipole moment and therefore has a large stabilizing energy in polar media constitutes the two-state theory of dual emission, although this theory is not directly related to DMABN. It is no wonder that many computational investigations have been carried out to confirm or reject the basic two-state model. The emphasis of these computational investigations has been on the following aspects:

(i) Is twisting the only important mode of relaxation or are there other equally if not more important modes shaping the ICT dynamics?
(ii) Are there only two or are there more than two states participating in the ICT state formation and the dynamics of charge transfer (CT)?
(iii) What role does the solvent or medium play in the ICT phenomenon? Is macroscopic solvent polarization the dominant process responsible for imparting enhanced stability to the ICT state or does specific solvation of the ICT state with the solvent molecules (e.g., hydrogen bonding in the excited state) have an equally important bearing on the phenomenon?

In addition to these questions (which have been sought to be answered by detailed computational calculation in the gas phase or in solution), several theoretical questions have been investigated in connection with the phenomenon of ICT. One of these concerns the connection, if any, between the photoinduced ICT state and molecular conductance on one hand and electrochemical ICT rate and molecular conductance on the other, assuming that the same bridging unit (the mobile π-electron network or the conduit) mediates in transporting the charge from the donor (D) to the acceptor (A) in the photoinduced ICT process (in the D-π-A system) and in the electrochemical charge transporter in which D and A are each replaced by metal electrodes M (the system here is M-π-M). Theoretical development in this area has sharpened our understanding of the basic nature and mechanism of the long-distance CT process witnessed in ICT molecules involved in a host of phenomena. A third significant area of theoretical investigation pertaining to the ICT phenomenon focuses on the electromagnetic (EM) consequences of the process of CT. If the charge (electron) indeed moves from the donor to the acceptor unit through the bridging unit and is accelerated, Maxwell's theory demands that the moving charge would emit EM radiation of appropriate wavelengths. If such emission indeed takes place and is detectable, it can be leveraged to understand the ICT process directly even if the system (D-π-A) is in vacuum. If the accelerated charge indeed emits radiation, it becomes a probe to understand and monitor the ICT dynamics directly. It turns out that such a process indeed takes place and has led to the development of what has become known as "terahertz" (THz) spectroscopy of ICT states. There are a few other theoretical aspects; of these, two deserve attention and are taken up briefly. Let us first take stock of the computational studies on the ICT molecules and the picture that emerges from them. The early theoretical work in this field has been critically examined and we refer the readers to the two authoritative reviews for a reference. Our emphasis is more on contemporary works to understand the present trend.

3.2 Computational Studies on ICT

Initial computational studies [5, 6] on the energy levels of DMABN as a function of the torsional coordinate (the reaction coordinate of the ICT "reaction") and the experimental findings led to the construction of approximate ground- and excited-state potential energy surfaces (PESs) of DMABN as displayed in Figure 3.2.

The structure of the PES lends credence to the basic two-state model of ICT fluorescence that has been proposed, although it has been criticized [7] on the ground that DMABN does not exhibit dual fluorescence at all in the gas phase, but requires a polar medium for doing so. It may be noted that polar macroscopic solvation can be accounted for by performing the calculations in the reaction field of the solvent and the solvent-modified PES also confirms the basic validity of the two-state model of TICT in DMABN and related fluorophores. A slightly different model appears to emerge from systematic theoretical studies undertaken by Sobolewki and Domcke [8] on a series of benzonitrile molecules, including DMABN at the Hartree–Fock (HF) configuration interaction (CI) among states singly excited (configuration interaction singles (CIS)) with respect to HF. Complete active space self-consistent field (CASSCF), CASPT2 – multiconfiguration second-order perturbation theory – takes the CASSCF wave function as reference function. Geometry optimization of the ICT state predicted the planar structure of all the aminobenzonitrile, 4-aminobenzonitrile (ABN), and

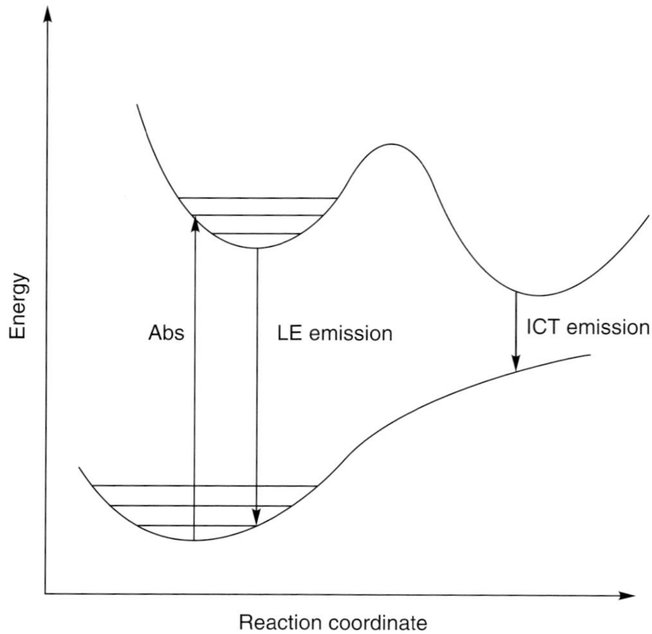

Figure 3.2 The potential energy diagram showing the intramolecular charge-transfer process.

DMABN in the ICT state with a bent —C≡N group [8]. The authors suggest that flow of electronic charge from the donor to the acceptor (—C≡N) π-system causes valence electrons on the nitrile carbon atom to undergo rehybridization (sp to sp^2) and consequently there is bending of the —CN moiety. It is this bending of the nitrile function and not the twisting of the amino function that is, according to the authors, responsible for stabilizing the ICT state. It turns out that the LE states in these molecules have strong singlet $\pi\pi^*$ character, while the ICT state has a dominant $\pi\sigma^*$ nature. They belong to two different irreducible representation of the C_s point group. Their potential energy curves are predicted to cross when examined along the —CN bending coordinate. The states may be expected to get coupled under the out-of-plane distortion produced by torsion or pyramidalization of —NH_2, with the crossing turning into an avoided crossing with the appearance of multidimensional conical intersection of the PESs. These theoretical investigations do not take into account the role of the polar medium, which has been experimentally claimed to be essential to observe the strongly redshifted ICT emission from DMABN in addition to the normal emission. Curiously, the proposed theoretical model of Sobolewki and Domcke is very much like the conical intersection model proposed by Rettig much earlier [3] for DMABN, except that the roles of bending and torsion are reversed.

Later, the same authors [9] carried out *ab initio* electronic structure calculations at the same level to characterize the ICT process in benzethyne (BE), 4-aminobenzethyne (ABE), and 4-dimethylamino benzethyne (DMABE). Full geometry optimization of the CT state predicted the lowest excited state to have a planar configuration with a *trans*-bent ethylene group in BE, ABE, and

DMABE [9]. The bending of the ethyne moiety appears to stabilize the CT states significantly in all the aforesaid molecules of the BE series and has roots in the rehybridization of the carbon atom of the ethyne group triggered by the transfer of electronic charge density to the ethyne acceptor group. More significantly, the authors found that the relative energies of the LE and relaxed excited states computed at the CASPT2 level predict exothermicity of the excited-state CT process in the ABEs even when isolated gas-phase molecules are considered. The observation was in sharp contrast with what was well known for DMABN – that there is no ICT emission unless the molecule interacts with a sufficiently polar environment. The ABEs were therefore theoretically projected as probes for experimentally observing photoinduced dual emission (LE + ICT) even in gas phase without the participation of polar solvent molecules. The claim was rejected by Zachariasse *et al.* [10] on the basis of their experimental findings. The "theory-experimental" controversy was finally settled when it was found experimentally [11] that the LE state in *p*-(dimethylamino)benzythene undergoes inter-system crossing (ISC) in a less polar environment, which is the most dominant non-radiative mode of deactivation while in media of higher polarity [$E_T(30) \geq 44$], the passage into the non-fluorescent CT state serves as the main non-radiative mode of deactivation. The exothermicity of the ICT formation process in DMABE observed theoretically in the gas phase therefore cannot guarantee dual emission.

Daum *et al.* [12] analyzed the gas-phase fluorescence excitation spectrum of 4-(diisopropylamino) benzonitrile (DIABN) and related compounds and established that they emit only from the ICT states. That is, there is no dual emission in the gas phase. It appears that the formation of the ICT state is fast and serves as an efficient channel of deactivation that stalls the LE emissions completely. DMABN in such a situation emits only from the LE state. Zachariasse *et al.* [13, 14] proposed a planar intramolecular charge-transfer (PICT) model in which the ICT state is assumed to have a planar disposition, while the LE state is theorized to retain a part of the pyramidality of the donor (dimethylamino) group present in the ground state (GS). In the PICT model, the energy gap $\Delta E(S_1, S_2)$ between two close-lying excited states S_1 and S_2 with $S_2 > S_1$, plays a crucial role in shaping the dynamics of the ICT state formation. A lower energy gap [small $\Delta E(S_1, S_2)$] favors fast formation of the ICT state. Thus, DIABN shows dual emission in hydrocarbon (alkane) solvents, but not in the gas phase. On the other hand, DMABN emits only from the LE state in such a situation as well as in the gas phase. In DMABN, $\Delta E(S_1, S_2)$ is high in the gas phase and also in hydrocarbon solvents, preventing ICT state formation and blocking dual emission. $\Delta E(S_1, S_2)$ in DIABN is much lower in the gas phase, favoring fast ICT formation in the gas phase and therefore allowing only ICT emission. In hydrocarbon solvents, the $\Delta E(S_1, S_2)$ gap in DIABN is slightly higher so that LE emission is not completely eliminated. In contrast with DMABN, the diisopropylamino group in DIABN is more bulky, which reduces the pyramidicity of the N-atom and prevents twisting.

The dynamics of photoinduced ICT state formation is an important feature of ICT molecules that must be clearly understood for efficient applications of solar energy for technological purpose on one hand and for exploiting the photosynthetic process of food production in plants, on the other. The primary

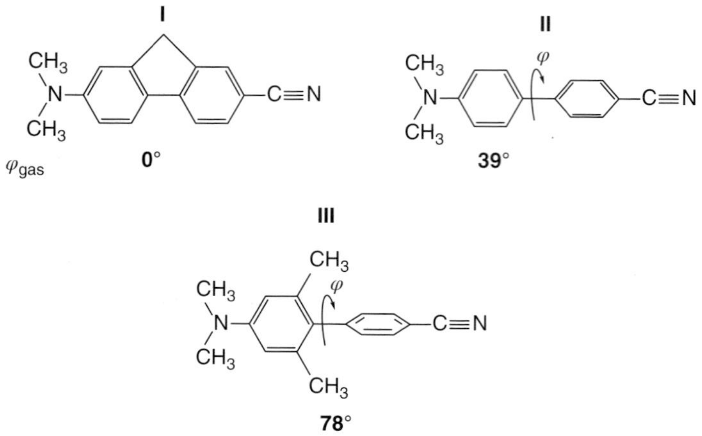

Figure 3.3 Chemical structures of the donor–acceptor biphenyls (I–III), investigated by Rettig and coworkers. The corresponding equilibrium ground-state twist angles (ϕ_g) obtained from AM1 calculations are also shown. (Maus *et al.* 1999 [15]. Reproduced with permission of American Chemical Society.)

step in the conversion of solar energy into electrical potential or chemical energy is photoinduced energy transfer or CT. To make optimal use of the photoconversion process, the finer details of primary steps involving forward and backward CT must be clearly understood. An important question concerning the evolution of the ICT state is whether the nuclear dynamics unleashed by absorption of light favors formation of a less or more conjugated structure. (We have already witnessed the TICT versus PICT debate in DMABN and related molecules.) Interesting examples of the evolution of nuclear dynamics in strongly coupled donor (D) and acceptor (A) systems are provided by biphenyls and D–A biphenyls (Figure 3.3) [15]. In unsubstituted biphenyls, we can neglect D–A interaction altogether. The GS (S_0) of biphenyl in solution has an intra-annular twist angle ϕ 15°–45°, implying a nonplanar structure. In the first excited state (S_1), the molecule assumes a planar structure. It has been established that the S_1 state is a short-axis ^1L_a state with rather a low rate of radiative (fluorescence) decay ($K_f \sim 0.01\,\text{ns}^{-1}$). Both steady-state and time-resolved fluorescence spectroscopy of a series of D–A-substituted biphenyls (I–III; Figure 3.3), supplemented by a ϕ-dependent CNDO/S calculations appear to suggest that the first excited singlet state ($^1S_1$) is an ICT singlet state of ^1L_a type that is emissive in nature.

The formation of this state is accompanied by migration of electronic charge density from the dimethylaminobenzene subunit to the cyanobenzene (acceptor) moiety irrespective of what the polarity of the medium or the intra-annular twist angle (ϕ) are. As observed in the D–A-substituted biphenyl (I, Figure 3.3) in which the D and A groups are restricted structurally, the more flexible D–A biphenyls (II), too exhibits only a weak dependence of fluorescence rate constant ($\langle k_f \rangle \sim 0.4$–$0.6\,\text{ns}^{-1}$) on the polarity of the medium, implying planarity in the excited state of II. The D–A biphenyl (III), which has a strongly pretwisted structure in the S_0 state, displays I- and II-like behavior in nonpolar solvents with $\langle k_f \rangle = 0.3\,\text{ns}^{-1}$, indicating that at least a partial relaxation of the excited state has taken place toward planarity. In media of higher polarity, $\langle k_f \rangle$ goes down to values as low as $0.03\,\text{ns}^{-1}$, implying full relaxation to planarity in the excited CT state of III. It is curious to note that III is associated with two relatively long fluorescent lifetimes (>200 ps) with a clear precursor–successor relationship, which suggests that a fast equilibrium between a more planar and a more twisted rotamer distribution takes place in the single ICT state.

A hemocyanine dye (Figure 3.4) has been theoretically probed to decipher the role of solvent polarity in the formation of the TICT state by rotation of one part of the molecule around different chemical bonds present in the molecule [16].

In Figure 3.4, ϕ_1, ϕ_2, ϕ_3, and ϕ_4 refer to the four different internal modes of rotational or torsional motion that can lead to TICT state formation. The ground-state properties were calculated by the AM1 method available in the GAMESS quantum chemical software. A cylindrical solvation model [17] was used to take care of the macroscopic solvation. The excited-state energies were computed by invoking the INDO/S method at the linear variational (CI) level of calculations in which all singlet configurations generated by exciting one electron from the top 10 highest occupied molecular orbitals (HOMOs; i.e., HOMO to HOMO-9) to 10 lowest unoccupied molecular orbitals (LUMOs; i.e., LUMO to LUMO+9). The excited-state energies E_i(exc) calculated by combining the INDO/S excitation energies with ground-state energy at AM1

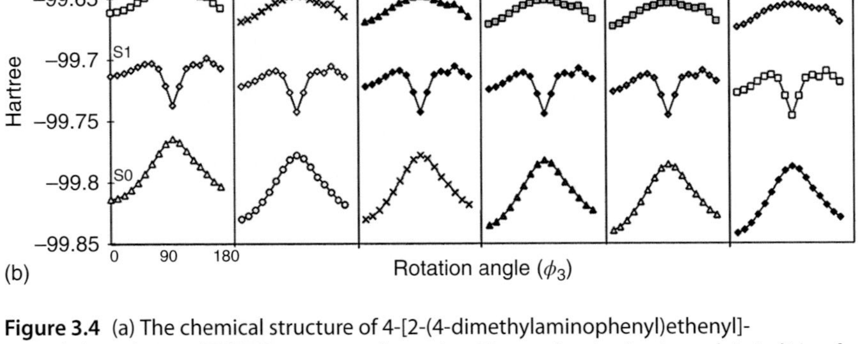

Figure 3.4 (a) The chemical structure of 4-[2-(4-dimethylaminophenyl)ethenyl]-1-methylpyridinium (HR) in its trans conformation. The angles ϕ_1, ϕ_2, ϕ_3, and ϕ_4 indicate four possible rotations inside the molecule (see text). (b) Energy of the ground (S_0), first (S_1), and second excited (S_2) states of HR as a function of rotation about the central double bond for different solvent dielectric constants. The numbers a|b|c on the top denote the bulk solvent dielectric constants, dielectric constants of the subshells, and the dielectric constants of the inner solute shells, respectively. (Copyright 2001 American Chemical Society; Reproduced with permission from Ref. [16].)

level of approximation and adding solvation energies give

$$E_i(\text{exc}) = (E_i - E_0)_{\text{INDO/S}} + (E_0)_{\text{AM1}} + E_{\text{sol}}$$

The calculations reveal [16] that the TICT state can be formed in polar solvents by rotating (ϕ_1) the dimethylamino group to the perpendicular position. The probability of such a process taking place is however rather small due to the high barrier to the rotation in the excited state. The rotation of the pyridyl ring (ϕ_4) and the aniline ring (ϕ_2) enhances CT in the S_1 state; the energy barrier to rotation of the pyridyl ring in the S_1 state is small, while the barrier to rotation of the aniline ring in the excited state practically vanishes in a polar solvent. There is a high energy barrier to rotation about the central double bond (ϕ_3) and the calculations ascribe the low photoisomerization yield to the high energy barrier to rotations about the relevant double bond.

Dreyer and Kummrow [18] theoretically analyzed the femtosecond transient infrared (IR) spectrum of DMABN and ABN using the *ab initio* CASSCF method in an effort to determine the molecular structures of the excited state involved in the formation of the ICT states of the aforesaid molecules. The *ab initio* CASSCF calculations were the first such in the given context, with all states under consideration being fully optimized. The calculations probed two nuclear configurations in the LE state: one with a novel pyramidal structure, the other being planar in disposition. The authors considered several pathways for the formation of the ICT state, namely, TICT, pseudo-Jahn-Teller ICT (PICT), and the rehybridized intramolecular charge transfer (RICT). The calculations provide enough evidence to suggest that the LE state is pyramidal in ABN but planar in DMABN. As to the ICT state formation, the calculations favor the TICT route in DMABN, but no compelling evidence is provided. The authors hope that eventual resolution of the yet-to-be resolved phenylamino stretching frequency would settle the issue finally. One notes here that the suggested picture may be appropriate only for gas-phase observations as solvent corrections have not been attempted.

An interesting ICT molecule (Figure 3.5) that has been investigated by time-resolved ultrafast visible (electronic) pump–vibrational (IR) probe-based transient absorption (TA) measurement [19] to shed light on the excited-state structure is the 4-dicyanomethylene-2-methyl-6-(*p*-(dimethylamino)styril)-4H pyran molecule (DCM) and its isotopomer DCM-d6. The experiments have been conducted in acetonitrile and dimethyl sulfoxide (DMSO) solvents. The measurements indicate that the dynamics of the ICT process involves two steps. The first step concerns charge separation with electronic charge density moving to the dicyanomethylene group. The second step involves a structural evolution of the dimethylamino group. Theoretical studies of vibrational spectrum and effects of isotopic substitution were used to draw a definite conclusion about what happens after charge separation. The vibrational spectrum computed by

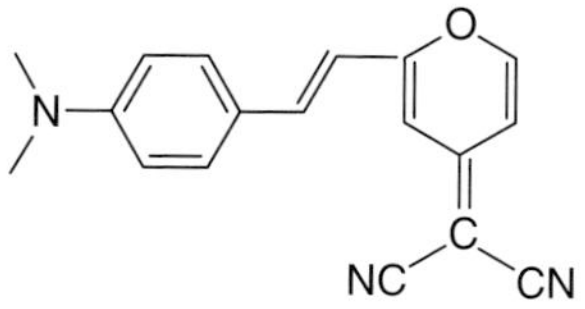

Figure 3.5 The molecular structure of DCM.

density functional theory (DFT) with B3LYP functional and STO 6-31G(d) basis set appears to suggest that the dimethylamino group twists out of the plane, stabilizing the charge-separated species formed in the first step. The basic idea of the TICT model is apparently vindicated in the ICT molecule (DCM).

Earlier, Marguet *et al.* [20] undertook electronic structure calculation on DCM by invoking the CS-INDO method and predicted the existence of an excited state with a high dipole moment that matches with the experimental data. The authors pointed out that in the excited state referred to, DCM assumes a configuration in which the dimethylamino group is perpendicular to the aromatic ring, again asserting the validity of the TICT mechanism. A recent and state-of-the-art electronic structure calculation on DMABN by Robb and coworkers [21] adds a new dimension to the ICT dynamics of DMABN. These authors have tended to conclude that optical excitation initially populates the S_2 state. Radiation-less decay then takes place via the seam of a conical intersection whereby both the LE and ICT states are directly populated, with equilibration eventually taking place on the S_1 surface. The idea of conical-intersection-mediated excited-state dynamics is rather new and rapidly growing into an area of research that can potentially revolutionize our understanding of organic photochemistry or chemistry of the excited state. Calculations by Robb and coworkers on DMABN have raised the hope that it would be soon possible to extend the calculations to investigate the evolution of vibrational spectra of molecules like DMABN or DCM or similar ICT molecules that display a complex interplay of electronic structure and nuclear dynamics.

Zhang *et al.* [22] recently exploited polarization-resolved UV pump–mid-IR probe spectroscopy backed up by time-dependent density functional theory (TDDFT) calculations to investigate the dynamics of CT-induced intramolecular rotation in julolidine malononitrile (JDMN, Figure 3.6).

The dynamics of photoisomerization has been probed with both symmetric and asymmetric C—N stretch of the malononitrile group. Their measurements revealed that the excited electronic state (S_1) relaxes with a time constant of 12.3 ps by rotation about both the C—C single (b) and C—C double bonds (c) of the malononitrile group with a branching ratio of 1 : 5. The isomerization about the "b" bond (Figure 3.6) leads to the formation of a metastable excited state. The rotational isomerization about the C=C bond (c in Figure 3.6), on the other hand, leads to excited-state quenching via a conical intersection between the S_1 and S_0 electronic states. The electronic and nuclear structures of the long-lived excited state as measured by pump–probe anisotropy measurements

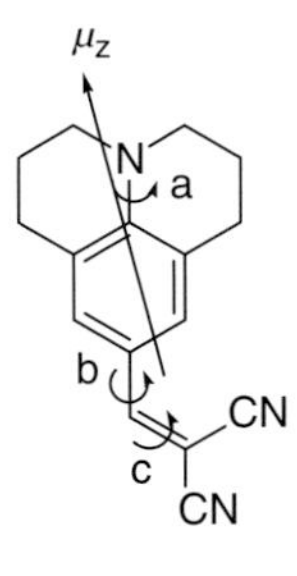

Figure 3.6 The chemical structure of *para*-julolidine malononitrile (JDMN). The $S_0 \rightarrow S_1$ electronic transition dipole and the torsional angles that are involved in the S_1 electronic excited-state relaxation dynamics are also shown.

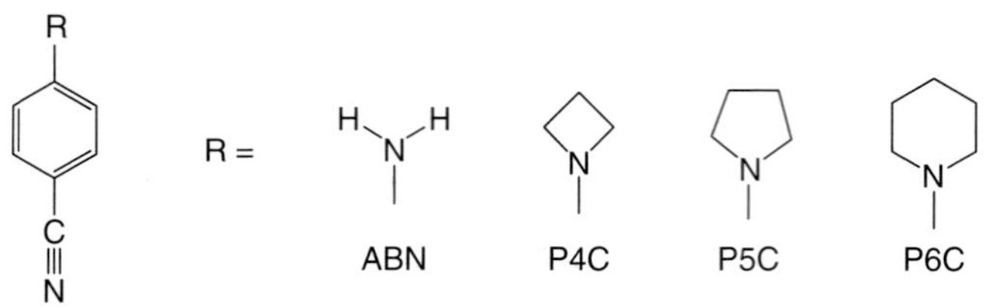

Figure 3.7 The chemical structures of the molecules studied by Dahl *et al.* (Dahl *et al.* 2005 [23]. Reproduced with permission of American Chemical Society.)

and TDDFT calculations using the CAM-B3LYP functional with 6-31G(d,p) basis sets tend to confirm that the isomerization about the malononitrile single bond (b, Figure 3.6) forms a TICT excited state with a dipole moment 6.2 Debye more than that of the S_1 state and originates from CT from the julolidine π-system to nitrile π-system, the two π-systems being decoupled by rotation about the C=C bond (c in Figure 3.6).

Solvent dependence of spectra and kinetics of excited-state CT in several alkylamino benzonitriles (Figure 3.7) was investigated both experimentally and theoretically by Dahl *et al.* [23]. While the experimental tools included steady-state absorption and emission spectroscopic measurements and characterization of emission delay kinetics, the theoretical tools used were *ab initio* quantum chemical calculations on the GS at the restricted Hartree–Fock (RHF)/6-31G(d,p)/MP2/6-31G(d) levels. Excited-state calculations were carried out at the semiempirical AM1/CI level. The effects of solvation were accounted for in some of the excited-state calculations using the CDSMO self-consistent reaction field (SCRF) approach. The ultimate goal of the investigation was to learn how to model the solution-phase reaction potentials of the solutes studied and to assess more completely how CT is controlled by the static and dynamic solvent effects. As the first step toward the realization of the goal, the authors assessed the solvent dependence of the steady-state spectra of the molecules and attempted to fit the main characteristics of the spectra to a rather simplified two-dimensional model of the solution-phase surface. The new feature of the work can be identified with the attempt to directly fit such a model to the experimental data with the help of theoretical calculations. The authors found that the dielectric continuum treatment describes the solvent dependence of the spectral frequencies reasonably well and produces state-specific dipole moments in good agreement with values predicted by semiempirical quantum chemical calculations. The model failed, however, to reproduce both frequency and intensity data.

Ma *et al.* [24] undertook picosecond time-resolved resonance Raman and TA spectroscopy combined with *ab initio* geometrical and vibrational analysis of ABN and DMABN molecules in the ground as well as excited states. The authors claim that both the molecules have planar structures in the LE states. The pyramidal conformation of the molecules in the GS is predicted to flatten out along the inversion coordinate, while the phenyl ring expands and the Ph—N bond shortens in the process. Partial transfer of electron density from the nitrogen lone pair orbital to the ring π* orbital was indicated by their observation of frequency downshift of the C—C stretching mode of the phenyl ring. A downshift of 30 cm^{-1} in the frequency of the C—N stretching mode was also observed. These shifts are corroborated by the changes in the pattern of charge distribution upon excitation. Picosecond time-resolved resonance Raman spectra of the ICT state (A-state) of DMABN, the methyl-deuterated analog (DMABN-d_6) and ^{15}N isotopic substitution (DMABN-^{15}N) were analyzed, leading to the identification of the frequency of the ν(Ph-N) mode. A 96 cm^{-1} downshift in the ν(Ph-N) rules out the PICT mechanism for the formation of the ICT state and firmly favors the TICT or electronic decoupling mode of the formation of the ICT state. A residual pyramidal character of the conformation of the amino group appears to be present in the A-state.

Peridinin is a light-harvesting carotenoid [25]. It is a pigment associated with the chlorophyll that occurs in the pyridinin-chlorophyll protein light-harvesting complex in dinoflagellates. Wagner *et al.* investigated the nature of the ICT state in peridinin on the basis of experimental and theoretical evidence. The evidence tends to support the theory that the ICT state of peridinin is an evolved state formed by an excited-state bond order reversal and solvent relaxation or reorganization. The ICT state evolves in less than 100 fs and possesses a large dipole moment ($\mu \sim 35$ D). The CT character evolves from a shift of electron density within the polyene chain. There appears to be no participation of molecular orbitals (MOs) localized in either of the β-rings in the CT process [26]. Rather, charges move from the allenic side of the polyene into the furanic ring region with concomitant reversal of bond orders in the central part of the polyene chain. These authors argue that the electronic properties of the ICT state are generated via mixing of the $1^1Bu_1^+$ ionic state with the lowest lying 2^1Ag^- covalent state. The state that forms from this ionic-covalent mixing is dominantly $^1Bu^s$-like in character and not only displays a large oscillator strength but also possesses an unusually high doubly excited character. The authors found that the two populations, one with the lowest ionic ICT state and the other with the lowest lying 2^1Ag^- covalent state, exist in equilibrium with each other in a majority of the solvents. The two populations are separated by a small barrier arising from solvent reorganization and cavity formation. In the reported calculations [25], the authors made use of the B3LYP/6-31G(d) method for predicting ground-state geometries. Excited-state geometries were generated by the CIS method. The authors claim that the CIS method provides good quality results for both gas-phase molecules and molecules in a solvent. The solvent environment was modeled by adopting the polarizable continuum model (PCM) as implemented in Gaussian 09 software. They found that a combination of CIS and PCM method provides a rather reliable method of predicting excited-state geometry in the presence of a medium. The CIS calculations with geometry optimizations were performed in an active space spanned by the eight highest filled MOs and eight lowest energy virtual (unoccupied) orbitals. The spectroscopic properties were, however, calculated using MNDO-P + SDCI (configuration interaction with singly and doubly excited states), the symmetry-adapted cluster configuration interaction (SAC-CI) method and equation of motion coupled cluster singles and doubles (EOM-CCSD) method. It turns out that the MNDO-SDCI is quite sound, albeit empirical, when it comes to the calculation of spectroscopic properties of carotenoids and, for that matter, of peridinin as well. The SAC-CI methods appear to be very efficient and almost optimal for calculating dipolar properties of excited states. They suffer, however, from occasional inability to predict the correct ordering of the ionic and covalent states. One must therefore exercise caution while using SAC-CI results for interpreting experimental data. The EOM-CCSD methods, on the other hand, are relatively new and predict transition energies as well as oscillator strengths with high accuracy. These methods work equally well in conjunction with PCM of macroscopic solvation.

It must be mentioned here that the majority of the calculations were performed on a suitable model chromophore in which the central polyene portion of the molecule was retained, but the cyclohexane as well as the epoxy cyclohexane rings were replaced with a methyl group and an allenic >CH_2 group, respectively. The resulting model molecule belongs to the C_s point symmetry and has a reduced size that makes high-quality calculation affordable from the point of view of computational labor or hardware requirement.

The IR and visible TA spectra of all trans-β-apo-8′ carotenal in a number of solvents differing both in polarity and polarizability have been carefully analyzed by Ragnoni *et al.* [26] at different excitation wavelengths. The authors sought to correlate the solvent dependence of the kinetics and band-shape changes in the IR with those of excited-state absorption bands in the visible spectra and were able to demonstrate that information obtained from the two spectral regions is complementary in nature. It is interesting to note that the collected data could be interpreted within the framework of a scheme of relaxation which postulates the so-called bright $^1Bu^+$ state as the major component of the ICT state. This state is dynamically stabilized, in polar solvents, by molecular distortion (twisting) and solvent relaxation. The detailed analysis of the solvent effects on the excited-state bands in the IR and visible spectra led the authors to conclude that both solvent polarity and solvent polarizability are important factors that must be taken into account to provide a complete rationalization of the excited-state relaxation dynamics of trans-8′-β-apo carotenal and understanding of the nature of the ICT state and the role that the ICT state plays in this molecule. The experimental observations reported by Ragnoni *et al.* [26] lend credence to the theory that wave functions of the S_1 and S_0 states at the FC excited-state geometry have a mixed ionic-covalent character. It appears that the extent of ionic-covalent mixing not only depends on the solvent polarity but also can get dynamically modified by the effects of solvent polarizability. The authors argue that the effects of different excitation wavelengths on the kinetics as well as spectral dynamics could be traced to photoselection of a subpopulation of partially distorted molecules. A more detailed analysis of the S_0, S_1 and S_2 PESs including the effects of non-adiabatic coupling and effects of solvation would be necessary before the full picture emerges.

Determining the structural features of the ICT state in a fluorophore is no doubt interesting. Overwhelming theoretical and experimental evidences available till now support the nonplanar (twisted) structure of the DA molecule like DMABN in the ICT states. A more complex and interesting question that needs to be addressed concerns the electronic pathway that the fluorophore follows as it passes from the initially excited $\pi\pi^*$ state to the ICT state. The two-state (LE + ICT) model no doubt appears very plausible, but lacks fool-proof experimental evidence. The dominant experimental support favoring the standard two-state model of ICT reaction comes from the time-correlated single-photon counting (TCSPC), which has a typical instrument response time of 20 ps or more. This response time is, however, substantially longer than the decay time (4.1 ps) of the fastest decaying component of the LE fluorescence. It is clear therefore that experimental measurements (temporal) with much higher time resolution would be necessary before one can confirm or reject the two-state dynamical model of the ICT process.

Femtosecond TA measurement [27] strongly suggests that the fully charge-separated ICT state is formed from the $\pi\sigma^*$ state. Thus, the well-known benzonitrile radical anion-like absorptions of the TICT state at 320 and 420 nm have a rise time of 4.1 ps that matches well with the decay time of the room temperature 700 nm $\pi\sigma^*$ state transient of DMABN in acetonitrile medium. The observed kinetics, as the authors claim, appear to be consistent with the sequential ICT process in which an initially excited $\pi\pi^*$ state (S_2, L_a-type) produces the TICT state via an intermediate electronic state arising from $\pi\sigma_{CN}{}^*$ configuration. That the ICT process in DMABN is more complex than envisaged in the popular two-state model is also evident from the observation that the decay time of the long-lived component of the LE fluorescence is not identical with the decay time of the ICT fluorescence. Gustavsson *et al.* [28], however, point out that the fluorescent ICT state may be different from the TICT state seen in TA.

The tenability of the two-state model has been probed from a purely theoretical point of view recently by Jarowski and Mo [29]. They computed and analyzed structural weights of the canonical resonance forms used in the valence bond configuration interaction (VB-CI) model, in the ground (neutral) and excited (ionic) states of a series of linear dipolar ICT chromophores carrying buta-1,3-diene-1,4-diaryl bridge. The calculations were done using block-localized wave function at B3LYP/6-311+G(d) level of approximation and provided, for the first time, a quantitative theoretical assessment of the validity of the two-state model. For comparison, a similar analysis was also made within the framework of an extended 10-state model. It turns out that both the 2-state and 10-state models predict surprisingly low structural weights of the VB-CT resonance form in the GS and substantially higher weights for the same form in the excited states. Individually, resonance forms were structurally optimized for clear assessment of the origin of bond-length alternation (BLA) in the bridging units. The authors [29] used a Wheland-energy-based weighting scheme. It is curious that the weighted average of the optimized bond lengths failed to reproduce the BLA correctly in the two-state model. The 10-state model, however, recovers the BLA feature quite accurately. Similar studies for DMABN and related molecules are not forthcoming yet. We anticipate, however, that similar calculations and analysis will soon provide more quantitative assessment of the validity of the two-state model of ICT phenomenon in DMABN and related ICT molecules.

The role of twisting and the $\pi\sigma^*$ state in the excited-state ICT process in DMABN have been theoretically examined closely using state-of-the-art electronic structure theory by Georgieva *et al.* [30] very recently. These authors made good use of a combination of multiple reference configuration interaction (MRCI) and second-order algebraic diagrammatic construction [ADC(2)] methods for deciphering the structural process leading to the dual emission of

DMABN in the gas phase and in acetonitrile medium. The solvent effects were accounted for on the basis of a conductor-like screening model. These authors invoked the MRCI method to estimate the extent of non-adiabatic interaction between two of the lowest $\pi\pi^*$ singlet excited state – that is, the S_1 (LE) and S_2 (CT) states of the traditional ICT theory and locate the minimum of the crossing seam (MXS). The ADC(2) method, on the other hand, was exploited by the authors to assess the role (if any) of the C-CN bent $\pi\sigma^*$ state in the photodynamics of the molecule in the excited state. The authors suggest that the initial photodynamics begins in the S_2 ($\pi\pi^*$) state with $—N(CH_3)_2$ twisting taking place and/or simultaneous pyramidalization of the ring carbon atom attached to the nitrogen atom. The $\pi\sigma^*$ state lies too high in energy in the gas phase to be of significance in shaping the dynamics. In solution, however, the bent $\pi\sigma^*$ state is stabilized enough to be energetically comparable to the LE state and to be able to influence the dynamics, although the C-CN bent will initially destabilize the $\pi\pi^*$ state. The authors found that the structure associated with the MXS corresponds to a twisting angle equal to 50° and that the S_2 to S_1 deactivation was not mediated directly by the twisting motion of the dimethylamino group. Not only that, they also concluded, on the basis of the theoretical analysis, that polar solvation does not play any significant role in S_2 to S_1 deactivation. When the molecule is constrained to have a structure-retaining C_s point symmetry, the C-CN bending has a strong stabilization effect (on the $\pi\sigma^*$ state) and leads to a quite significant amount of CT. The structure is, however, not a true minimum on the PES and moves over into the LE state when the C_s-symmetry constraint is removed and the molecule is allowed to relax further. The results seem to indicate that the dynamics is dominantly shaped by the non-adiabatic interactions between the CT state and the LE state, and the main source of dual fluorescence can be traced to the twisted ICT and the LE state. With the refinement of theoretical techniques of computing electronic structure and dynamics in the presence of strong non-adiabatic effects, and enhancement of the computing hardware, many more similar calculations are expected to be undertaken in the coming years and a comprehensive understanding of the photodynamics in ICT molecules will emerge.

The bapo molecule has been the object of yet another study throwing some light on the excited-state nuclear dynamics of carotenoids, which are known to play a dual role in photosynthesis [31]. They are efficient light harvesters and act as photoprotective radical quenchers. The exact non-radiative mechanisms that shape the functionality of carotenoids have hitherto remained unclear. The exact number of ultrafast relaxation channels that help these molecules travel down from the optically bright S_2 ($1^1B_u^+$) state to the optically dark S_1 ($2\,A_g^-$) state has not yet been established with any degree of certainty. Thus, it is difficult to say if intermediate dark states of uncharacterized or unknown electronic structure take part in the process. It is also unknown if the ground-state conformers of the molecules lead to a separate manifold of states. Interestingly, it appears more or less certain that in the polar environment, carotenoids can access an ICT state supposedly associated with the >C=O group held in conjugation with the polyene chain. A number of investigators [32, 33] have naturally focused attention on the ICT state with a view to ascertaining if it is related to the S1 manifold or is an independent electronic state. The role of nuclear motion that mediates the S_2–S_1 coupling also has been investigated in several experimental and theoretical works on β-carotene. There appears to be a concern that the $S_2 \leftrightarrow S_1$ crossing is driven by a conical intersection that links the S_2 and S_1 states.

Olivier and Fleming [34] studied the role of coupled electronic and nuclear motion through the conical intersections in the ultrafast relaxation of bapo – a model aldehyde containing the carotenoid moiety. The authors did a combined theoretical and experimental investigation. TDDFT calculations in conjunction with ultrafast electronic absorption, one- and two-dimensional vibrational spectroscopies in cyclohexane (nonpolar medium) and acetonitrile (polar medium) solution enabled the authors to establish a direct correlation between the intrinsically coupled electronic and vibrational degrees of freedom, which are assumed to play an important role in driving relaxation of bapo from the bright S_2 to the dark S_1 state. Line shapes of some features in two-dimensional electronic-vibrational (2DEV) spectra allowed the authors to make some definitive assignments of the excited-state vibrations of bapo in acetonitrile. Studies of anisotropy firmly establish that the excited-state dynamics in bapo does not involve a *trans–cis* isomerization, thereby dismissing a previously made hypothesis. It has been demonstrated that the electronic and vibrational line shapes, for specific vibrational modes, remain correlated well beyond the decay time of the S_2 state. The observation tends to suggest that transfer of the molecules from the S_2 state to the S_1 state is impulsive and perhaps involves a conical intersection located in the vertical FC region. The work once again emphasizes the need to construct detailed and accurate *ab initio* PESs including non-adiabatic effects and carry out complex electronic and molecular dynamics on the constructed PES to understand clearly the mechanisms that led to the emergence of stepwise functionalities of the carotenoids. Fujiwara *et al.* [35] carried out a computational TDDFT study on the electronic pathway leading to ICT in di-tertiarybutyl aminobenzonitriles and 2,4,6-tricyano aniline. Their calculations provide evidence for ultrafast ICT state formation in *meta-* and *para-*di-tertiarybutyl aminonitriles through the following sequential switching of states:

$$\pi\pi^* \rightarrow (L_a) \rightarrow \pi\sigma^* \rightarrow \text{ICT}$$

in the presence of conical interaction involving three closely lying excited states. No such evidence in favor of LE → ICT state formation in 2,4,6-tricyanoaniline (TCA) was confirmed by their theoretical calculations. These calculations, however, predicted that 2,4,6-tricyano-*N*,*N*-dimethyl aniline (TCDMA) but not 2,4,6-tricyanoaniline possesses two ICT states which have the typical quinoidal structures associated with the ICT states, both lying below the initially photoexcited S_1 ($\pi\pi^*$) state. The CC2 calculations further predicted the existence [35] of two conformers labeled, respectively, as ICT (Q) and ICT (AQ) with "Q" standing for quinoidal and "AQ" for anti-quinoidal character. These two conformers were predicted to interconvert rapidly. The lower energy ICT(Q) state appears to get populated from the unstable ICT(AQ), which is responsible for the observed time-resolved fluorescence as well as excited-state absorption from the mixed S_1 ($\pi\pi^*$)/ICT states of TCDMA. In both TCDMA and TCA, the $\pi\sigma^*$ state is located much higher in energy, thus precluding the $\pi\sigma^* \rightarrow$ ICT formation pathway that has been predicted to be operative in *meta-* and *para-*di-tertiatybutyl aminonitriles.

It appears now that a consensus about the energetics and structural dispositions of ICT molecules has been slowly emerging, although much of it is static in nature. Very little seems to be known about what happens during the very early stages of the non-adiabatic dynamics in the prototypical and most extensively studied ICT molecule like DMABN even in the gas phase where no interaction of the probe with the solvent molecules would complicate the dynamical picture due to operations of the multiple timescales. As we have noted earlier in this chapter that there has been so far a broad agreement on one aspect of the problem of ICT in DMABN – that in the gas phase the molecule is initially photoexcited to the second excited state (S_2 in the standard nomenclature), which possesses ICT character. Fuβ *et al.* [36] carried out multiphoton ionization experiment on DMABN and on the basis of the popular assumption that the twisting of the dimethylamino group dominates the S_2/S_1 crossing event, these authors proposed that their data would be consistent with DMABN reaching the S_2/S_1 crossing (following the initial photoexcitation) in less than 100 fs. CASSCF-level calculations [33] earlier predicted that the S_2/S_1 minimum energy conical intersection would be energetically accessible over a rather wide range of twist angle (τ) close to 180°. Subsequent CASSCF calculations using different active spaces, however, clouded the entire issue of dimethylamino group twisting by predicting conical intersection geometry that ranged from a half twisted dimethylamino group to a completely twisted group with $\tau = \pi/2$.

In a near-complete overturning of the emerging picture, recent studies [37] on the first excited-state dynamics of DMABN using the surface-hopping trajectory method with ADC(2) and LR-TDDFT seem to suggest rather curiously that the twisting of the dimethylamino group is not of much relevance in the much important S_2/S_1 non-adiabatic transition.

Todd Martinez's group [38] very recently invoked their *ab initio* multiple spawning method combined with GPU-accelerated LR-TDDFT theory to study the ultrafast decay dynamics of DMABN following the initial photoexcitation into the S_2 state. It transpires that there is almost complete non-adiabatic transfer of population from the S_2 state to the S_1 state in about 50 fs without exciting any significant torsion of the dimethylamino group at all. It is only after the nuclear wave packet has reached the S_1 surface and started relaxing and acquired the LE character that significant torsion of the dimethylamino group is noticed. The authors do not, however, follow the long-time adiabatic equilibriation on the S_1 surface as their electronic structure calculation method is not adequate for accurately predicting the relative energies of the S_1-LE and S_1-ICT minima. The LR-TDDFT method overstabilizes the S_1-ICT minimum, thereby vitiating long-time dynamics. It is reassuring to note that the ultrafast S_2/S_1 non-adiabatic dynamics produced by the AIMS/LR-TDDFT method matches well with what was recently predicted by Kochman *et al.* [39] (see also Ref. [40] by Du *et al.*) using a different non-adiabatic scheme (e.g., TSH) and an electronic structure calculation method, ADC(2). The important points made by these authors are that (i) no significant twisting of the dimethylamino group is required for the S_2/S_1 non-adiabatic (radiation-less) transition to occur and (ii) the pyramidalization coordinate fluctuates around what is expected for a planar geometry. Their findings echo what has been recently established by Martinez group on the basis of their AIMS/LR-TDDFT calculations. What happens to the dynamics and relative ordering of the important energy levels when DMABN is in solution remains to be seen.

ICT in donor–acceptor (D-A) biaryls has been the subject of a number of studies in view of the technological relevance. Planar or moderately twisted D-A biaryls sustain full charge separation in the S1 state and are good candidates for electron transfer (ET)-initiated photocatalysis. Naturally, these molecules have been the target of a number of experimental as well as theoretical investigations. Let us consider electron donor–acceptor biaryls like the pyridinium and pyrilum phenolates [38]. These molecules are very good models for studying photoinduced ET. Barbara *et al.* carried [41–43] out the first-ever ultrafast spectroscopic measurement on betaine-30 by a one-color pump–probe experiment. While confirming the ultrafast non-radiative deactivation of the S1 state, they suggested that a rather strong entanglement exists between the non-radiative deactivation process and the solvent relaxation. The dynamics was described invoking a two-state model in which the GS is assumed to be zwitterionic and the excited state to be just weakly polar. The analysis was done within the framework of a hybrid model – the so-called Sumi–Marcus theory [44], in which the high-frequency intramolecular modes are handled quantum mechanically. Experiments done subsequently tended to suggest that the dynamics also involved substantial reorganization of the intramolecular modes. Time-resolved and TA experiments strongly indicated that the decay of the optically populated excited state back to the GS was mediated by a "dark" intermediate state which appears to be rather decoupled from the GS. The molecule in the dark state was speculated to have a twisted geometry, with the donor and the acceptor rings being at a dihedral angle of $\tau = \pi/2$. The authors felt that the molecule returns to the GS in the twisted geometry and subsequently relaxes to the planar form on the ground PES.

Fedunov *et al.* [45] very recently developed a theoretical model for simulating the transient electronic absorption spectra of donor–acceptor dyads undergoing ultrafast CT in solution based on what has become known as the multichannel stochastic point transition approach. The method treats the reorganization of the high-frequency intramolecular modes quantum mechanically and the low-frequency intramolecular as well as the solvent relaxation modes classically. They assumed the relaxation of the slow modes to be exponential with time constants borrowed from experimental data. The excited-state dynamics is obtained by simulating the time distribution of population function of each quantum state after electronic excitation and following electronic and vibrational transitions with the multichannel stochastic method. The model was invoked to simulate the TA of pyrylium phenolate in acetonitrile solution. The authors claim that a very good agreement with the experimental findings was achieved when a three-state model comprising optically excited state, the dark intermediate state with twisted geometry and large CT character, and the GS was adopted. The method has the capability to disentangle the contributions to the ultrafast spectral dynamics originating from population changes and relaxation processes. The authors note that for accurately describing the fast decay of the excited state and fast GS recovery, large reorganization energy associated with the slow modes is required for enabling transitions from the excited to the dark state as well as from the dark state to the GS to take place. The simulation suggests that the electronic transition takes place in sub-picosecond timescale and the observed spectral dynamics emerges only from the structural relaxation of pyrylium phenolate on the ground surface after a lapse of 2 ps. The solvent relaxation appears to be ultrafast with no effect on the spectral dynamics beyond 2 ps or so.

Monomethine dyes [46] are ICT chromophores that have attracted a lot of attention in view of their technological relevance. These dyes display fluorescence decay timescales of subpicoseconds, while the GS recovery time ranges between 1 and 10 ps. These molecules exhibit twisted motion (Figure 3.8) in the excited state, which appears to be environment sensitive. The environment can be tailored to cause suppression of twisting in the excited state, thereby causing very significant enhancement of fluorescence quantum yield. The excited-state twisting in monomethine dyes has been often characterized as an environment-controlled excited-state process devoid of any intrinsic barrier to twisting. There is thus the need for developing a theoretical model that can lead to a better understanding of the process of controlling of fluorescent emission by selecting the appropriate medium, which in turn can lead to the development of better sensors. It is believed that twisting the methine bond in the excited state lowers the potential energy, leading to charge-localized or charge-separated molecular states from which internal conversion becomes rather facile due to reduction of the adiabatic energy gap. Electronic structure calculations [47, 48], at least in several cases, have predicted the existence of low-energy conical intersection seams with sizable twist, facilitating ultrafast deactivation of the photoexcited states. The conical interactions themselves stem from crossing of diabatic states with different patterns of charge localization, and simulations carried out tend to suggest that there is significant coupling between the CT and solvation processes [49], which appears to affect the decay rate rather significantly. The coupling between twisting and CT has also been speculated for some time, but explicit inclusion of such couplings in a theoretical model has been lacking. Oslen and McKenzie [46] proposed a two-state model Hamiltonian which explicitly includes coupling between the twisting displacements and CT in the GS as well as excited state of a monomethine dye molecule. The model Hamiltonian (H) has been carefully parameterized against standard quantum chemical calculations on a number of protonated states of green fluorescent protein chromophores. The states were chosen so as to sample different regimes of detuning from the cyanine resonant limit. The model "H" is constructed on

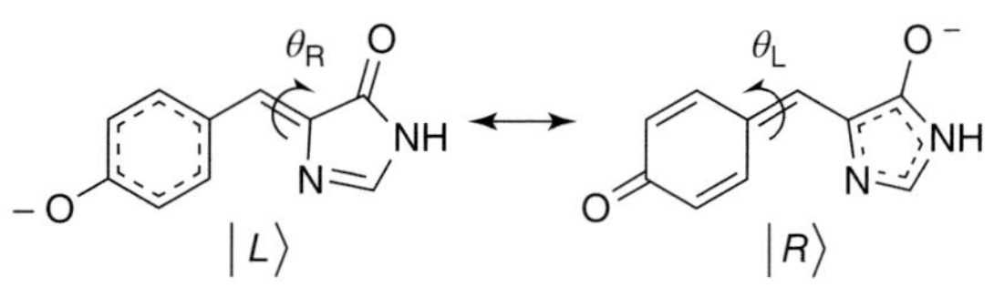

Figure 3.8 The diabatic states |*L*> and |*R*> used to define the model Hamiltonian. (Oslen and McKenzie 2012 [46]. Reproduced with permission of American Institute of Physics.)

the basis of diabatic states $|L>$ and $|R>$ and parameterized as follows:

$$< L|H(\theta_L, \theta_R|L) = \frac{\delta}{2} + 2\gamma_R \sin^2 \theta_R + 2\gamma_L' \sin^2 \theta_L \tag{3.1}$$

$$< R|H(\theta_L, \theta_R|R) = -\frac{\delta}{2} + 2\gamma_L \sin^2 \theta_L + 2\gamma_R' \sin^2 \theta_R \tag{3.2}$$

$$< L|H(\theta_L, \theta_R|R) = \frac{1}{2}\varepsilon \cos \theta_L \cos \theta_R \tag{3.3}$$

In these equations, γ_L and γ_R are the energies associated with twisting about the double and single bonds, ε being the hopping strength.

With the Hamiltonian thus defined, the adiabatic states and their energies are completely specified by a dimensionless parameter,

$$\lambda(\theta_L, \theta_R) = \frac{\delta + 2(\gamma_R - \gamma_R')\sin^2\theta_R - 2(\gamma_L - \gamma_L')\sin^2\theta_L}{\varepsilon \cos\theta_L \cos\theta_R} \tag{3.4}$$

Also

$$\lambda(\theta_L, \theta_R) = \cot 2\phi(\theta_L, \theta_R) \tag{3.5}$$

The adiabatic gap energy turns out to be

$$\Delta E(\theta_L, \theta_R) = [\{(\varepsilon \cos\theta_L \cos\theta_R)^2 + (\delta + 2(\gamma_R - \gamma_R')\sin^2\theta_R - 2(\gamma_L - \gamma_L')\sin^2\theta_L\}^2]^{1/2} \tag{3.6}$$

The mixing angle ϕ completely defines the transformation to the adiabatic representation. It is needless to mention that it is an extended version of the Mulliken–Hush model. The model, as the authors claim, provides a simple yet realistic description of the evolution of the CT character along the two twisting channels (θ_L, θ_R) associated with the methine bridge in the excited state. The model predicts qualitatively different responses in the three regimes that can be identified by their relationships with the resonant cyanine limit. The regimes referred to differ in the presence or absence of twist-dependent polarization reversal and the existence of conical intersection. The authors predict that the selective biasing of one twisting channel (θ_L or θ_R) over the other can be accomplished by the application of a diabatic biasing potential. The biasing succeeds only if the parameters are carefully chosen from a narrow range of values near the cyanine limit. The simple effective valence bond Hamiltonian provides a detailed picture of the phenomena originating from the coupling between the CT state and the twisting modes. The authors do not, however, attempt to include the influence of the medium on the twisting in their theoretical model. Had it been done, it would have shed light on the possible role of solvents in the excited-state relaxation process along with interplay of twisting and solvent reorganization or macroscopic solvent polarization.

Although the exact role of twisting in the TICT fluorescence and related phenomena is still an open question, the emission properties (TICT emission) are known to be potentially environment dependent – an attribute that makes the TICT-based fluorophores ideal sensors for solvents, microviscosity, and different chemical species [50]. It is interesting that several TICT-based materials have been recently discovered to become fluorescent upon aggregation. Various recent studies in organic optoelectronics, nonlinear optics, and solar energy conversion have exploited the idea of TICT to control the electronic-state mixing and coupling on the CT state. Some of these applications are discussed in Chapters 5 and 6 in a aggregation.

ICT, as we have already emphasized, is a ubiquitous phenomenon playing a vital role in processes like photosynthesis, redox reaction, xerography corrosion, and so on. The prevailing technique of probing such an important phenomenon relies on indirect means. Since the ICT takes place most often in donor–acceptor-based systems, a traditional technique of probing the CT has been to monitor transient emission of EM radiation or its absorption by the donor/acceptor fragment of the ICT system and measure how the local environment affects the emission or absorption, that is, how frequency and intensity of radiation and its polarizations are modulated. The indirect measurements have shed considerable light on the ICT process or phenomenon, – but a more direct means of following ICT would certainly be desirable. The ICT process, we may note, involves movement of charge from one end of the molecule to another. An accelerated charge (or a time-varying current), we know from classical electrodynamics, emits EM radiation. The moving charge involved in the ICT is therefore expected to emit transient EM waves (in the THz domain) and through the emitted waveform "broadcast" the dynamics of the underlying CT process. THz spectroscopy came to life in the late 1980s when it became possible to generate and propagate nearly single cycle pulses of far-IR radiation through free space and subsequently detect them in time domain. Schmuttenmaer group [51] successfully measured ICT-linked emitted field amplitudes with sub-picosecond temporal resolution. We note here that the pulse generation from moving charge requires that fields emitted by all the molecules in the sample are constructively added up. That requires (i) all the molecules appropriately oriented, (ii) all the molecules to be coherently excited, and (iii) if all the molecules are axially oriented, the CT must induce a change in the polarization along the axis of orientation. The techniques used by Beard *et al.* [51] are related to the method of THz generation from biased semiconductors and the transient dc photocurrent techniques. The THz generation from biased semiconductors exploits the generated EM pulses to probe the underlying ultrafast carrier dynamics and measure transient photoconductivity in a variety of semiconductors like gallium arsenide (GaAs), GaAs grown at low temperatures, nanocrystalline colloidal TiO_2, and CdSe quantum dots and other quantum well structures [52]. In all such experiments, optical pump–THz probe configuration was used. The transient photocurrent technique actually measures the charge in polarization of a solution following photoexcitation. When the photoexcited molecules reorient themselves in a static electric field (dc field), a displacement current is generated, the measurement of which can lead to ground and excited dipole moments of the molecule along with lifetimes of the long-lived excited states (the time-resolution of the thermal dissipation probe (TDP) technique is low). Another related technique has been referred to as coherent infrared emission (CIE) interferometry. Here, vibrations are coherently excited by an ultrafast electronic absorption and the emitted EM field is detected by interferometry. It is needless to mention that only such vibrations that are coupled to the electronic excitation will contribute to the emitted field. In a way, the technique adopted is akin to the stimulated resonance Raman scattering (SRS) experiment. As in SRS, here too, a low-frequency oscillation coupled to an electronic transition is responsible for the emitted field. In the experiments of Beard *et al.* [51, 53, 54], it is the ICT that couples, while in CIE the intramolecular vibrations take the place of CT. In what follows, the theoretical model developed to extract the dynamics is briefly outlined (for details, see Beard *et al.* [51, 53, 54]).

The EM waveform emitted during the CT process is detected in the time domain through free-space electrooptic sampling and then modeling is done as follows. If we neglect propagation effects of the field through the medium, the emitted amplitude ε_i^{Ω} in the far field is equal to the second derivative of the time-dependent polarization:

$$\varepsilon_i^{\Omega} = \frac{\partial^2 P(t)}{\partial t^2} \tag{3.7}$$

where Ω is the emitted frequency. For describing the CT dynamics correctly, the nonnegligible propagation effects must be taken care of. The authors do that by solving the Maxwell's equation in the time domain coupled to a phenomenological model for the time-dependent polarization $[P(\varepsilon)]$. The experimental data is then given a nonlinear least-square fit to the model for extracting the dynamics of the CT. Thus, the relevant equation for a transverse field in one dimension under zero free current condition is given by

$$\frac{\partial^2 \varepsilon(z,t)}{\partial z^2} - \frac{1}{c^2}\frac{\partial^2 \varepsilon(z,t)}{\partial t^2} = \mu_0 \frac{\partial^2 P(z,t)}{\partial t^2} \tag{3.8}$$

In this equation, c is the velocity of light in vacuum. $\varepsilon(z, t)$ is the electric field at "z" at time "t", while $P(z, t)$ is the induced polarization at "z" at the same instant. The polarization P can be partitioned into its linear and nonlinear components; it is only the third-order nonlinear term $P(3)$ that contributes to the generated EM pulse. The linear term is responsible for the dissipation and modulation of the generated signal as it propagates through the medium. Equation 3.8 then becomes

$$\frac{\partial^2 \varepsilon(z,t)}{\partial z^2} - \frac{1}{c^2}\frac{\partial^2}{\partial t^2}\int R^{(1)}(t-t')\varepsilon^{\Omega}(z,t')\partial t' = \mu_0 \frac{\partial^2 P^{(3)}(z,t)}{\partial t^2} \tag{3.9}$$

$R^{(1)}(t)$ is the linear response of the solvent molecules to the field $\varepsilon(z, t)$.

Equation 3.9 describes the generation and propagation of the generated EM transient through the solution. The linear term is known from measurement of static THz pulse propagation through the solvent without excitation. By expressing the third-order nonlinear polarization in terms of the third-order time-domain response function, the authors [55] have solved Equation 3.9 coupled with Equation 3.7 by the finite-difference time-domain (FDTD) method. The FDTD calculation provides the generated field in the near-field regime. A near- to far-field transformation is performed by taking the first derivative of the calculated field. A nonlinear least-square fit is performed for extracting the forward and backward ET rates.

The application of the technique to elucidate and understand ICT dynamics, especially during the early stages, has revealed interesting information [56]. The details are discussed later on in this chapter.

3.3 Theoretical Treatment of ICT Rates

The ET process, as we have mentioned repeatedly, is central to all redox reactions and crucial in all electrochemical events including corrosion. The evidence of ET lies in the redistribution of electronic charge density and it is natural to try to estimate the rate at which the CT takes place and understand the factors that shape the rate. The CT event may be interatomic or intermolecular or intramolecular and may be triggered thermally as well as photochemically. So far, we focused on photochemically induced ICT, primarily from the electronic structural point of view. The problem of calculating ICT rates, however, requires us to consider the CT event dynamically. For an isolated molecule, the problem is straightforward. The situation encountered in solution is far more complicated as several timescales appear in the description. To understand the nature of the difficulties, let us consider a simple CT event, say $M \rightarrow M^+$ occurring in solution. If we assume that the atom "M" was in equilibrium with its surrounding (the medium), the rapid ET causes a redistribution of electron density on a timescale that the solvent molecules typically take to reorganize themselves in a manner that is consistent with the altered electron density distribution. The solvent molecules were polarized in response to the initial electronic charge distribution and the polarization must change to respond to the altered charge distribution. This leads to the appearance of solvent reorganization energy (an energy deficit) associated with the CT process. Energy conservation demands that ICT can take place only if the extra energy is supplied to the system. In photochemical ET, the photon energy balances the energy deficit, while no such balancing is possible in thermal ET. In thermal ET, the mechanism therefore must be very different, as was first noticed by Marcus who cogently argued that fast ET takes place only if the much slower moving solvent molecules surrounding the probe attain a configuration at which the initial (pre-transfer) and the final electronic states (post-transfer) of the probe become electronically degenerate. Adopting a simple model invoking the Landau–Zener theory with the additional assumption that solvent reorganization is fast compared to the reaction rate and the dynamical interaction with the solvent can be neglected leads to the adiabatic ET (charge transfer) rate constant [57]

$$k_{\text{ad}} = \frac{\omega_s}{2\pi} \mathrm{e}^{-\beta \overline{E}_{\text{A}}} \tag{3.10}$$

The term ω_s is defined as

$$\omega_s = \left(\frac{k}{m}\right)^{1/2} \tag{3.11}$$

In Equation 3.10, $\overline{E}_{\text{A}}$ is the barrier height on the adiabatic PES which the transferring electron (charge) must cross over (Figure 3.9). In the non-adiabatic limit, the rate constant k_{nad} is more complex and turns out to be

$$k_{\text{nad}} = \sqrt{\frac{\pi \beta k}{2}} \frac{|V_{12}|}{\hbar |\Delta f|_{X=X^*}} \mathrm{e}^{-\beta E_{\text{A}}} \tag{3.12}$$

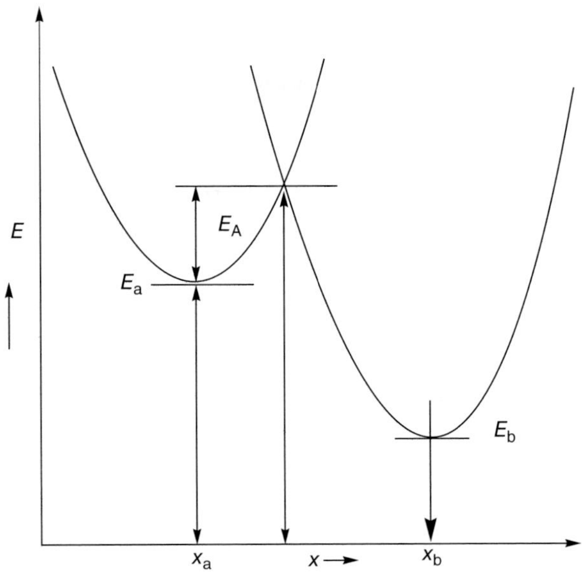

Figure 3.9 A simple model depicting energetics of electron transfer on two intersecting parabolic (harmonic) potential energy surfaces.

where

$$\Delta f = k(X_2 - X_1) \tag{3.13}$$

V_{12} is the non-adiabatic coupling strength between the two surfaces involved in the transition, while k is the harmonic force constant characterizing the PESs.

The simple model does not use the energetic and dynamic properties of the solvent molecules surrounding the probe. Nor does it provide any quantum mechanical insight into the process that takes place when the two electronic states involved in the transition come close to each other.

The parabolic PESs so far have been described by a rather abstract coordinate "X", which is taken to somehow represent the nuclear configuration of the solvent molecules surrounding the probe (solute). The continuum dielectric model of the polar solvent was used by Marcus to provide, for the first time, a physical meaning to the coordinate "X". Marcus attributes this to two solvent timescales – the fast one associated with electronic and the slow one to the nuclear response. The solvent dielectric response function is modeled by the static response (ε_s) and the fast electronic component ($\varepsilon_e = n^2$). The basic condition relevant to the Marcus theory is that the electronic charge redistribution or ET or CT takes place only in fixed nuclear positions – a restatement of the FC principle in the context of continuum dielectric theory of electronic transitions. Using the results of equilibrium electrostatics for the simple case of a spherical ion of charge "q" and radius "R" held in an equilibrium with an infinite dielectric medium characterized by the dielectric constants ε_s and ε_e, the free energy difference $\Delta G_{a \to b}$ associated with transition from the charge state "a" to charge state "b" of the ion is calculated to be

$$\Delta G_{a \to b} = \frac{(q_a - q_b)q_0}{\varepsilon_s R}\theta - \frac{(q_a - q_b)^2\theta^2}{2\varepsilon_s R} \tag{3.14}$$

where

$$q_b = q_a + (q_b - q_a)\theta \tag{3.15}$$

q_b and q_a are the charges on the ion in the state b and a, respectively. The free energy change $\Delta G_{a\to b}$ then leads to the definition of the "potential energy surfaces" for fluctuations of the nuclear polarization of the solvent about the equilibrium state "a".

$$W_a(\theta) \equiv \Delta G_{a\to b} = \frac{(q_b - q_a)^2}{2a}\left(\frac{1}{\varepsilon_c} - \frac{1}{\varepsilon_s}\right) \tag{3.16}$$

The parabolic surface $W_a(\theta)$ is, however, a free energy surface (not a PES) and $e^{-\beta W_0(\theta)}$ specifies the probability that the system in the equilibrium state "a" (charge q_a) at a temperature T will be found to have a value of nuclear polarization that is characteristic of another equilibrium state "b" (charge q_b) due to dielectric fluctuation.

The simple model quantifies the harmonic force constant appearing in Equation 3.10, but not the "mass" associated with dielectric fluctuations, appearing in Equation 3.10. The non-adiabatic ET rate constant, on the other hand (Equation 3.13), requires only the non-adiabatic coupling constant as an external input. One may still question the use of free energy surface in the Landau–Zener rate expression. One can make use of the Fermi golden rule to obtain an estimate of the non-adiabatic ET rate.

$$k_{a\to b}(E_{ab}) = \frac{2\pi}{\hbar}|V_{ab}|^2 F(E_{ab}) \tag{3.17}$$

where the thermally averaged FC factor is given by

$$F(E_{ab}) = \frac{1}{Q_a}\sum_i e^{-\beta E_{a,i}} \sum_f |< \chi_{ai}|\chi_{bj} >|^2 \delta(\Delta\varepsilon + E_{ai} - E_{bj}) \tag{3.18}$$

where Q_a represents the nuclear partition function in the electronic state a, E_{ai}s are the vibrational energy levels supported by the nuclear PES in the electronic state "a" measured from the electronic origin, and χ_{ai}s are the nuclear wave functions. E_{bj}, χ_{bi}s are similarly defined for state "b". The use of the Fermi golden rule is an approximation that is expected to work well in the weak coupling non-adiabatic limit and the FC approximation should not be adequate for the electronic transition that is being described. We must note that the influence of the solvent on the electronic rate process must be contained in the FC factor and this influence includes the effects of slow dielectric response. The qualitative trends predicted by the simple theoretical model described have been experimentally confirmed, thereby validating, to an extent, the basic tenets of the ET rate theory, including the dielectric continuum model used for the medium surrounding the probe or the solute and the postulated linearity of dielectric response. The success of the model in predicting the rate of cross-exchange reaction ($A + B \to A^- + B^+$) is well documented [58]. An interesting prediction of the theory concerned the observed dependence of the rate of ET on the energy gap $E_{ab} = E_b - E_a$.

The rate expression turns out to be

$$k_{ab,na}(\Delta E) = \frac{1}{\hbar}\sqrt{\frac{\pi}{E_A(k_B T)}}|V_{ab}|^2 e^{-[(E_{ab}-E_A)^2/(4E_A(k_B T))]} \tag{3.19}$$

The term $k_{ab},_{na}(\Delta E)$ in Equation 3.19 clearly suggests that k_{ab} would increase as the E_{ab} value increases from zero; once the $E_{ab} = E_x$ threshold is crossed, the rate would begin to decrease, resulting in the so-called inverted Marcus regime – a feature that has been substantiated experimentally [59]. A second aspect of the ET rate theory focuses on the dependence of the predicted (observed) rates on the distance "R_{DA}" separating the two centers (donor D and acceptor A) between which the ET is assumed to take place. The ET takes place in the transition state (the PES of Figure 3.5b), where a nuclear fluctuation is supposed to have brought the system into a nuclear configuration for which the diabatic electron energies of the donor and acceptor have become equal. The rate of ET can then be computed from the tunneling rate between the two centers (D and A), the tunneling being mediated by the coupling between the diabatic states, which depends on the donor–acceptor separation (R_{DA}). The rate turns out to be

$$k \sim e^{-\beta' R_{DA}} \tag{3.20}$$

where

$$\beta' = \frac{2}{\hbar}\sqrt{2m_e I_D} \tag{3.21}$$

where m_e is the mass of the electron and I_D is the ionization energy of the donor. The predicted rate constant suffers from the fact that even for small donor ionization energy (4–5 eV), the rate drops very quickly to zero as R_{DA} increases, which seems to suggest that ET can take place only if the donor–acceptor separation is small. The experimental observation is just the contrary – ET does take place between the D and A, even when they are separated by a large distance. The contradiction can be avoided by assuming that "bridge-assisted" long-range ET can take place (Figure 3.10). The bridge provides centers (unoccupied molecular levels) on the path of the electron where it can reside for a while before moving to the next well. The model predicts that the rate constant decreases exponentially with the distance L separating the D and A.

$$k \sim e^{-\beta' L} \tag{3.22}$$

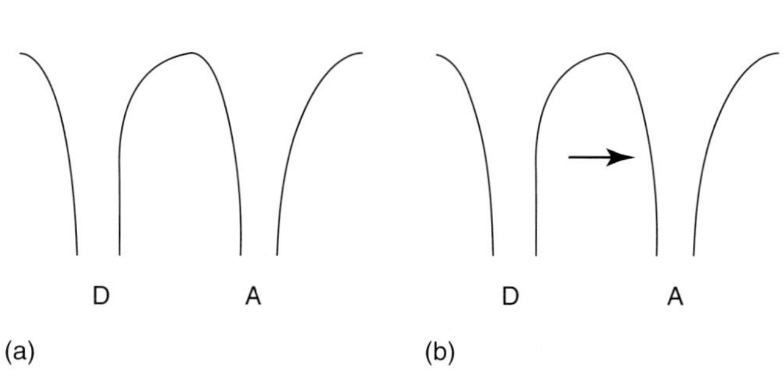

Figure 3.10 Model potential for bridge-assisted long-range electron transfer.

where

$$\beta' = -\frac{2}{b}\ln\left|\frac{V_B}{E_{D/A} - E_B}\right| \tag{3.23}$$

The bridge-assisted ET process referenced to is a coherent quantum mechanical process dominated entirely by tunneling and is useful for short bridges. An alternative model is one in which the electron is thermally activated into the bridge where it can move down the bridge by a sequence of ET steps, each characterized by a Marcus-type ET rate. One expects, therefore, that a crossover from the observed exponential dependence of tunneling ET rate operating in short bridges, to bridge-length-independent behavior may take place as the bridge length increases sufficiently and ET by hopping takes over. Indeed, experimental observation confirms such a transition [60]. It has also been observed [61, 62], as can be theoretically anticipated, that at low temperatures the ET rate is dominated by tunneling contribution which gives place to Arrhenius rate (exponential) at higher temperatures as thermally assisted transfer becomes dominant. It would be interesting to explore if a mechanism-independent model of computing ICT rate can be developed. The ICT system can be classified more generally into donor-bridge-acceptor (D-B-A), electrode-bridge-acceptor (M-B-A), and electrode-bridge-electrode (M-B-M) types so that electrochemical CT rates can be accommodated within the ambit of a general theory of ICT rates in D-B-A (also referred to as D-π-A) type of molecular system. In the M-B-A or M-B-M systems, the electrochemical CT rates or molecular conductance are the relevant kinetic quantities which are expected to be mathematically linked with the ICT state, if identical bridges act as conduits between the donor (M) and acceptor (M). In fact, the ICT rate in a D-B-A molecule is expected to be linearly correlated with molecular conductance in M-B-M systems at zero bias, if one assumes the same molecular bridge and "super-exchange" CT mechanism to be operative [63, 64]. Even for hopping-mediated ICT across fairly long bridges with high tunneling barrier, one expects the linear correlation to persist [65, 66]. Lewis *et al.* [63] exploited Fermi's golden rule to establish a relation between the ICT rate in a D-B-A molecule, the electrochemical constant in the M-B-A system, and molecular conductance in M-B-M junctions. Traub *et al.* [64] predicted the resistance of M-B-M structures assuming that multiple states on the electrodes are available. It turns out that for high-barrier molecular bridges and ET by hopping, tunneling CT rate (D-B-A) and molecular conductance (in M-B-M systems) are linearly correlated. The experimental data, however, testifies against the linear correlation. The cause of a breakdown of the linear correlation is not far to seek. Even if the molecular bridge (B) has the same configuration and electronic structure, the molecule–environment (bath) interactions and energy difference between the donor and the electrode charge states could easily cause a deviation from a linear correlation between ICT rate and molecular conductance.

Venkatramani *et al.* [67] developed a theoretical framework to compute ICT rates and molecular conductance without assuming any specific mechanism of charge transport. Their model leads to a nonlinear relation between the CT rate and molecular conductance. The deviation from linearity arises mainly from the difference in energy barrier to the CT process in the two types of systems and the basic difference in the decoherence rates. The authors demonstrate how a decrease in the CT energy barrier and an increase in the decoherence rate shifts the position of the crossover point from exponential to softer dependence of ICT rates and molecular conductance on the bridge length (i.e., the distance over which the charge is effectively transferred). The authors make use of a reduced density matrix formalism and adopt a phenomenological approach to handle the system – bath coupling for calculating the CT kinetics. To compute the steady-state currents, they exploit the Landauer–Butiker method based on Green's function. The most interesting aspect of their treatment is its regime-independent applicability and inclusion of decoherence effect of the bath. To be specific, the authors predict a power-law relationship between the ICT rates and molecular conductances. The nonlinear rate–conductance relation is traced to have its origin partly in the differences in charge-transport barrier height and partly in the bath-induced decoherence rates encountered in the two types of experiments. The theoretical model also successfully explains the rather queer correlation between electrochemical kinetics (in M-B-A systems) and molecular conductance (in M-B-M systems). Having described the predictions of the model, we now briefly outline the theoretical apparatus used by the authors that represent the bridge by a tight-binding Hamiltonian H_B with N identical units with equal on-site energy (E_B) and equal nearest-neighbor interaction (t_B). Thus, H_B reads

$$H_\mathrm{B} = \sum_{m=1}^{N} |m > E_\mathrm{B} < m| + \sum_{m=1}^{N-1} \left(|m > t_\mathrm{B} < m+1| + |m+1 > t_\mathrm{B} < m|\right) \tag{3.24}$$

The CT rate is calculated from the reduced density matrix (P) of the donor-bridge-acceptor system coupled to a bath. P is computed by solving the relevant quantum-Liouville equation

$$\frac{dP}{dt} = -\frac{i}{\hbar}(L_\mathrm{s} + L_\mathrm{SB})P \tag{3.25}$$

In Equation 3.25, L_S is the real system Liouvillian $[H_{S,P}]$, while L_SB is a complex system–bath interaction Liouvillian that accounts for the decay of the reduced density matrix elements. The system Hamiltonian H_S reads

$$\begin{aligned} H_\mathrm{s} = H_\mathrm{B} + \sum_{m\varepsilon_\mathrm{DA}} \left(|m > \varepsilon_m < m|\right) + \left(|\mathrm{D} > t_\mathrm{D1} < 1| + < 1\,|t_\mathrm{D1} < \mathrm{D}|\right) \\ + \left(|\mathrm{A} > t_\mathrm{NA} < N| + < N\,|t_\mathrm{NA} < \mathrm{A}|\right) \end{aligned} \tag{3.26}$$

The structure of $\mathrm{D_S}$ indicates that the donor with energy ε_D is coupled to the first bridge unit with coupling strength t_D1; similarly, the acceptor A is coupled to the last bridge unit with a coupling constant t_A1. L_SB is modeled purely phenomenologically by taking

$$(L_\mathrm{SB})_{mn.mn} = i\left[\frac{\gamma_m}{2} + \frac{\gamma_n}{2} + \gamma_{mn}(1 - \delta_{mn})\right] \tag{3.27}$$

The term $\gamma_m/\hbar$ represents the population relaxation rate from the site "m", while $\gamma_{mn}/\hbar$ is the pure dephasing rate that leads to the loss of coherence between the sites m, n.

At $t = 0$, the charge is assumed to be entirely localized on the donor (D), which amounts to taking $\rho_{DD} = 1$ and $\rho_{AA} = 0$. The CT (D → A) rate (k_{CT}) is then calculated by computing the quantity

$$k_{CT} = \int_0^{\infty} \rho_{AA}(t)dt \bigg/ \!\! \int_0^{\infty} t\rho_{AA}(t)dt \tag{3.28}$$

which is the CT rate constant K_{DA}. $\rho_{AA}(t)$ is obtained by solving the Liouville's Equation 3.28. For further details, we refer the readers to the excellent paper by Venkatramani *et al.* [67]. The authors show that in cases where the bridge states are energetically distant from the donor–acceptor states, the non-adiabatic CT rates are given by k_{CT} evaluated at resonant donor–acceptor energies and weighted by the Marcus activation factor. These terms are, at worst, weakly dependent on bridge length or bridge properties.

The theoretical models presented have been applied by several research groups to study the rates of CT in several ICT-based molecules. Recently, Manna *et al.* [68] used a combined quantum chemical and molecular dynamics simulation approach to investigate the photoinduced CT process in a carotenoid-porphyrin-fullerene (C_{60}) dyad (Figure 3.11). The authors reported

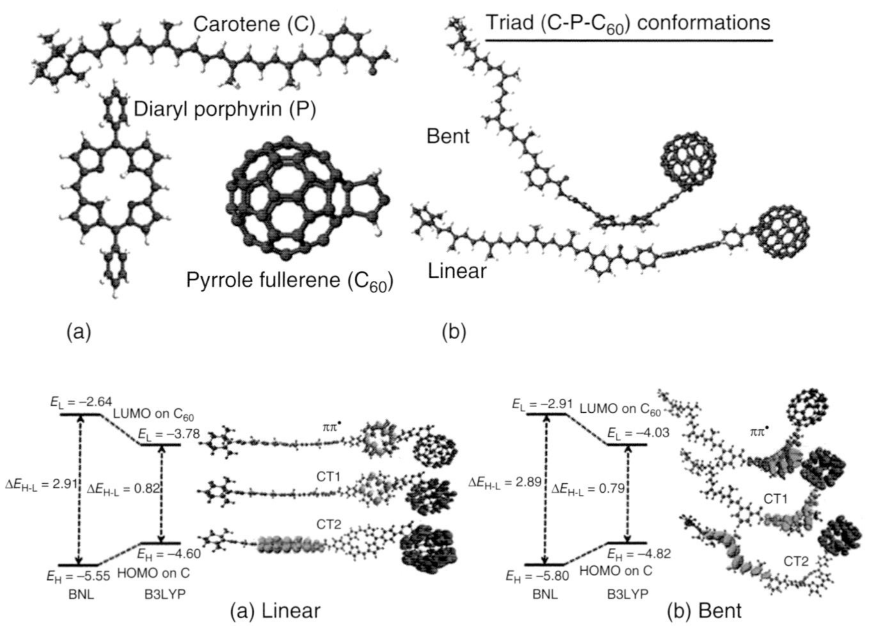

Figure 3.11 In the upper panel: (a) The molecular structures of three chemical moieties of the triad: carotene (C), diaryl-porphyrin (P), and pyrrole-fullerene (C60). (b) The linear and bent conformations of the C-P-C60 dyad. In the lower panel: the HOMO and LUMO energies of linear and bent conformation of the dyads (in eV). The nature of the different electronic excitations (ππ*, CT1 and CT2) are also shown. (Manna *et al.* 2015 [68]. Reproduced with permission of American Chemical Society.)

that the aforesaid dyad is quite flexible and can take different shapes, depending on the nature of the medium. Semiclassical Marcus theory [69] was used to calculate the rate of CT processes in these systems. The authors found that the linearly extended conformation of the dyad, where the donor (carotenoid) and the acceptor (fullerene) are separated by about 50 Å, enabled charge separation through a sequential CT process. They also considered a representative bent conformation, which in spite of being entropically favored (as found in their simulation studies), diminishes the CT process, although the donor and the acceptor are spatially closer than its linear counterpart. Previous studies [70, 71] on the sub-picosecond timescale following the excitation of the porphyrin system predicted a semi-separated CT state (CT1) that involves the bridging P as the donor, generating the C–P^+–$C60^-$ state. Later on, simulation studies on the larger timescale based on a semiempirical approach confirmed the subsequent generation of the fully charge-separated CT2 state (C^+–P–$C60^-$) on the hundreds of picoseconds timescale from the initially generated CT1 state [72]. Their computed rate of CT at the linear conformation matches fairly well with measured values. The authors assigned the reduced CT in the bent conformation to a significantly reduced rate for the transfer process between the shorter and longer range CT states, that is, CT1 to CT2. Therefore, the results presented by the authors encourage developing strategies to populate linear conformations over the bent ones in the C–P^+–C60 dyad for achieving efficient charge separation.

3.4 Experimental Methods for Studying ICT Process

As discussed in the preceding section, numerous techniques have been devoted to explore the mechanism of ICT process in several molecules. Initially, the steady-state electronic absorption and emission spectroscopy was used to study the aforesaid process; advent of several state-of-the-art spectroscopic techniques has enriched this field. Some of the common and recent techniques used to understand the ICT process are discussed in this chapter.

3.4.1 Steady-State UV–Visible Absorption and Emission Spectroscopy

Steady-state electronic spectroscopic techniques have been used widely to study the ICT process in several molecules. As mentioned in Chapter 2, some of the ICT molecules exhibit dual emission, while some show single emission from either the LE or ICT state only. Atsbeha *et al.* [73] recently studied the absorption and emission spectra of DMABN in several solvents. They found that DMABN shows dual emission in 1,4-dioxane, DCM, ethanol, and acetonitrile, while a single emission peak with λ_{max} at 342 nm was observed in cyclohexane. Their results suggest the redshift of λ_{max} due to increase in polarity is much higher for the ICT species than for the LE species, as expected from the higher dipole moment of the former. Druzhinin *et al.* studied the effect of photoexcitation of DMABN and related molecules (Figure 3.12), namely, 4-(*N*-pyrrolidinyl)aminobenzonitrile (P5C) and 4-(*N*-piperidinyl)

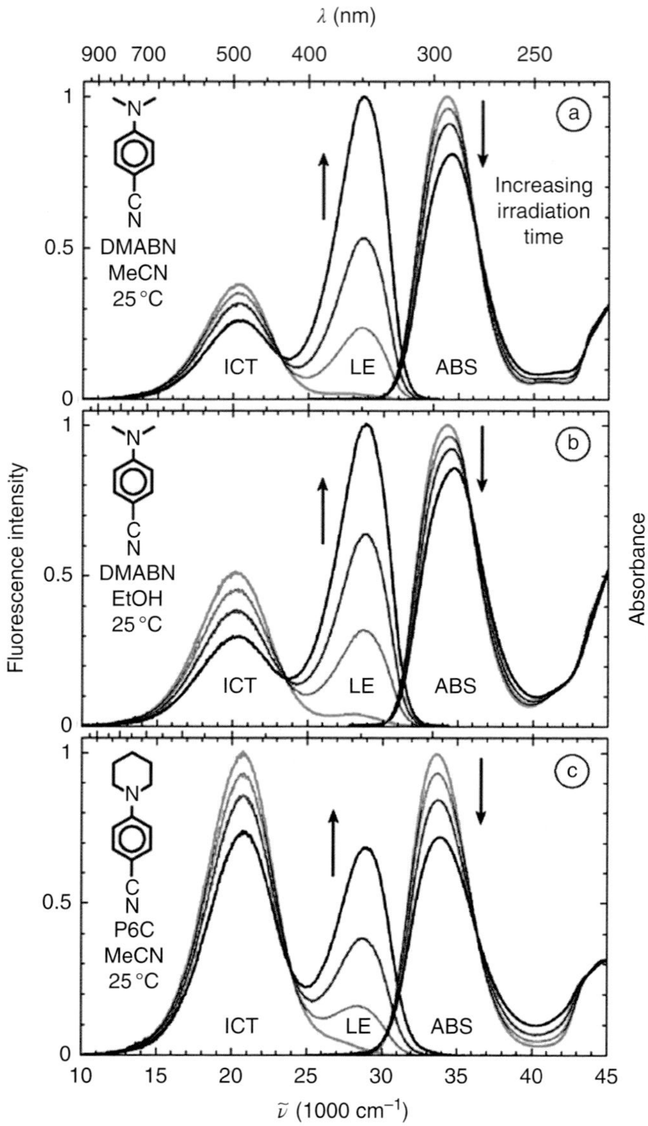

Figure 3.12 Absorption and fluorescence (dual emission from LE and ICT states) spectra of (a) DMABN in acetonitrile, (b) DMABN in ethanol, and (c) 4-(*N*-piperidinyl) aminobenzonitrile (P6C) in acetonitrile at 25 °C in arbitrary units. The changes in the spectral intensities due to photoproduct formation with increasing duration of controlled photoexcitation at 285 nm are also shown. (Zachariasse *et al*. 2005 [74]. Reproduced with permission of American Chemical Society.)

aminobenzonitrile (P6C) in acetonitrile [74]. They reported that DMABN forms a stable photoproduct 4-(methylamino)benzonitrile (MABN) that has one methyl group of DMABN replaced by a hydrogen atom. The emission of the photoproduct MABN occurs in the same range of LE state of DMABN. As the LE emission DMABN gets quenched substantially in acetonitrile while that of MABN does not, even a small amount of this photoproduct can influence the LE intensity. As the lifetime of the fluorescent excited state of MABN is comparable to the nanosecond decay component of LE state of DMABN, the photoproduct formation also enhances the relative intensity of the nanosecond decay time compared to the picosecond decay component in bi-exponential LE emission decay of DMABN in acetonitrile. This photoproduct formation therefore leads to misinterpretation of the kinetic data obtained from photostationary and time-resolved fluorescence studies of DMABN in polar solvents. The authors found that the main photoproducts of P5C and P6C are 4-cyano-N-phenylpyrrole (PP4C) and ABN, respectively.

3.4.2 Time-Resolved Ultrafast Spectroscopic Techniques

TCSPC technique with picosecond resolution is used to measure the lifetime of the excited state species. The common technique to obtain time-resolved emission decay in a TCSPC setup is to excite the system at the absorption maximum of the molecule and the emission is monitored with time generally at the emission maximum of the molecule. The excited-state lifetime of ICT species generally ranges between a few nanoseconds to a few microseconds. The formation of ICT species is generally faster than the instrument response time of a typical TCSPC setup, and other spectroscopic instruments with better temporal resolution are needed to study the fast photophysics during the ICT process. Recently, Chen *et al.* studied the ICT process [75] and excited-state dynamics of trans-4-(N-arylamino)stilbenes with aryl = phenyl (p1H), 4-methoxyphenyl (p1OM), or 4-cyanophenyl (p1CN) in solvents with varying polarities and viscosities. In polar solvents ACN, DMF, and DMSO, they found the emission curves of p1CN display a biexponential decay, analysis of which leads to a component of short lifetime that is less than their TCSPC instrument response and component with longer lifetime of 0.93–1.15 ns in the emission wavelength 380–500 nm. They also found a single exponential decay with a lifetime of 0.78–1.2 ns in the emission wavelength 520–600 nm in the same solvents. They assigned the shorter component to conversion to TICT, the long one in the blue region to fluorescence decay of the PICT state, and that in the red region to the decay of the TICT state. The authors used femtosecond TA spectral measurements to study the fast process during ICT in detail.

Ultrafast spectroscopy is used to study the dynamics of ICT process in several molecules. Due to the development in the field of laser spectroscopy, it is now possible to measure processes happening in femtosecond timescale and even faster. Some of the examples are discussed here.

Palit *et al.* [76] studied the relaxation dynamics of an ICT probe, namely, bis-[(4-dimethylamino)-phenyl]methaniminium chloride (also known as Auramine) using the time-resolved absorption technique with sub-picosecond resolution in conjunction with quantum chemical calculations. Their results suggest that the molecule reaches the LE state following photoexcitation with 400 nm light. The LE to ICT process in Auramine involves the twisting of the dimethylamino groups in this molecule. The authors claimed that the relaxation dynamics from LE state follows at least two transient states (they call them

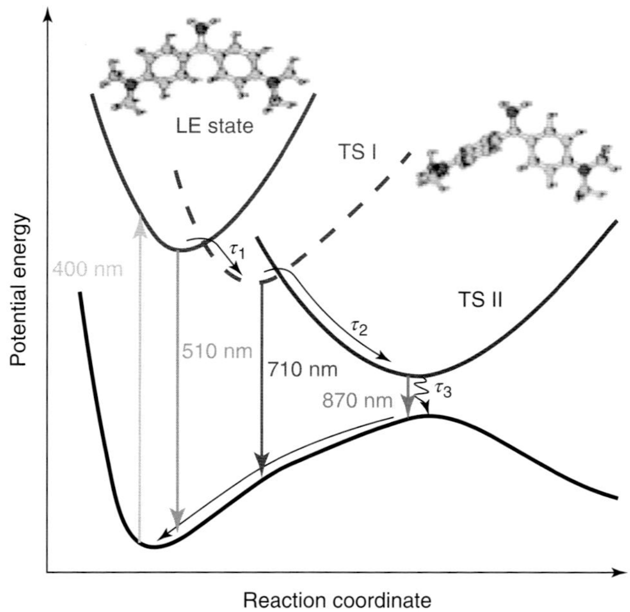

Figure 3.13 The potential energy surfaces of Auramine, as proposed by Palit and coworkers. The wavelength of the corresponding transitions (absorption, emission) are shown in the figure. (Palit *et al.* 2011 [76]. Reproduced with permission of American Chemical Society.)

TS I and TS II) that are geometrical conformers of each other. These transient states are formed consecutively following the decay of the LE state. A schematic representation of excited-state photophysics of auramine is shown in Figure 3.13.

The same group [77] studied the role of coupling between the electron donor (D) and acceptor (A) groups in ICT dynamics of isomeric *N*,*N*-dimethylaminochalcone derivatives, namely, DMAC-A and DMAC-B (Figure 3.14) using femtosecond TA spectroscopy in conjunction with DFT calculations. The coupling between the D and A units are important in view of the fact that it decides whether a TICT or a PICT state will be formed (discussed in Chapter 2). In DMAC-A, the distance between the *N*,*N*-dimethylaniline (donor) and the carbonyl (acceptor) groups is one ethylene unit shorter than that of DMAC-B. The authors claim that the twisting of the donor group is feasible in DMAC-A in the S_1 state, leading to a TICT state; while due to the large energy barrier along the twisting coordinate(s) of DMAC-B, it forms a PICT. The authors explained these results on the basis of strong donor–acceptor coupling in DMAC-A due to the shorter distance between the D and A groups. Nearly two times lower quantum yield of fluorescence and shorter lifetime of the S_1 state of DMAC-A than those in DMAC-B also support the conclusion.

Thioflavin T, a benzothiazole dye, is an ICT molecule used to detect the amyloid fibrils, a material responsible for several protein-misfolding diseases. This molecule undergoes an ICT process from the photoexcited LE state to form a nonfluorescent TICT state [78]. The emission of this molecule is found to be

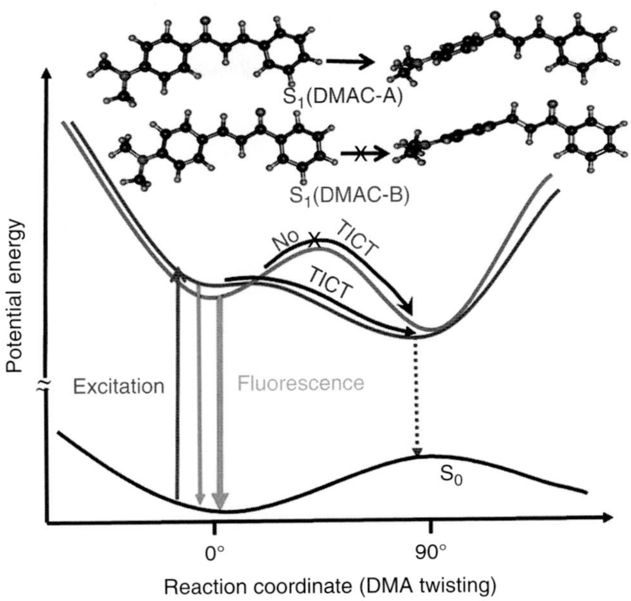

Figure 3.14 The potential energy surfaces of DMAC-A and DMAC-B. The authors found that the TICT process is feasible in DMAC-A, while no ICT reaction was detected in DMAC-B. The possible changes in structure of the aforesaid molecules due to ICT process are also shown. (Palit *et al.* 2015 [77]. Reproduced with permission of American Chemical Society.)

dependent on the viscosity of the medium. The twisting dynamics of thioflavin T in the excited state was studied using sub-picosecond and nanosecond time-resolved spectroscopic techniques. The authors found that twisting around the central C—C bond between two aromatic moieties to be barrierless and this process dominantly contributes to the ultrafast dynamics of the first excited (S_1) state [79].

Wasielewski *et al.* [80] studied the ultrafast intramolecular electron transfer (IET) in two covalently bound chromophore–catalyst ET dyads (Figure 3.15) using the femtosecond TA spectroscopic technique. In the aforesaid dyads, the dyes are derivatives of strong photooxidant perylene-3, 4:9,10-bis(dicarboximide), acronymed PDI, whereas the molecular catalyst is the Cp*Ir(2-phenylpyridine)Cl metal complex. The femtosecond TA measurements were carried out by exciting the molecules under investigation with 550 nm pulses with 150 fs duration that selectively excite the PDI unit. The authors reported that photoexcitation of the PDI units in these dyads leads to the reduction of the chromophore to PDI^- in less than 10 ps with quantum yield of unity. Substitution of bis(CF_3)phenyl groups to the bay position of PDI increases the rate of the ET as it makes the dyad a better photooxidant and also more soluble in polar, water-miscible organic solvents. A rapid charge recombination in these dyads that occurs through a biexponential decay of the PDI^- unit is reported, suggesting two ion-pair populations, probably formed due to structural changes at the iridium center upon oxidation. The relative abundance of these ion pairs are reported to be solvent dependent.

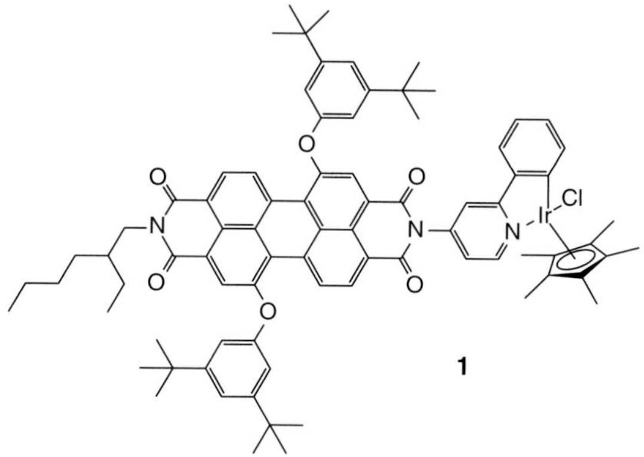

1

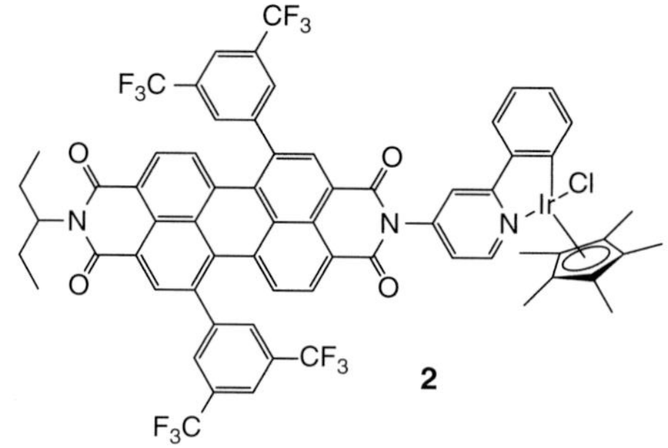

2

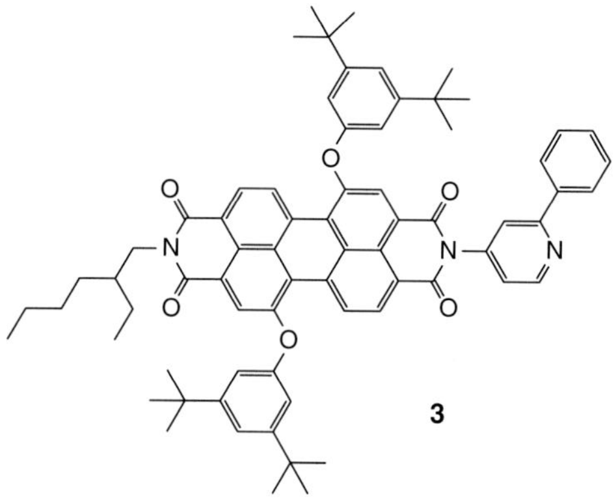

3

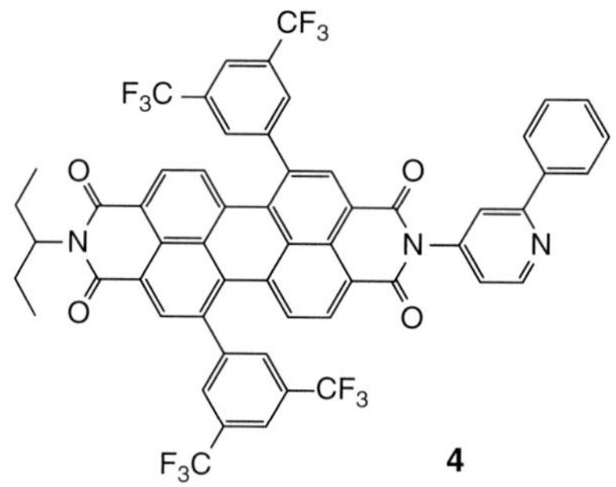

4

Figure 3.15 Molecular structure of PDI/Ir complex dyads (1 and 2) and their respective ligands (3 and 4). (Wasielewski *et al.* 2012 [80]. Reproduced with permission of National Academy of Sciences.)

The same group [81] reported an ultrafast photodriven intramolecular ET from a zinc porphyrin to a readily reduced diiron hydrogenase model complex.

Karunakaran and Das [82] reported the cascade of processes involved in the excited-state relaxation process in the cyano- and mono/di methoxy-substituted diphenyl acetylene derivatives. In these ICT molecules, either the monomethoxy or the dimethoxy group functions as the donor and the cyano group functions as acceptor (acronymed MA1 and DA1, respectively). Using femtosecond pump–probe spectroscopy and nanosecond laser flash photolysis, the authors studied the ultrafast dynamics and intermediates involved in the ICT process in these molecules. In acetonitrile, a polar solvent, laser irradiation excites the DA1 molecule to the FC state where the dipole is induced and the solvent configuration is almost the same as in the GS. Due to interaction between the electric field and permanent dipoles of the solvents, the solvent molecules start to reorient themselves to reach a new equilibrium position. This initial dominant solvent relaxation dynamics occurs due to coupled libration and translational motion of the solvent molecules. This state is called the "hot" state. The process was followed by the combined intermediate and final parts of the solvation process that involves damped rotation and diffusive rotational and translational motions, leading to the formation of the LE (Sol) state (Figure 3.16). In a polar solvent (acetonitrile), the authors proposed formation of a PICT state, and this state is further relaxed to form a TICT state of higher dipole moment than that of the GS. The TICT state has charge-separated radical cation and anion species that recombine to form the triplet state. In cyclohexane, a nonpolar solvent, the emission of an ICT molecule mainly comes from the LE state. The authors found that the LE and PICT states are the major intermediates for formation of the triplet state of DA1 in cyclohexane. These results led the authors to conclude that in acetonitrile, excited-state dynamics of MA1 and DA1 is controlled by ICT dynamics along with ultrafast solvation dynamics, whereas in cyclohexane it is primarily controlled by the solvent dynamics.

Abe *et al.* [83] studied the ICT process in 2-pyridin-1-yl-1*H*-benzimidazole (SBPa), a betaine pyridinium dye, using a combined steady-state, femtosecond photoionization (in the gas phase) and femtosecond TA (in solution) spectroscopic measurements in conjunction with quantum chemical calculations. Their results predicted and showed a negative solvatochromism in the CT absorption band of this molecule due to an S_0 to S_2 vertical transition. The emission in this molecule occurs due to a weakly solvatochromic S_1 to S_0 transition. Their quantum chemical studies predicted a two-step CT process in this molecule. The underlying photophysics of SBPa is shown in Figure 3.17 using both diabatic and adiabatic representations. Their experimental results indeed support the two distinct processes in this molecule. Photoexcitation leads this molecule to the second excited (S_2) state, followed by ultrafast deactivation to the first excited (S_1) state. The authors found that the CT process in this molecule is dependent on the solvent polarity and viscosity in aprotic media, while in proteic media the hydrogen bond formation plays an important role. The authors compared the results to that obtained from pretwisted analog of SBPa that led them to discard the idea of formation of a highly twisted ICT state. The process occurring in SBPa after reaching the S_2 state at the FC geometry involves relaxation to the

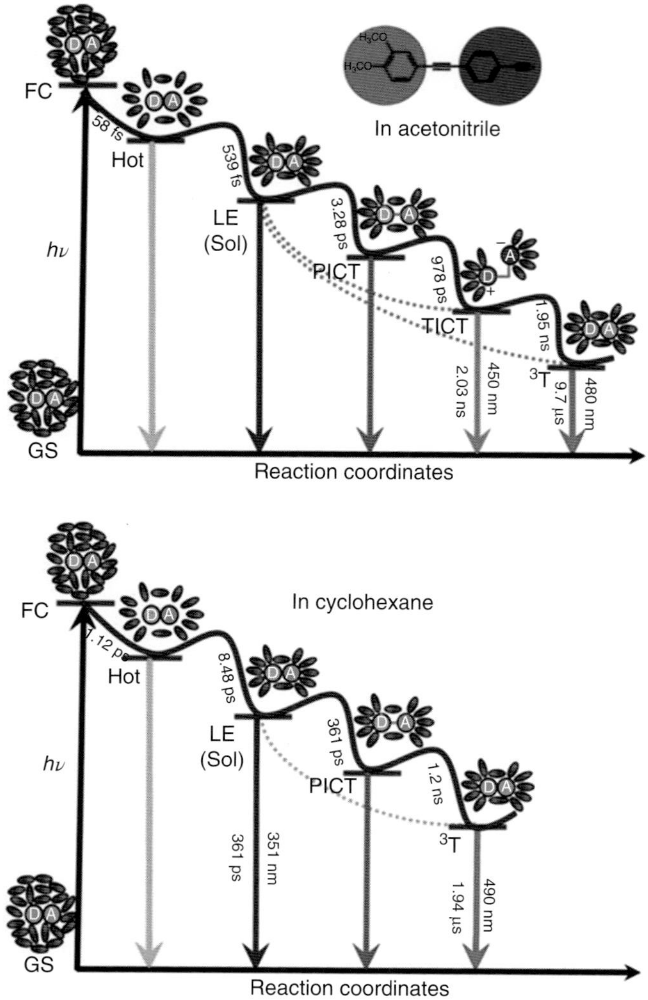

Figure 3.16 The excited-state deactivation pathway of DA1 in acetonitrile and cyclohexane involving the solvation and ICT relaxation (see text for details). (Karunakaran *et al.* 2016 [82]. Reproduced with permission of American Chemical Society.)

S_2(CT) state in about 100 fs, which is accompanied by reorganization of the electronic configuration. The system then undergoes internal conversion from S_2(CT) to S_1(CT) in a timescale of about 300 fs to 20 ps, depending on the nature of the media. The authors found a competitive process that leads the molecule to the S_1(E) state from the S_2(CT) state, and this S_1(E) is reported to be the only emissive state in this molecule. In the gas phase, the authors reported the S_2(CT) to S_1(E) transition to be only deactivation route in SBPa.

3.4.3 Raman/Resonance Raman Spectroscopy

Resonance Raman spectroscopy has been used to study the ICT process in several molecules. Interested readers may go through a concise review that

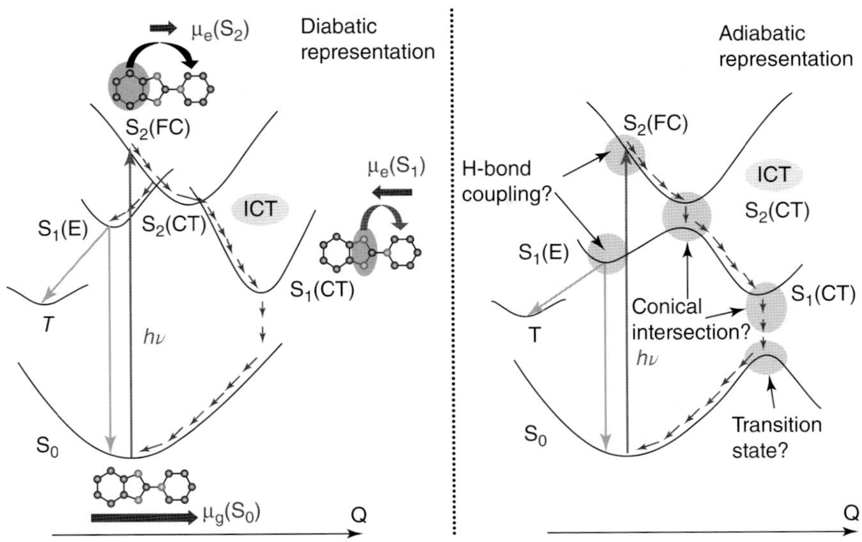

Figure 3.17 Diabatic and adiabatic representations of the photophysics of SBPa upon photoexcitation. (© PCCP Owner Societies; Reproduced with permission from Ref. [83].)

summarizes the early studies of resonance Raman spectroscopy to explore the CT process [84]. As mentioned earlier, picosecond time-resolved resonance Raman spectroscopy (TR^3) has been used to study the structure of the ICT state (A-state) of the DMABN [85]. The authors have used Kerr gate of 3 ps resolution to reject the intense fluorescence of DMABN that earlier hindered the use of TR^3. They have used 267-nm pump and 330-nm probe pulses with 5–10 and 1–5 μJ per pulse, respectively. To identify the frequency of the phenyl-amino stretching mode [ν(Ph-N)] of DMABN, they have also used the methyl-deuterated analog DMABN-d_6 and ^{15}N isotopic substitution (DMABN-^{15}N). Their results suggest that in the ICT state the transferred electron is delocalized over the whole benzonitrile group that finally leads to lengthening the C≡N bond. A downshift of the ν(Ph-N) by about 96 cm^{-1} in the ICT state from that in the GS indicates a notable decrease in the n–π interaction and electronic decoupling of the dimethylamino and benzonitrile groups in DMABN. This result matches with the minimum overlap principle of the TICT state and therefore rules out the possibility of formation of the PICT state that needs a strong electronic coupling between the donor and acceptor subgroups. Similar results were obtained by Hamaguchi *et al.* [86] who had used nanosecond time-resolved IR spectroscopy to study the structure of the excited singlet and triplet states of DMABN in several polar and nonpolar solvents. In a polar solvent (butanol), they observed two excited-state species, out of which the 2.2 ns component was assigned to the TICT singlet state, while the other one having an oxygen-sensitive excited-state lifetime was attributed to TICT triplet state. In a nonpolar solvent (hexane), only one species was found in the excited state, which they assigned to a non-CT triplet state. Their studies showed a large downshift of the C≡N stretching frequency by 120 cm^{-1} as it changes from 2216 cm^{-1} in the GS to 2096 cm^{-1} in the TICT singlet state. The authors also proposed that the anionic benzonitrile

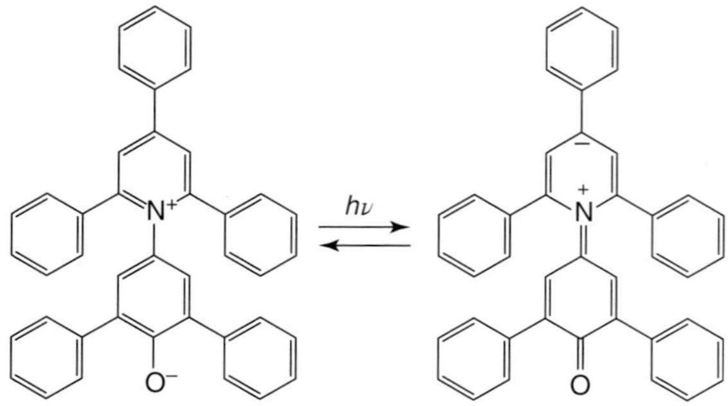

Figure 3.18 The ICT process in betaine-30. (Kovalenko *et al.* 2001 [87]. Reproduced with permission of American Chemical Society.)

group gets electronically decoupled from the dimethylamino group due to CT in DMABN, supporting the TICT mechanism.

Betaine-30, a canonical CT complex, is a well-studied ICT probe that undergoes a fast ET following photoexcitation [41–43]. The photoexcitation leads to transfer of electron density from the oxygen atom of the phenoxide ring to the pyridine ring, as shown in Figure 3.18. The charge-transferred first excited state (S_1) rapidly decays via a back electron transfer (bET) process. This bET process in betaine-30 was subjected to several studies that concluded that this process is several order faster than that was predicted by the Marcus theory. It was predicted that in addition to the solvent reorganization and thermal fluctuations, a coupling between high-frequency vibrational modes and CT coordinate must occur to drive the bET process. Kovalenko *et al.* [87] studied the spectral behavior of betaine-30 in polar solvents using pump-supercontinuum probe technique after about 50 fs excitation at 532 and 634 nm. They monitored the spectral evolution of stimulated emission and excited-state absorption to monitor the photoinduced solvation and ultrafast intramolecular rearrangement of the probe molecule in the excited state.

Frontiera *et al.* [88] used femtosecond-stimulated Raman spectroscopy (FSRS) in conjunction with DFT calculations to study the bET of betaine-30. They used the 4.4 W fundamental output of a femtosecond amplifier to generate the pump, continuum probe, and actinic pulses needed for their experiments. The authors claimed that this system has the capability to probe the excited-state vibrational features in the femtosecond timescales that are required to study the instantaneous structural changes of betaine-30 following photoexcitation (Figure 3.19).

The betaine-30 molecule remains in the D^+A^- state in the GS [41–43], unlike the normal ICT molecules that attain this form in the excited state. Upon photoexcitation, betaine-30 obtains a DA structure and therefore the dipole moment of this molecule is higher in the GS than in the excited state. Frontiera *et al.* [88] found that a large fraction of the population oscillates in the excited (DA) state and transiently pre-resonantly leads to the higher lying S^* state. The authors found that the population then undergoes bET in about 2 ps in the Marcus inverted region. An additional analysis of the frequency and amplitude kinetics of this molecule obtained from their FSRS experiments led the authors to conclude that a new relaxation pathway is available in this molecule through a low-lying electronic excited state.

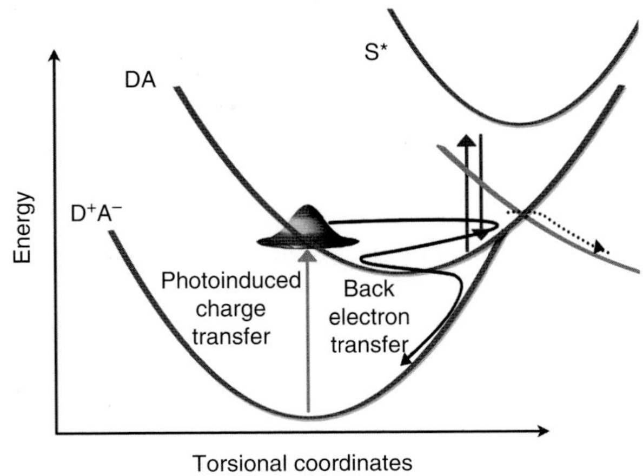

Figure 3.19 Pictorial depiction of the photophysics of betaine-30 as obtained by Frontiera *et al.* (see text for details). (Copyright 2016 PCCP Owner Societies; Reproduced with permission from Ref. [88].)

3.4.4 THz Spectroscopy and ICT Dynamics

Since the time James Clerk Maxwell wrote down his famous equations, which were verified by Heinrich Hertz experimentally, it is well known that any accelerating charge generates EM radiation [89]. This is in fact the basis for light sources in free-electron lasers and cyclotrons. The generation of EM radiation is also reported to be the basis for generation of THz pulses by photoconductive antennae [90]. THz pulses are about half-cycle pulses of EM radiation that are usually a few hundred femtoseconds in duration. A photoconductive antenna can be created by applying an electric field across the face of a semiconductor wafer (e.g., GaAs), as photoexcitation leads the electrons to the conduction band and accelerated by bias field, which generated the EM radiation. As the ICT process involves movement of charge, it should also generate an EM transient. The temporal information embedded in the emitted waveform is expected to be linked directly to the underlying pulse generation dynamics. Schmuttenmaer and coworkers [51] were probably the first to report the direct observation of IET process in betaine-30 by measuring the EM waveform broadcast by the IET process itself. They have shown that the rates of both the forward and backward CT processes can be measured on a timescale of 0.1–10 ps. The advantage this method provides that one need not to be worried on the emissive nature of the ICT molecule, as several ICT-based molecules showed poor emission quantum yield. To generate an EM pulse, the field emitted by all individual chromophores must be added together constructively. There are two fundamental requirements to achieve this goal. Firstly, the molecules must be oriented and, second, they must be excited coherently. If these molecules are oriented in only one direction, to generate the EM pulse, the CT process must induce a change in polarization along the orientation axis. The authors successfully oriented the dipolar molecules, at least partially, in a static electric field before coherent photoexcitation. They used short pulses of about 100 fs duration to photoexcite the molecules. The signal strength is expected to increase linearly with the degree of orientation, number of photoexcited molecules, and change in the dipole moment on photoexcitation. The signal strength is also expected to depend on the cosine of the angle between the axis of molecular orientation and that of the excited state dipole. The authors argued that the sign of the radiative field depends on whether the change in dipole moment is positive or negative. Therefore, it is possible find out the direction of CT directly from the information obtained in these experiments. Betaine-30, also known as Reichardt's dye, is an interesting ICT molecule, in which an electron is being already transferred from the pyridinium donor (D) group to the phenolate acceptor (A) group in the GS [41–43]. ET in the excited state upon photoexcitation at about 800 nm leads the molecule from $S_0(D^+\text{-}A^-)$ GS to S_1(D-A) excited state. Therefore, it is expected that the first excited state of this molecule will have a smaller dipole moment than the GS. The values of dipole moment in betaine-30 in the GS and excited state are reported to be 15 D and −6 D, respectively [91]. Later on, the authors [54] reported the studies of ICT process in 4-dimentylamino-4′-nitrostilbene (DMANS) using the same technique (we call it THz spectroscopy). The authors found that the pulse generated from betaine-30 is opposite in polarity, compared to the EM pulse generated from the biased GaAs, although their applied fields are in the same direction (Figure 3.20). This fact indicated that the direction

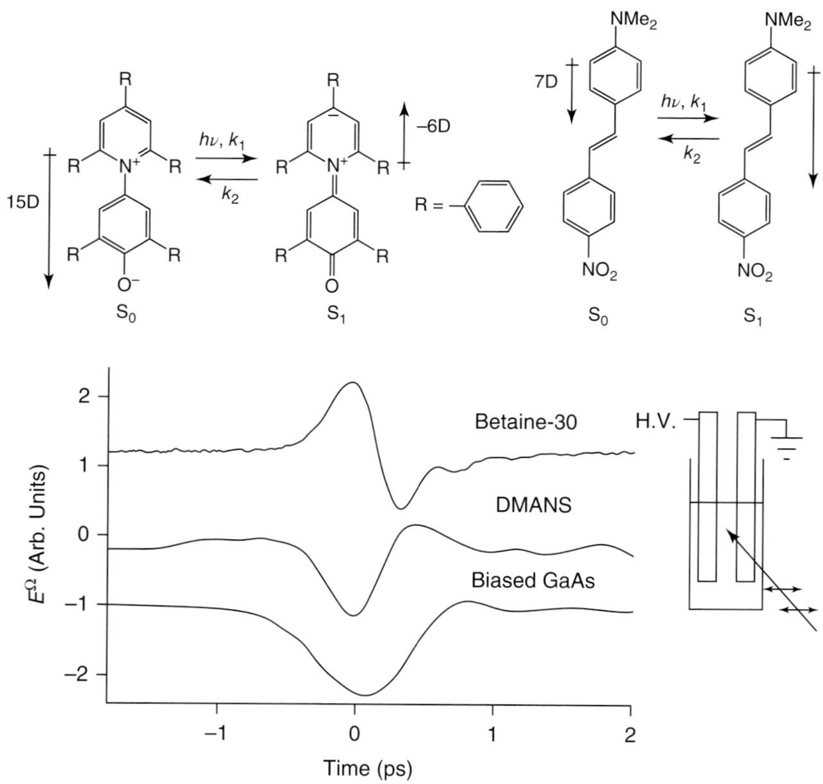

Figure 3.20 The upper panel shows the photoinduced charge-transfer process of betaine-30 (left) and DMANS along with their dipole moments in the ground and excited states. Lower panel shows the electric fields generated via charge transfer in DMANS and betaine-30 compared to that generated from photoexcitation of biased GaAs, while the direction of the applied field is the same in all the cases. The experimental configuration is also shown (right). (Beard *et al.* 2002 [54]. Reproduced with permission of American Chemical Society.)

of CT in betaine-30 is in the direction opposite to the applied field. Upon photoexcitation of 400 nm, DMANS showed the same general characteristics as betaine-30, but the polarity was opposite. The authors explained these results on the basis that DMANS shows higher dipole moment in the excited state (31 D) than that in the GS (7 D), which in the case of betaine-30 is opposite. The authors used the equation of time-dependent polarization as shown in Equation 3.29 to obtain the rates of forward (k_1) and backward (k_2) ET processes.

$$P(t) = \frac{N_0 k_1}{(k_2 - k_1)} \left[\exp(-k_1 t) - \exp(-k_2 t)\right] \times \left(\mu_e^2 - \mu_g^2\right) \frac{E_1}{3k_B T} \tag{3.29}$$

where μ_g and μ_e are the dipole moments of the molecule in the GS and excited state, N_0 is the initial density of excited molecules, E_1 denotes local electric field, k_B is the Boltzmann constant, and T is the absolute temperature.

The authors evaluated the values of k_1 and k_2 for betaine-30 to be 3.37 and 0.67 ps^{-1}, respectively. As this method is independent of the nature of the donor and acceptor as this technique measures the movement of charge that generates the signal, it can be expected that the method is applicable to study the CT processes in other systems like self-assembled layer structures.

References

1 Lippert, E., Luder, W., Moll, F., Nagele, W., Boos, H., Prigge, H., and Seibold-Blankenstein, I. (1961) *Angew. Chem.*, **73**, 695.

2 Lippert, E. *et al* (1962) in *Advances in Molecular Spectroscopy* (ed. A. Mangini), Pentagon Press, Oxford.

3 Rettig, W. (1986) *Angew. Chem. Int. Ed.*, **25**, 791.

4 Grabowski, Z.R., Rotkiewicz, K., and Rettig, W. (2003) *Chem. Rev.*, **103**, 3899.

5 Lippert, E. *et al* (1987) *Adv. Chem. Phys.*, 68.1.

6 Majumdar, D. *et al* (1991) *J. Phys. Chem.*, **95**, 4324.

7 Zachariasse, K.A. (2000) *Chem. Phys. Lett.*, **320**, 8.

8 Sobolewki, A.L. and Domcke, W. (1996) *Chem. Phys. Lett.*, **250**, 428.

9 Sobolewki, A.L. and Domcke, W. (1996) *Chem. Phys. Lett.*, **259**, 119.

10 Druzhinin, S.I., Ernsting, N.P., Kovalenko, S.A., Lustres, L.P., Senyushkina, T.A., and Zachariasse, K.A. (2006) *J. Phys. Chem. A*, **110**, 2955.

11 Chattopadhyay, N., Serpa, C., Pereira, M.M., de Melo, J.S., Arnaut, L.G., and Formosinho, S.J. (2001) *J. Phys. Chem. A*, **105**, 10025.

12 Daum, R., Druzhinin, S., Earnst, D., Rupp, L., Schroeder, J., and Zachariasse, K.A. (2000) *Chem. Phys. Lett.*, **341**, 272.

13 Zachariasse, K.A. *et al* (2001) *Chem. Phys. Lett.*, **320**, 8.

14 Zachariasse, K.A. *et al* (2000) *Chem. Phys. Lett.*, **323**, 351.

15 Maus, M. *et al* (1999) *J. Phys. Chem. A*, **103**, 3388.

16 Cao, X. *et al* (1998) *J. Phys. Chem. A*, **102**, 2739.

17 Jayaram, B. *et al* (1990) *J. Phys. Chem.*, **94**, 4666.

18 Dreyer, J. and Kummrow, A. (2000) *J. Am. Chem. Soc.*, **122**, 2577.

19 Tassle, A.J.V., Prantil, M.A., and Fleming, G.R. (2006) *J. Phys. Chem. B*, **110**, 18989.

20 Marguet, S. *et al* (1993) *Chem. Phys.*, **160**, 265.

21 Gomez, I. *et al* (2005) *J. Am. Chem. Soc.*, **127**, 7119.
22 Zhang, W., Lan, Z., Sun, Z., and Gaffney, K.J. (2012) *J. Phys. Chem. B*, **116**, 11527.
23 Dahl, K. *et al* (2005) *J. Phys. Chem. B*, **109**, 1563.
24 Ma, C. *et al* (2002) *J. Phys. Chem. A*, **106**, 3294.
25 Wagner, N.L., Greco, J.A., Enriquez, M.M., Frank, H.A., and Birge, R.R. (2013) *Biophys. J.*, **104**, 1314.
26 Ragnoni, E. *et al* (2015) *J. Phys. Chem. B*, **119**, 420.
27 Lee, J.K., Fujiwara, T. *et al* (2008) *J. Chem. Phys.*, **128**, 164512.
28 Gustavsson, T. *et al* (2009) *J. Chem. Phys.*, **131**, 031101.
29 Jarowski, P.D. and Mo, Y. (2014) *Chem. Eur. J.*, **20**, 17214.
30 Georgieva, I. *et al* (2015) *J. Phys. Chem. A*, **119**, 6232.
31 Boda, S. *et al* (2009) *Proc. Natl. Acad. Sci. U.S.A.*, **106**, 12311.
32 Zigmantas, D. *et al* (2004) *Phys. Chem. Chem. Phys.*, **6**, 3009.
33 Copczynski, M. *et al* (2007) *J. Phys. Chem. A*, **111**, 2257.
34 Olivier, T.A.A. and Fleming, G.R. (2015) *J. Phys. Chem. B*, **119**, 11428.
35 Fujiwara, T. *et al* (2014) *Int. Symp. Mol. Spectrosc.* doi: 10.15278/isms.2014.FC10
36 Fuβ, W. *et al* (2007) *Photochem. Photobiol.*, **9**, 1151.
37 Amatatsu, Y. *et al* (2005) *J. Phys. Chem. A*, **109**, 7225.
38 Curchod, B.F.E., Sisto, A., and Martinez, T.J. (2017) *J. Phys. Chem. A*, **121**, 265.
39 Kochman, M.A. *et al* (2015) *J. Chem. Theory Comput.*, **11**, 1118.
40 Du, L. *et al* (2015) *J. Chem. Theory Comput.*, **11**, 1360.
41 Levinger, N.E., Johnson, A.E., Walker, G.C., Akkeson, E., and Barbara, P.F. (1992) *Chem. Phys. Lett.*, **196**, 159.
42 Akkesson, E. *et al* (1992) *J. Phys. Chem.*, **96**, 7859.
43 Walker, G.C. *et al* (1992) *J. Phys. Chem.*, **96**, 3728.
44 Sumi, H. and Marcus, R.A. (1986) *J. Chem. Phys.*, **84**, 4894.
45 Fedunov, R.G., Plotnikova, A.V., Ivanov, A.I., and Vauthey, E. (2017) *J. Phys. Chem. A*, **121**, 471.
46 Oslen, S. and McKenzie, R.H. (2012) *J. Chem. Phys.*, **137**, 164319.
47 Tolbert, L.M. *et al* (2012) *Acc. Chem. Res.*, **45**, 171.
48 Ediz, V. *et al* (2008) *J. Phys. Chem. A*, **112**, 9692.
49 Weigel, A. *et al* (2012) *Phys. Chem. Chem. Phys.*, **14**, 11150.
50 Sasaki, S. *et al* (2016) *J. Mater. Chem. C*, **4**, 2731.
51 Beard, M.C., Turner, G.M., and Schmuttenmaer, C.A. (2000) *J. Am. Chem. Soc.*, **122**, 11541.
52 Smith, P.R. *et al* (1988) *IEEE J. Quantum Electron.*, **24**, 255.
53 Beard, M.C. *et al* (2002) *J. Phys. Chem. A*, **106**, 7146.
54 Beard, M.C., Turner, G.M., and Schmuttenmaer, C.A. (2002) *J. Phys. Chem. A*, **106**, 878.
55 Schmuttenmaer, C.A. *et al* (1759) *Chem. Rev.*, **104**.
56 Ulbrichtetal, R. *et al* (2011) *Rev. Mod. Phys.*, **83**, 543.
57 Nitzan, A. (2006) *Chemical Dynamics in Condensed Phases: Relaxation, Transfer and Reactions in Condensed Molecular Systems*, Oxford University Press, Chapter **9**, 14–16.

58 Bennet, L.E. (1973) *Prog. Inorg. Chem.*, **18**, 1.
59 Miller, J.R. *et al* (1984) *J. Am. Chem. Soc.*, **106**, 3047.
60 Giese, B., Amaudrut, J., Kohler, A.K., Spormann, M., and Wesseley, S. (2001) *Nature*, **412**, 318.
61 Segal, D.A., Nitzan, A. *et al* (2000) *J. Phys. Chem. B*, **104**, 3817.
62 Segal, D.A. *et al* (2000) *J. Phys. Chem. B*, **104**, 2790.
63 Lewis, F.D. *et al* (1997) *Science*, **277**, 673.
64 Traub, M.C. *et al.* (2007) *J. Phys. Chem. B*, **111**, 6676.
65 Nitzen, A. *et al* (2001) *J. Phys. Chem. A*, **105**, 2677.
66 Nitzen, A. *et al* (2002) *Isr. J. Chem.*, **42**, 163.
67 Venkatramani, R. *et al* (2014) *Faraday Discuss.*, **174**, 57.
68 Manna, A., Balamurugan, D., Cheung, M.S., and Dunietz, B.D. (2015) *J. Phys. Chem. Lett.*, **6**, 1231.
69 Marcus, R.A. (1993) *Rev. Mod. Phys.*, **65**, 599.
70 Rozzi, C.A. *et al.* (2013) *Nat. Commun.*, **4**, 1.
71 Spallanzani, N., Rozzi, C.A., Varsano, D., Baruah, T., Pederson, M.R., Manghi, F., and Rubio, A.J. (2009) *Phys. Chem. B*, **113**, 5345.
72 Rego, L.G.C., Hames, B.C., Mazon, K.T., and Joswig, J.-O. (2014) *J. Phys. Chem. C*, **118**, 126.
73 Atsbeha, T. *et al* (2010) *J. Fluoresc.*, **20**, 1241.
74 Druzhinin, S.I., Galievsky, V.A., and Zachariasse, K.A. (2005) *J. Phys. Chem. A*, **109**, 11213.
75 Chen, I.C. *et al* (2016) *Phys. Chem. Chem. Phys.*, **18**, 28164.
76 Palit, D.K. *et al* (2011) *J. Phys. Chem. A*, **115**, 8183.
77 Palit, D.K. *et al* (2015) *J. Phys. Chem. A*, **119**, 11128.
78 Stsipura, V.I. *et al* (2007) *J. Phys. Chem. A*, **111**, 4829.
79 Ghosh, R. and Palit, D.K. (2014) *ChemPhysChem*, **15**, 4126.
80 Wasielewski, M.R. *et al* (2012) *Proc. Natl. Acad. Sci. U.S.A.*, **109**, 15651.
81 Wasielewski, M.R. *et al* (2010) *J. Am. Chem. Soc.*, **132**, 8813.
82 Karunakaran *et al* (2016) *J. Phys. Chem. B*, **120**, 7016.
83 Abe, J. *et al* (2012) *Phys. Chem. Chem. Phys.*, **14**, 1945.
84 Myers, A.B. (1996) *Chem. Rev.*, **96**, 911.
85 Kwok, W.M. *et al* (2001) *J. Phys. Chem. A*, **105**, 984.
86 Hamaguchi *et al* (1995) *J. Phys. Chem.*, **99**, 7875.
87 Kovalenko *et al* (2001) *J. Phys. Chem. A*, **105**, 4834.
88 Frontiera *et al* (2016) *Phys. Chem. Chem. Phys.*, **18**, 20290.
89 Jackson, J.D. (1962) *Classical Electrodynamics*, John Wiley & Sons, Inc, New York.
90 Fattinger, C. and Grischkowsky, D. (1989) *Appl. Phys. Lett.*, **54**, 490.
91 Liptey, W. (1969) *Angew. Chem. Int. Ed.*, **8**, 177.

4

Medium Effect on ICT Process: Theory and Experiments

4.1 Introduction

It is now a well-known fact that solvents play a major role in determining the electronic structure and properties of a molecule [1–31]. The rate of a reaction occurring in solution can be dramatically different compared to that in the gas phase or in a different kind of solvent. Therefore, the rate of a reaction occurring in solution can be tuned by altering the properties of the medium. Using its equilibrium solvation, a solvent can increase or decrease the energy of the reactants compared to that of the transition state of that reaction, affecting the energy barrier of the reaction and hence its speed. It can also use the dynamic or nonequilibrium solvation for the molecules in which the electronic charge is distributed during the reaction. As the dipole moments of intramolecular charge-transfer (ICT) molecules are notably different in the ground and excited states due to the difference in charge distribution in these states, the interaction of the molecule with the solvents in the ground and excited states could be changed dramatically. Therefore, ICT molecules are ideal systems for studies on both equilibrium and dynamic solvation. In this chapter, we discuss the effect of the properties of the medium on the ICT process and their implications. The effect of polarity of the medium and hydrogen bonding on the ICT process in several molecules are mainly emphasized.

4.2 Some Theories and Models of Solvation

We have mentioned in Chapter 2 that ICT processes are mostly studied in solution. The properties of the solvents (polarity, viscosity, etc.) play a significant role in the aforesaid process. It is now known that a solute interacts with a solvent mainly in the two following ways, depending on the nature of the solute and solvents. Firstly, the solute can form hydrogen bonds with the solvents, known as *specific* or *microscopic* solvation. This gives rise to clusters of specific sizes and stoichiometries. This cluster formation occurs only when the solute and solvents have hydrogen bond donating and accepting groups. On the other hand, the solvent can dissolve the solute through its dielectric continuum. The latter one is non-stoichiometric and known as *macroscopic* solvation. In polar protic solvents

Intramolecular Charge Transfer: Theory and Applications,
First Edition. Ramprasad Misra and S. P. Bhattacharyya.

like ethanol, methanol, and water, an ICT molecule could be solvated using both microscopic and macroscopic solvation. It has been reported that both microscopic and macroscopic solvations play crucial roles in determining the structure and properties of a solute in solution, although their relative contributions can vary from system to system. Therefore, the structure and reactivity of a molecule may differ in the gas phase from what is seen in the solution, mainly due to differences in the electronic arrangements in the gas and solution phases. The differences in properties are prominent when we move from nonpolar solvents to polar solvents. To understand the solvation of a molecule in a medium, several groups have studied the formation of microclusters with the solvents [16, 32–40]. Studies on clusters with specific size provide us with unprecedented details about the interaction of the molecule with the solvent in the bulk, which otherwise could be extremely difficult to understand. Although the studies on photophysics of ICT molecules are carried out mostly in solution because of availability of spectroscopic instruments, the studies on formation of size-selective molecular or ionic clusters have become a hotly pursued area in molecular science in the past few decades with the advent of state-of-the-art spectroscopic techniques like nozzle jet, resonance two-photon ionization (R2PI), and so on [16, 32–40].

In solutions, a molecule is generally surrounded by a few hundred to several thousand solvent molecules. Therefore, it needs huge computational effort to treat the solvent effect on a molecule explicitly. In this chapter, we review some of the theories and models of solvation that are relevant for studies on ICT molecules. Interested readers may go through a comprehensive review [41] on implicit solvent models by Cramer and Truhler. One of the early models to calculate the solvation energy that primarily dictates the change in absorption energy of a molecule was proposed by Kirkwood [42]. With an approximation of spherical molecules, the solvation energy $[w(\varepsilon)]$ can be expressed as

$$w(\varepsilon) = \frac{1}{4\pi\varepsilon_0}\left[\frac{Z^2e^2}{2a}\left(1-\frac{1}{\varepsilon_s}\right)+\frac{|\mu|^2}{a^3}\left(\frac{\varepsilon_s-1}{2\varepsilon_s+1}\right)\right] \tag{4.1}$$

where Z and e are the molecular charge and elementary charge, respectively, and μ is the electric dipole moment. The terms ε_0 and ε_s stand for the dielectric constants of vacuum and the medium under investigation, respectively, and "a" denotes the radius of the molecule. If the absorption of a molecule involves the transition from the S_0 state to the S_n state, the change in solvation energies (Δw) of the aforesaid states can be written as

$$\Delta w = w(S_n) - w(S_0) = \frac{1}{4\pi\varepsilon_0 a^3}\left[|\mu_{S_n}|^2 - |\mu_{S_0}|^2\right]\left(\frac{\varepsilon_s-1}{2\varepsilon_s+1}\right) \tag{4.2}$$

where μ_{S_n} and μ_{S_0} stand for the electronic dipole moments of the S_0 and S_n states, respectively, in the gas phase. Therefore, one can expect that the difference in dipole moment between the absorbing excited state(s) to that with the ground state dominantly dictate the shift in absorption maxima of a molecule in a particular solvent.

Nakano *et al.* [43] reported the solvent-independent absorption spectra of the unsubstituted boron-dipyrromethene (BODIPY) dye and explained their results obtained from quantum chemical calculations using the Kirkwood model of solvation. In the gas phase, their calculations predict the dipole moments of 4.50 and 4.78 D in the S_0 and S_1 state of this molecule. Their results show that the Δw varies between 0.002 and 0.005 eV that corresponds to change less than 1 nm in the absorption wavelength. The authors claim that the negligible change in structure is responsible for the high quantum yield of fluorescence of these molecules. In the Bayliss–McRae model of solvatochromism [1], the energy (frequency) of the absorption of a solute in the gas phase and in solvent depend on the refractive index, bulk static relative permittivity, as well as the ground- and excited- state dipoles. Gutierrez and coworkers [44] proposed a solvatochromism model to account for the dipolar solute–polar solvent interactions qualitatively. Their model is based on the variation in dipole moment and the orientation in the excited state with respect to those in the ground state in the framework of the Franck–Condon excitation, considering explicit solute and solvent molecules. Considering an ideal case of interaction, the dipole moment of the solvent molecule (μ_s) belonging to the first solvation cell can be considered oriented according to the orientation of μ_0, as shown in Figure 4.1. According to the authors, the classical interaction potential between two dipoles can be expressed as

$$V_{12} = \frac{\mu_1 \times \mu_2}{r^3} - 3\frac{(\mu_1 \times r)(\mu_2 \times r)}{r^5} \tag{4.3}$$

where μ_1 and μ_2 are the dipoles and r is the vector between these two dipoles. For the expression one can expect that V_{12} will decrease as the value of r increases. Under certain approximations, the authors proposed a crude model that shows that the interaction potential between the excited state S_n and the ground state S_0 depends on the variation in angle (ϕ) between the dipoles.

Several quantum chemical studies are devoted to understand the effect of solvation on the absorption and emission properties of ICT molecules [45, 46]. Many

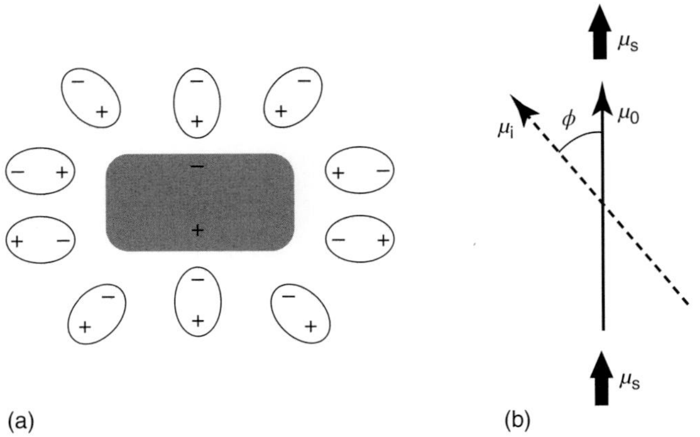

Figure 4.1 (a) Polar solvent cage around a dipolar solute, as proposed by Gutierrez and coworkers. (b) Model of dipole–dipole interaction between a polar molecule (solute) and two polar solvent molecules. (Gutierrez *et al.* 2005 [44]. Reproduced with permission of Royal Society of Chemistry.)

of these studies deal with the bulk solvation of a molecule, followed by calculations of desired properties. Some of these studies are discussed in this chapter.

One of the very popular ways to study the effect of bulk solvation on the absorption and fluorescence energies is using the general solvent effect. In this theory, it is considered that the fluorophore is a dipole immersed in a continuous medium of uniform dielectric constant. It should be noted here that formation of the ICT state may need the rotation of group(s) and/or change of structure of the fluorophore. In several cases, the formation of the ICT state is not fully representable within the framework of the general solvent effect. Within the limit of general solvent effect, Lippert–Mataga analysis [47, 48] is reported to provide a reliable means to examine the specific solvent effect on the photophysics of several probe molecules. This theory connects the Stokes' shift to properties of the solvent [49, 50]. To derive the Lippert–Mataga equation, it was considered that the energy difference between the ground and excited states of a molecule is dependent on the refractive index (n) and dielectric constant (ε) of the solvent. While deriving the Lippert–Mataga equation, it was also considered that the change in the energy of fluorophore depends on the change in its dipole moment and there is no chemical interaction between the probe and solvent molecules. The Lippert–Mataga equation can be expressed mathematically as follows:

$$\upsilon_A - \upsilon_F = \frac{2}{hc}\left[\frac{\varepsilon - 1}{2\varepsilon + 1} - \frac{n^2 - 1}{2n^2 + 1}\right]\frac{(\mu_E - \mu_G)^2}{a^3} + C \tag{4.4}$$

where υ_A and υ_F are the wavenumbers (expresses in cm^{-1}) of absorption and emission of the probe molecule, respectively. The term h represents the Planck's constant, c is the speed of light, and "a" stand for the radius of the probe molecule. μ_E and μ_G are the dipole moments of the molecule in its ground and excited states, respectively, and C is a constant. In the right side of Equation 4.4, the first term inside the bracket considers the effects of both the reorientation of the solvent dipole and redistribution of electrons in the solvent molecule on the spectral change of the probe molecule, whereas the second term accounts for the redistribution of electrons only. So the difference between these two terms accounts for the spectral shifts due to reorientation of solvent molecules and is known as *orientation polarizability* ($\Delta f(\varepsilon, n)$), which is sometimes written as Δf. The parameter $\Delta f(\varepsilon, n)$, which is also known as *Lippert solvent polarity parameters,* can be mathematically expressed as shown in Equation 4.5.

$$\Delta f(\varepsilon, n) = \frac{(\varepsilon - 1)}{(2\varepsilon + 1)} - \frac{(n^2 - 1)}{(2n^2 + 1)} \tag{4.5}$$

The Lippert–Mataga plot is constructed by plotting the Stokes' shift, that is, the difference in absorption and emission energies (in cm^{-1}) of the probe in a solvent, against the corresponding orientation polarizability, $\Delta f(\varepsilon, n)$. Usually, a linear Lippert–Mataga plot is expected for molecules that do not have any specific interaction with the solvents. Nonlinearity in this plot indicates that the solute and solvent molecules either interact through specific solvation or react chemically. The Lippert–Mataga analysis proved to be quite useful for analysis of solvent effect, at least qualitatively [49–52]. Zhu *et al.* [53] used the Lippert–Mataga analysis (Figure 4.2) to find out the difference in

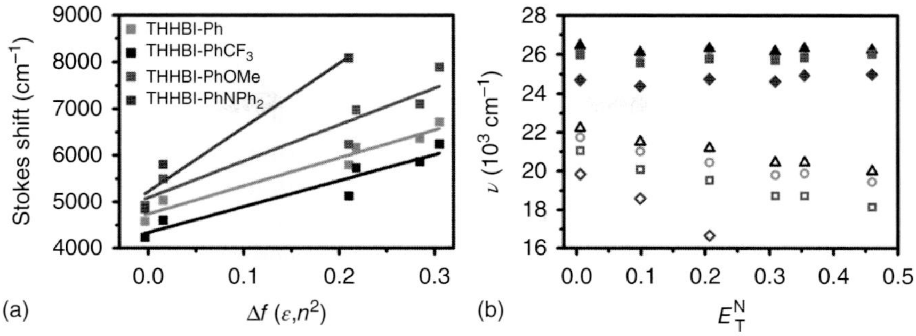

Figure 4.2 (a) The Lippert–Mataga plot for THHBI molecules (see text). (b) The change in absorption and emission maxima of the molecules with Reichardt's solvent parameters. (Zhu *et al.* 2016 [53]. Reproduced with permission of Nature Publishing Group.)

ground- and excited-state dipole moments ($\Delta\mu_{EG}$) of several tetrahydro[5] helicine-based imide (THHBI) derivatives. Their studies reveal that the values of $\Delta\mu_{EG}$ are about 13 D for THHBI-Ph and THHBI-PhCF_3, while the same is 15.3 D for THHBI-PhOMe and 24.3 D for THHBI-PhNPh_2 (see Figure 2.17 for the structure of these molecules).

We have studied the effect of solvents on the absorption and emission properties of 3-(phenylamino)-2-cyclohexen-1-one (PACO), an ICT probe, using a combined experimental and theoretical studies [54]. Our results show that the Lippert–Mataga plot for this probe is nonlinear, indicating strong interaction between the probe and solvent molecules. The similarity in the decrease in the excited-state lifetime of this molecule in polar protic solvents, like ethanol and methanol and in the presence of a dilute acid, led us to surmise that the probe molecule abstracts proton from the solvents in the excited state. A similar observation was reported by Ramkumar and Kannan [55] who found the Lippert–Mataga plot for two pyrazoline core fluorescent materials to be nonlinear, indicating hydrogen bonding between the solute and solvent molecules.

The Lippert–Mataga equation was derived assuming that the fluorescence arises from the state directly reached upon photoexcitation; however, this is not always the case. As the emission often takes place from a relaxed state, the modified treatment proposed by Liptay may be more appropriate to estimate the excited-state dipole moment (μ_E) of a solute. According to this modification, the fluorescence maxima (ν_f) are correlated to solvent polarity parameter ($\Delta f'$) as [56, 57]

$$\upsilon_f = -\frac{2\mu_e'(\mu_e' - \mu_g)\Delta f'}{hca^3} + C \tag{4.6}$$

The solvent polarity parameter ($\Delta f'$) can be expressed as

$$\Delta f' = \frac{(\varepsilon - 1)}{(2\varepsilon + 1)} - 0.5\left[\frac{(n^2 - 1)}{(2n^2 + 1)}\right] \tag{4.7}$$

Although Liptay's modified equation is used by several groups [58, 59] to extract the excited-state dipole moment of a solute, the Lippert–Mataga analysis seems to be more popular for this purpose.

For studying the solvent effect on photophysics of a probe quantitatively, the Kamlet–Taft analysis is quite popular among the scientific community [60, 61]. The Kamlet–Taft analysis provides one with the opportunity to separate the role of solvent polarity and hydrogen bonding abilities on the absorption and emission maxima of a probe. To do this, the change in the spectral properties of a molecule (e.g., absorption, (ν_{abs}) and emission (ν_{emm}) maxima/stretching frequency) due to change in medium is modeled with the change in polarizibalities (π^*), hydrogen bond donor abilities (α), and hydrogen bond acceptor abilities (β) of the solvents. The Kamlet–Taft equation can be expressed as shown in Equation 4.8.

$$\nu = \nu^0 + s\pi^* + a\alpha + b\beta \tag{4.8}$$

where ν^0 is the maximum frequency of absorption or emission (or infrared (IR) stretching frequency) of the probe in gas phase, while ν is the corresponding quantity in the presence of the solvent. Sometimes, the corresponding values in a nonpolar solvent are used as ν^0 when measurement in the gas phase is difficult. The s, a, and b are the coefficients of the corresponding solvent parameters. The values of solvent polarizabilities (π^*), hydrogen bond donor abilities (α), and hydrogen bond acceptor abilities (β) of a large number of solvents can be found in the literature.

The Reichardt solvent parameter, $E_T(30)$ values (in kcal mol^{-1}) of solvents are derived from the negative solvatochromism of pyridinium N-phenolate betaine dye [62]. The aforesaid parameter is expressed as

$$E_T(30) = hc\nu_{max}N_A = \frac{28\,591}{\lambda_{max}} \tag{4.9}$$

where h, c, and N_A represent the Planck's constant, velocity of light, and Avogadro's number, respectively. ν_{max} (in cm^{-1}) and λ_{max} (in nm) denote the maximum frequency and maximum wavelength of absorption of the π–π^* charge-transfer (CT) band of the aforesaid dye. Marcus [63] used linear solvation energy relationship for $E_T(30)$ and showed that it can be related to the Kamlet–Taft solvent parameters, as shown in Equation 4.10.

$$E_T(30) = 30.2 + 12.99(\pi * -0.21\delta) + 14.45\alpha + 2.13\beta \tag{4.10}$$

From this equation it is evident that the $E_T(30)$ parameter takes the polarity and hydrogen bonding abilities of solvents into account.

Catalan [64] proposed a solvatochromism scale based on four empirical solvent parameters, namely, polarizability (SP), dipolarity (SdP), acidity (SA), and basicity (SB) of the solvents. The author separated the nonspecific solvent effect into the dipolarity and polarizability terms. The solvatochromic scale can be mathematically expressed as

$$\mathrm{A} = \mathrm{A}_0 + b\mathrm{SA} + c\mathrm{SB} + d\mathrm{SP} + e\mathrm{SdP} \tag{4.11}$$

where A is the solvent-dependent physiochemical property of a molecule in a solvent, while A_0 is the same property in the gas phase. The b–e are the regression coefficients that are the measure of different parameters on solvation-dependent properties of a solute.

Cole and coworkers [65] used the Kamlet–Taft analysis as well as the model proposed by Catalan to rationalize the effect of solvation on the absorption and emission properties of several coumarin dyes. The authors found the fitted results to be useful to explain the solvation effect on the aforesaid dyes. The authors also pointed out the failure of the empirical solvent parameters in terms of the quality of the fitting or invalidity of some of the predicted interactions that indicate some other types of solute–solvent interactions present in the solution, such as significant conformational changes or aggregation of the solute in the solution.

Renger *et al.* [66] derived an expression to understand the solvatochromic shifts of nonpolar solutes in a nonpolar medium. Their studies reveal that a higher excited state will give rise to a large solvatochromic redshift. They also claimed that the solvatochromic shift does not depend on the oscillator strength of the transition, contrary to some other theories, but it reflects the change in dispersive solute–solvent interactions between the ground and excited states of the solute, which can be determined by the anisotropy of intramolecular electron correlation.

4.3 Effect of Solvent Polarity, Viscosity, and Temperature in the ICT Process

It is now well established that the ICT process depends on the properties of the media. The polarity and viscosity of the solvent as well as the temperature affect the ICT process of a molecule. For this reason, ICT molecules are being used as fluorescent sensors for probing the microscopic polarity and microviscosity of many biological systems and well as thermal sensors at the cellular level. The applications of ICT probes for the aforesaid purpose are discussed in Chapter 5. In this section, we discuss some of the representative examples on the studies on solvent effect on the ICT process.

Hicks *et al.* [67] reported one of the very first studies on the effect of solvent polarity on the ICT process of 4-*N*,*N*-dimethylaminobenzonitrile (DMABN) that is still relevant. They have shown that the polarity of the medium plays a major role in the isomerization dynamics of molecules that show a large change in dipole moment during this process. In the aforesaid process, the structural changes are generally described as the motion of a particle on a one-dimensional potential energy curve with the initial and final states separated by a barrier. The authors mentioned three regions of solute–solvent couplings in the barrier-crossing theories that also explain the experimental findings of structural changes of molecules in the excited state in solutions. In the weak-coupling limit, the rate of barrier crossing is increased with the increase in this coupling (e.g., collision frequency and viscosity). The solvent collision in this region assists the solute in gaining energy so that it can reach the top of the barrier. In the intermediate region, the rate first approaches the maximum, followed by a decrease in rate with increase in coupling. In the high-coupling region, as the authors have mentioned, the motion becomes diffusive. The particle takes many small forward and backward steps in this region to cross the barrier to reach the final structure. In the so-called Smoluchowski limit, the rate of barrier crossing becomes inversely proportional to solvent coupling. Hicks *et al.* [67] showed that the barrier height for the twisted intramolecular charge transfer (TICT) process (E_a) decreases linearly with the increase in solvent polarity parameter $E_T(30)$, as shown in Equation 4.12.

$$E_a = E_a^0 - A[E_{30}(30) - 30] \tag{4.12}$$

where E_a^0 is the activation energy in an alkane solvent with $E_T(30)$ value equal to 30 kcal mol^{-1} and A is measure of how strongly the barrier height changes with change in solvent polarity.

Although initially proposed for DMABN, this theory was found to be useful to explain the polarity-dependent behavior of many other ICT molecules [68]. As decrease in solvent polarity restricts the ICT process, according to this model, the lifetime of the ICT emission is expected to increase monotonically with decrease in solvent polarity. Das *et al.* [69] used this model to explain the effect of solvent polarity on the spectral response of two anilino-naphthalene sulfonates, namely, 2-tolouidino-6-naphthalene-sulfonate (TNS) and 1-anilino-8-naphthalene-sulfonate (ANS). The authors expressed the overall rate of TICT process in a particular medium (k_T), as shown Equation 4.13.

$$k_T = k_T^0 \exp(-E_a/RT) \tag{4.13}$$

where R is the universal gas constant and T is the absolute temperature.

When Equations 4.12 and 4.13 are combined, one can see that the k_T increases exponentially with the increase in the solvent polarity parameter, $E_T(30)$.

The authors argued that the rate of radiative processes (k_R) is not much affected by solvent polarity and considering the rate of non-radiative processes (k_{NR}) as sum of TICT rate (k_T) and rate of other processes (k_X), one can write

$$k_{NR} = k_T + k_X \tag{4.14}$$

If the rate of ICT process is very high, as expected in polar solvent, $k_T \gg k_R + k_X$. Under this approximation

$$1/\tau_f \approx k_T^0 \exp[-E_a^0 + A\{E_{30}(30) - 30\}] \tag{4.15}$$

Then one gets

$$\ln\left(1/\tau_f\right) = B + AE_T(30) \tag{4.16}$$

where B is a constant. So, one can expect that the plot of $\ln(1/\tau)$ versus $E_T(30)$ to be a straight line. The authors found that in TNS the $\ln(1/\tau)$ versus $E_T(30)$ plot is indeed a straight line at high polarity, which indicated that at this polarity the ICT process is the main non-radiative pathway of this molecule. However, at low polarity region, the aforesaid plot deviates from the linearity. To rationalize this behavior, the authors studied the effect of addition of external heavy atom (ethyl iodide) on the lifetime of TNS as heavy atoms enhance the intersystem crossing (ISC), decreasing the radiative lifetime. They found that with addition of ethyl iodide, the lifetime of TNS is decreased. Therefore, the authors conclude that

ISC is an important channel of excited-state deactivation, especially at the low polarity region. The authors also found that for ANS the lifetime and quantum yield of fluorescence decrease monotonically with the rise in solvent polarity and ethyl iodide has negligible effect on these parameters. Therefore, they conclude that for ANS the ICT process is the main excited-state deactivation pathway, irrespective of solvent polarity.

Chang and Cheung [70] proposed a model to account for the solvent polarity effect on the non-radiative rates of several TICT molecules. They showed that the model fits well for rhodamine B dissolved in a series of ethanol–water mixtures as well as for 8-anilino-1-naphthalene sulfonate in the same medium. They proposed a model for evaluating the solvent polarity corrected non-radiative rate (k_{cor}) as

$$k_{cor} = k_{NR}\left[\frac{\beta[E_T(30) - 30\,\text{kcal mol}^{-1}]}{RT}\right] \tag{4.17}$$

where β is a dimensionless parameter that determines how strongly the activation barrier will vary with the solvent polarity. From a negative value of β for rhodamine B, the authors claimed that the energy barrier from the planar to the twisted state in this molecule decreases with increasing solvent polarity. The authors also noted that the solvent viscosity has a role in shaping the non-radiative rate constant of these molecules. Usually, with increase in solvent viscosity, the non-radiative deactivation rate of these molecules is decreased. The authors used solvents that have a narrow range of viscosity to avoid any influence of viscosity in their studies.

Theodorakis and coworkers [71] studied the effect of solvent polarity and viscosity of the emission properties of several ICT molecules with the aim of separating the polarity and viscosity effect on the fluorescence properties of these molecules. The probe molecules used in their studies are 9-(dicyanovinyl)-julolidine (DCVJ) and 9-(2-cyano-2-hydroxy carbonyl)-vinyl julolidine-triethyleneglycol ester (CCVJ-TEG) (Figure 4.3) that are known as *molecular rotors* (see Chapter 6 for a further discussion on molecular rotors). They compared their results to that of DMABN, a well-known ICT probe, and three coumarin derivatives, namely, ((7-amino-4-methylcoumarin-3-acetyl)amino)hexanoic acid (abbreviated as AMCA), coumarin 1, and coumarin 6. Molecular rotors are usually fluorescent molecules that show TICT and their emission behavior depends on the rotation of a group or groups present in these molecules. As increase in the viscosity of the medium can restrict the rotation of group(s) in these molecules, the quantum yield of fluorescence of these molecules depends on the viscosity of the medium. Generally, in molecular rotors the excited states deexcite non-radiatively from the TICT state and do not show dual emission. To understand the difference in probe sensitivity toward solvent polarity and viscosity, the authors carried out steady-state absorption and fluorescence spectral measurements of the aforesaid dyes in several solvents. Their results show that the Stokes shifts as well as emission intensity of DMABN strongly depend on the polarity of the medium. The intensity of the locally excited (LE) peak of DMABN increases

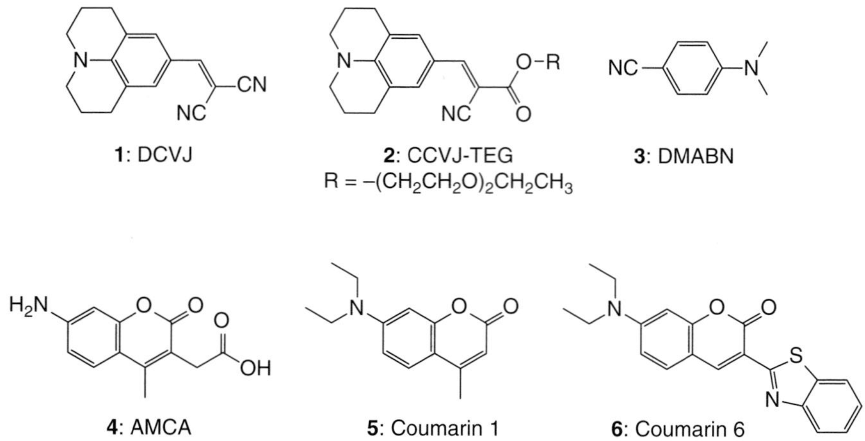

Figure 4.3 The chemical structure of DCVJ, CCVJ-TEG, DMABN, AMCA, coumarin 1, and coumarin 6. (Haidekker *et al.* 2005 [71]. Reproduced with permission of Elsevier.)

notably, whereas the ICT peak increases slowly with increase in viscosity while keeping the polarity of the medium almost unchanged. Therefore, the authors attributed these changes in emission of DMABN to medium viscosity. As these molecules showed changes in emission behavior with change in both polarity and viscosity of the medium, the authors conclude that it is difficult separate the effects from these parameters. Emission from coumarin dyes showed strong dependency on the solvent polarity, while they show very weak dependency on the viscosity of the medium. The emission intensity of molecular rotors showed a strong increase with increase in viscosity. The authors showed that the change in emission intensities of DCVJ and CCVJ-TEG depend on the viscosity of the medium and can be quantified as shown in Equation 4.18.

$$I_{\text{Peak}} = C\eta^{x} \tag{4.18}$$

where η is the viscosity of the medium, while C and x are constants.

The authors claimed that the power-law relationship better describes the emission intensity with change in viscosity if one can eliminate (or minimize) the effect of polarity on it. This is an important task, because to use as a molecular rotor the effect of polarity and viscosity on the emission properties of a probe must be separated. As the emission intensities of molecular rotors depend on the ability to form TICT states in the viscous medium, the Stokes shift of the probe is not much important, although molecules with smaller variation of the Stokes shift are preferred for studies of microviscosity. The results presented by the authors could be useful in designing molecular rotors for studying microviscosity of several important systems.

Darling and coworkers [72] studied the effect of temperature and medium viscosity on the ICT emission of 4-*p*-dimethylaminostyrylpyridinium salt in several protic solvents. The authors varied the viscosity of the medium by adding sucrose, glycerol, or poly(ethylene glycol) to pure water. They mentioned that it is not possible to separate the polarity and viscosity effect of a medium on the ICT process completely. Therefore, the authors tried to minimize the effect of polarity while studying the effect of viscosity on the aforesaid probe. They argued that the dynamics of relaxation of molecules in solution where the rotating solute exerts high friction with the solvent can be modeled using Debye–Stokes–Einstein model. So, at a particular temperature T, the orientation relaxation time of the solvent molecule (τ_{or}) can be expressed as

$$\tau_{\text{or}} = C(\eta/T) \tag{4.19}$$

where C is the geometry-dependent rotational friction coefficient of the probe (solute) molecule and η is the shear viscosity of the medium.

Assuming that contribution from deactivation channels other than CT to be negligible, the authors showed that the quantum yield of fluorescence of the probe molecule is directly proportional to the viscosity of the medium and inversely proportional to the temperature of the medium. In many ICT-based molecules, the increase in quantum yield of fluorescence (ϕ) is reported in the viscous media, and one must be careful about the approximations involved while expecting a linear relationship between the values of ϕ and viscosity of the solvents.

Palit and coworker [73] reported the ultrafast twisting dynamics of thioflavin-T, a molecular rotor, mainly known for sensing amyloid fibrils. They found that the reaction time for LE to ICT conversion which is about 0.73 ps in acetonitrile, increases to about 3.2 ps in propylene carbonate, a solvent with higher viscosity than acetonitrile. The authors mentioned that the rate of decay of the excited state that is undergoing a barrier-crossing process through twisting of the dimethylaniline group about a single bond or conformational relaxation through rotational motion about a double bond can be modeled on the basis of change in viscosity and temperature of the medium. The aforesaid decay rate (k) can be fitted to a power-law function as shown in Equation 4.20.

$$k = Z\eta^{-\alpha} \exp\left({}^{E_\alpha}/_{RT}\right) \quad (0 < \alpha < 1) \tag{4.20}$$

$$\ln(k) = -\alpha \ln(\eta) + \ln(Z) - \frac{E_\alpha}{RT} \tag{4.21}$$

where Z is the pre-exponential factor and E_α is the activation energy of the process. T and R denote the absolute temperature and universal gas constant, respectively, while α is a constant.

Therefore, at a given temperature, in a series of solvents with similar properties in which the E_α can be assumed to be constant, a linear relation between $\ln(k)$ and $\ln(\eta)$ can be observed. In fact, the $\ln(k)$ versus $\ln(\eta)$ plot for thioflavin-T was found to be linear, and from there the value of α was extracted to be about 0.70. Some possible reasons, as the authors mentioned, for the origin of the fractional viscosity dependence could be breakdown of the Stokes–Einstein relationship, specific solute–solvent interaction, the multidimensionality of the potential energy surfaces (PESs) and time-dependent friction. The authors discarded the possibility of breakdown of the Stokes–Einstein relationship for thioflavin-T on the ground that the solute is larger than the solvent molecules. The multidimensionality of the PES of this molecule was reported to be one of the reasons for fractional viscosity dependence in this molecule.

Nath and coworkers [74] studied the viscosity effect of the medium on the ultrafast bond twisting dynamics of thioflavin-T, an amyloid fibril sensor. The authors carried out steady-state as well as time-resolved fluorescence measurements of the probe molecule using a femtosecond fluorescence up-conversion instrument in acetonitrile, ethylene glycol and their mixtures. The authors assigned the ultrafast decay of the excited state to the ultrafast torsional motion around the central C—C bond of the molecule. They found the decay traces to follow nonexponential kinetics. Therefore, they fitted the decay traces using triexponential function and average lifetimes of the excited state of thioflavin-T in different media were calculated. Their studies showed that the average lifetime of the probe in acetonitrile to be 0.61 ps and that in ethylene glycol to be 17.64 ps. The authors explained the spectral behavior on the basis of viscosity of the medium. As ethylene glycol is much more viscous than acetonitrile, the excited-state decay is expected to be slower in the latter. Recently, Stsiapura *et al.* [75] reported the effect of solvent polarity on the non-radiative decay rate of thioflavin-T. They reviewed earlier studies on the viscosity-dependent fluorescence quantum yield (ϕ) of thioflavin-T and concluded that although the values of ϕ of this molecule can be explained well using the linear function $\phi = a + b(\eta/T)$ in the temperature range of 293–323 K, the values of parameters a and b vary significantly with the change in the molecular properties of the solvents. Moreover, the total data set of ϕ for all the solvents studied does not obey the unified linear dependence, indicating solvent properties other than viscosity play their role in shaping the values of ϕ. The authors found a correlation between the ratio of non-radiative to radiative decay rates (k_{NR}/k_R) to the polar properties of the solvents.

Dixon *et al.* [76] studied the effect of binding of calf thymus DNA in the intramolecular electron transfer process of a ruthenium(II) polypyridine naphthalene diimide, where a RuII(bpy)$_2$(bpy-CONH—)$^{2+}$ electron donor covalently attached to a naphthalene diimide (NDI) acceptor unit through a -$CH_2CH_2CH_2$— linker. The binding of DNA with the aforesaid probe (Ru-NDI) was confirmed through hypochromism in the absorption spectra as well as using circular dichromism spectroscopy. Their results show that the lifetime of the excited-state species of Ru-NDI remain almost the same even after binding with DNA, while their relative amplitudes are changed markedly.

It is worth mentioning that as several ICT probes are being used as a molecular rotor and fluorescent sensors, one must be careful about assigning the non-radiative decay to the ICT process in a molecule. Escudero [77] used quantum chemical calculations at the time-dependent density functional theory (TDDFT) as well as complete active space self-consistent field (CASSCF) level of theory to shed light on the origin of fluorescence quenching in several well-known photoinduced electron transfer (PET) molecules. The author noted that in studies of fluorescent sensor and switches, the fluorescence quenching is automatically assigned to the PET process in those molecules, leading to erroneous results. Even if the PET process in a molecule is thermodynamically feasible, the process may not occur as it might not be fast enough to compete with other deactivation channels. The author proposed a mechanism, which he called dark-state quenching, to account for the photophysics of several fluorophores that can bind proton, cation, and anion. The dark states, that is, states that do not absorb or emit radiation, are often elusive to experiments and theoretical investigations are sometime useful to understand the nature of these states. The analysis done by the author, as he claimed, goes beyond the common strategy of analyzing the frontier molecular orbital energy diagram and finding the thermodynamic feasibility of the PET process. Instead, he analyzed the PES of lowest energy excited states in the molecules under investigation. The author chose donor–acceptor (D–A) systems that do not show fluorescence on photoexcitation due to deactivation through the PET process (OFF state), while in the presence of analytes the PET process is blocked and the fluorescence is activated (ON state). Figure 4.4 depicts some of the compounds considered by the author for his studies and their respective potential energy diagrams, calculated at the TD-PBE0/6-31G* level of theory. The vertical absorption and emission energies (in eV) along with their oscillator strengths are also presented in the figure. The author found the amino group of compound **1** to be coplanar with the phenyl ring and the lone pair orbital of the amino group is delocalized within the highest occupied molecular orbital (HOMO) of the phenyl ring. The S_0 to S_1 absorption in this compound has oscillator strength (f) about 0.115. Once photoexcited to the S_1 state, the system evolves along a torsional motion involving the amino group, leading to the $(S_1)_{min}$ geometry. At this geometry, the author found that the amino group is twisted with respect to the phenyl ring; thus, it corresponds to a TICT species. The lone pair of the amino group is no longer delocalized over the phenyl ring in this state and the oscillator strength becomes close to zero ($f = 0.000$), that is, the state becomes a dark state. The negligible oscillator strength at the TICT minimum makes the radiationless deactivation pathway preferred over radiative deactivation. On the other hand, the compound **2** (the protonated species of **1**) shows emission from the LE minimum that has oscillator strength of 0.055. Therefore, compound **1** shows turn-on fluorescence on acidification. The author reported the increase of fluorescence quantum yield of compound **3** from almost zero in metal-free system upon addition of potassium salt. In compound **5**, which is an "ON" species, the photophysics is dominated by $\pi\pi^*$ state of the BODIPY moiety. Contrarily, in compound **6**, the author found the S_1 state of ICT becomes a dark state. Therefore, the author concluded that PET process is always not responsible for the quenching of fluorescence in these probes.

4.4 Studies of Solvation in Some ICT Molecules

Media play a pivotal role in shaping the structure and properties of ICT probes. There are several reports of using CT-based organic molecules for studying solvent effect, using different experimental and theoretical tools. Fayer and coworkers [78], Fleming and coworkers [2], Maroncelli and coworkers [79–81], Barbara and coworkers [82–84], Huppert and coworkers [85–87], Agmon [88], and many other research groups have studied the solvent effect on ICT probes. Zewail, Eisenthal, and coworkers [89] reported the studies on the solvent effect on ICT molecules. They used a systematic approach to understand the solvation

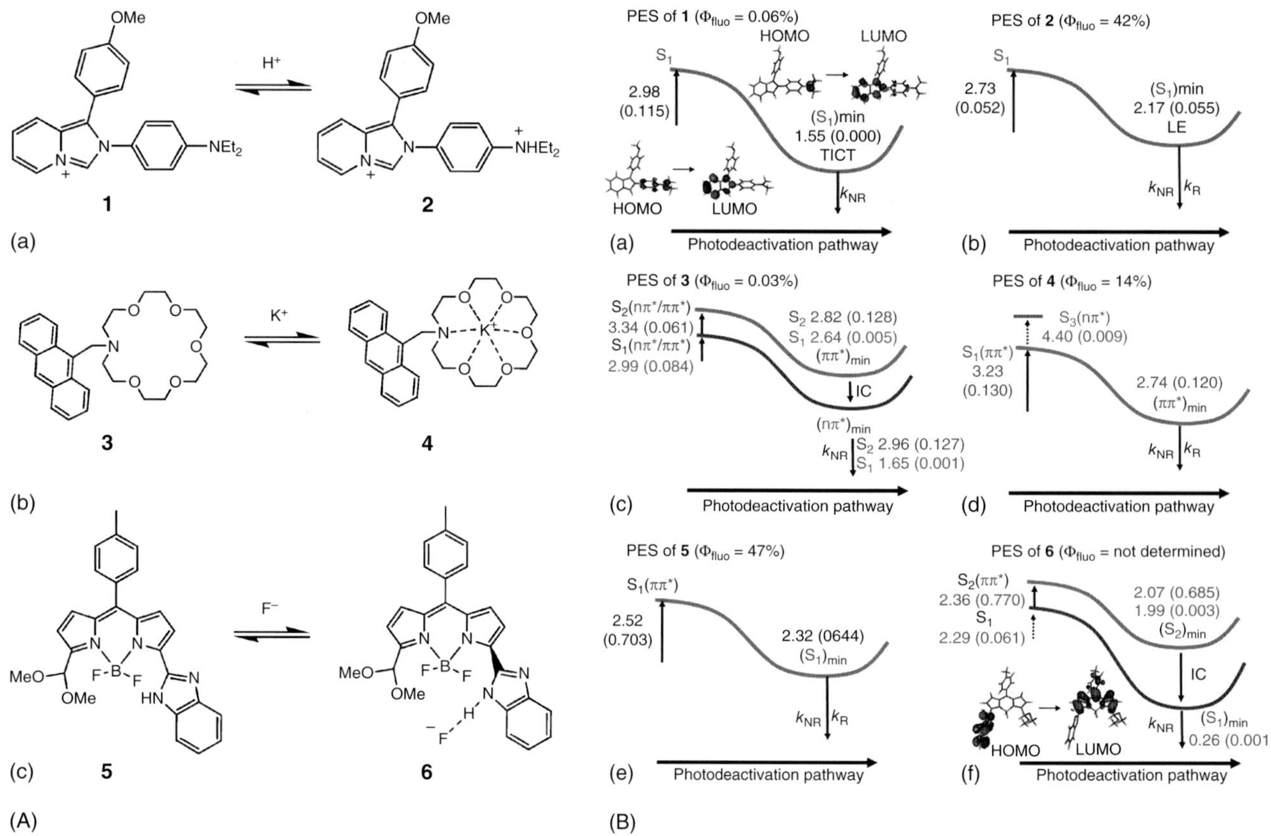

Figure 4.4 Chemical structures of compounds (**1–6**), studied by Escudero (A). Schematic representation of PES of compounds (**1–6**). Energies are presented in eV, oscillator strengths of the transitions are presented in parenthesis and experimental values of ϕ are also reported (B). (Escudero 2016 [77]. Reproduced with permission of American Chemical Society.)

of DMABN in the gas phase in supersonic jet expansion and thermalized vapor. They studied the ground- and excited-state vibrational spectra of the gas-phase DMABN as well as its stoichiometric clusters with water, ethanol, acetone, and ammonia. These studies concluded that 1 : 1 solvent–solute clusters were not enough to cause local perturbation to the molecule for charge separation.

N-Pyrrolobenzonitrile (PBN), a congener of DMABN, is one of the most studied molecules for solvent effects on the ICT [90]. In PBN (Figure 4.5), the dimethylamino group of DMABN is replaced by a pyrrole ring, which is considered a better electron-donor group than the dimethylamino group due to possession of six π-electrons. Therefore, it is expected that PBN will exhibit a more efficient ICT process than that in DMABN. Rettig *et al.* [90] were the first to report the dual emission and experimental studies of the ICT process in PBN.

The PBN absorbs at around 286 nm at room temperature and this peak is almost insensitive to the solvent polarity. The authors assigned the absorption of PBN dominantly to a ^1L_a-type state (in Platt's notation) with a small contribution from the ^1L_b-type state. They also mentioned that depending on the substitution, the role of ^1L_b and ^1L_a as S_1 and S_2 state can be interchanged. Dual emission from PBN was observed in nonpolar solvent *n*-hexane at room temperature and also at low temperature (frozen ethanol). The broad emission band observed with a maximum at around 480 nm in polar solvents at room temperature was assigned to a polar CT state. In frozen ethanol at 153 K, PBN shows two peaks with weaker peaks at around 320 and 430 nm, respectively. The relatively weaker peak at about 320 nm was assigned to the LE (S_2) state of this molecule. The authors also found that the relative intensities of the LE and ICT peaks as well as the solvatochromic shifts are strongly dependent on the temperature and the excitation wavelengths. The authors compared the results of PBN, PBAEE, and DPBN (Figure 4.5) with DMABN, a prototype for ICT systems. They also note that the compound $PBMes_2$ has comparable donor/acceptor capabilities with the molecules under investigation, although it differs in the acceptor

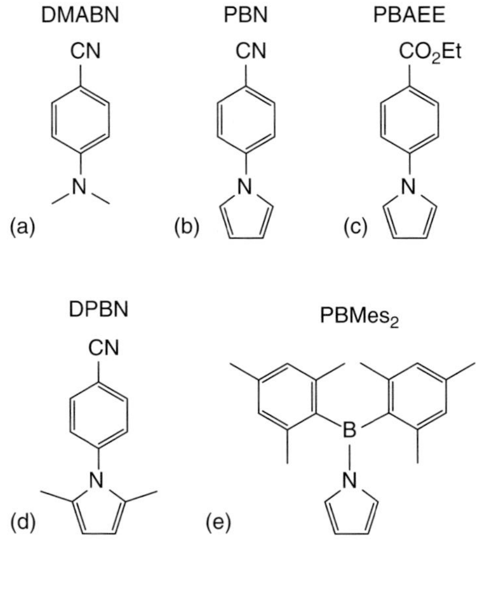

Figure 4.5 The structure of DMABN, PBN and related molecules. (Rettig *et al.* 1998 [90]. Reproduced with permission of American Chemical Society.)

strengths. The Lippert–Mataga analysis revealed the change in dipole moments between the ground and excited states of PBN and DPBN were 19.2 and 18.8 D, respectively. With their ground-state dipole moment about 3.2 D, they estimated the excited-state dipole moments of these molecules to be around 22 D. The fluorescence maxima were notably Stokes shifted for PBN, PBAEE, and DPBN with respect to the absorption maxima more than 10 000 cm^{-1}. They found that the dependence of the bathochromic shift of the emission maxima with solvent Lippert polarity parameter (Δf) were quite similar for all the three probes, although the Stokes shift of DPBN was much stronger than that of PBN. From these results the authors surmise that the CT state in these compounds are quite similar in nature. The larger Stokes shift in DPBN was assigned to stronger stabilization of the CT state due to enhanced electron-donating capability of the dimethyl-pyrrolo moiety and also to difference in emissive geometry. The longer radiative lifetimes of these molecules of about 130, 120, and 700 ns for PBN, PBAEE, and DPBN, respectively, indicated the possibility of a forbidden emission. The difference in fluorescence rate constants of pretwisted derivative DPBN with PBN and PBAEE also led to the conclusion that the structures of their emitting states were different from that of DPBN, being presumably more twisted than the others. The authors also concluded that in PBN, the CT state may not relax toward 90° that could have led it to full orbital decoupling. The relative intensities of the LE and ICT bands depend on the temperature and excitation wavelengths, indicating an energy barrier for the ICT emission to occur.

Belau *et al.* [91] studied the jet-cooled fluorescence and resonance-enhanced multiphoton ionization (REMPI) excitation spectra of pyrrolobenzene (PB) and PBN in the UV range. They observed only a single emission peak of PBN at around 310 nm in the gas phase upon excitation up to 2200 cm^{-1} above the 0–0 band. The observed fluorescence of PBN with a long lifetime (17 ± 2 ns) was assigned to the S_1 (L_b-type) state, while the absorption spectrum was assigned to the S_2 state of L_a type. The authors conclude that the energy gap between the S_1 and S_2 state is small, of about 400–500 cm^{-1}. This indicated that efficient radiationless coupling between these two states is possible, leading to population and emission from the S_1 state. Parusel [92] reported the density functional theory/ multiple reference configuration interaction (DFT/MRCI) study on the excited-state CT process of PBN. The author considered C_2 symmetry for PBN and found two CT states (2B and 3A) and two allowed $\pi\pi^*$ states (1B and 2A) relevant for the TICT process in this molecule. Based on the results, the author suggested that the redshifted fluorescence of PBN originated from the 1B-TICT state. Very recently, Dreuw and coworkers [93] reported the solvent polarity effect on the barrier height of TICT reaction of PBN. The authors used *ab initio* calculation, including, third-order algebraic diagrammatic construction of the polarization propagator [ADC(3)] level of theory in order to explain the experimental finding on PBN, including effect of solvent, temperature, and excitation wavelength on its emission properties. The effect of solvents on the transition energies of relevant excited states of PBN has been taken care of using conductor-like polarizable continuum model (C-PCM) and conductor-like screening model (COSMO). The authors noted that

PBN has C_2 symmetry in the ground state, and to reach an orthogonal conformation a small torsional barrier of about 0.13 eV need to overcome. Their calculations predicted that the S_1 state of PBN was a local $\pi\pi^*$ state with low CT character with *B* symmetry (they named it 1B), which can be assigned to the LE state (L_b-type) observed in the experiments. The S_2 state of A symmetry was named 2A, while they found the lowest-lying CT state has *B* symmetry (called 2B). The authors conclude that in the gas phase, photoexcitation led PBN to the bright 2A($\pi\pi^*$) state, where the molecule twists only a few degrees until an intersection with the close-lying 1B (LE) state is reached (Figure 4.6). After passing this 2A/1B intersection, the molecule relaxes to the 1B (LE) state minimum from which the deactivation to the ground state occurs, leading to emission from the LE state only. However, in polar solvents, this picture changes completely. In polar medium, both the 2B and 2A states get stabilized. This results in decrease in the energy barrier along the twisting coordinate as the polarity of the solvent increases. In polar solvents, the barrier along the twisting coordinate vanishes and twisting is facilitated. The authors found that at 63° the 2B state becomes the lowest in energy. Therefore, the occurrence of the redshifted emission in PBN is linked to the ability to reach the 2B TICT minimum.

Using a combined experimental and theoretical approach, we probed the effect of solvation on 3-(phenylamino)-2-cyclohexen-1-one (PACO), an ICT probe in

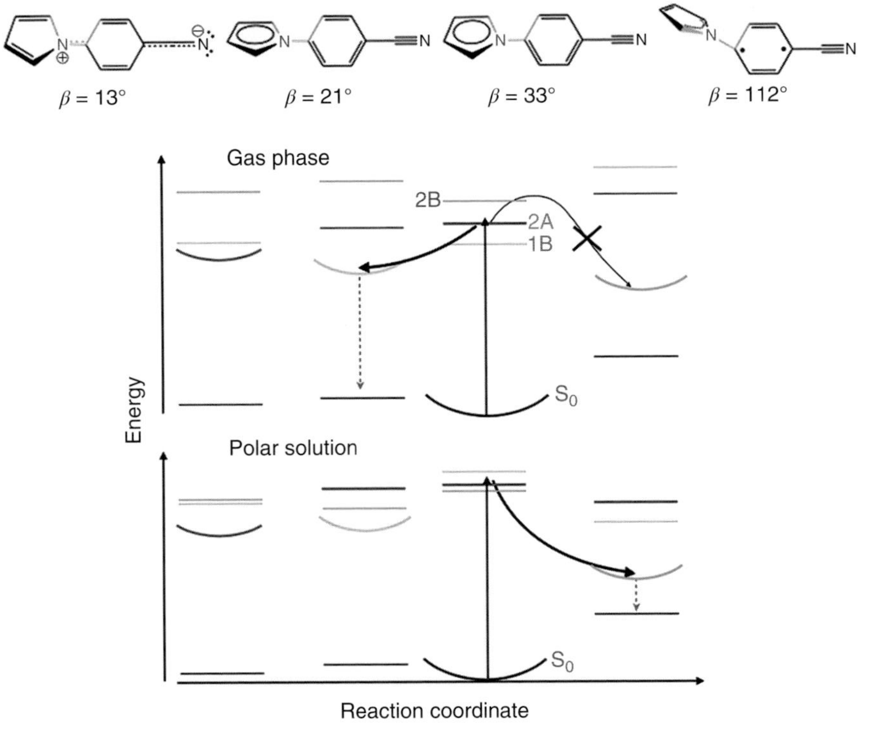

Figure 4.6 Schematic representation of photochemical reaction of PBN in gas phase and polar solvents. (Bohnwagner *et al.* 2016 [93]. Reproduced with permission of American Chemical Society.)

water [94]. The formation of microscopic clusters of PACO with water molecules was studied theoretically and it was found the probe molecule forms 1 : 3 molecular clusters with water (Figure 4.7). The macroscopic or bulk solvation was taken care of using the self-consistent reaction field (SCRF) model. We theoretically calculated the absorption spectra of PACO in the gas phase, 1 : 3 PACO-H_2O cluster in the gas phase, PACO macrosolvated in water and 1 : 3 PACO-H_2O cluster macrosolvated in water. Comparison with experimentally recorded absorption spectrum of PACO in water reveals that both specific and bulk solvation are important to rationalize the behavior of this molecule in aqueous solvent. Yun *et al.* [95] designed a novel ethylenediamine (EDA)-based chemosensor, namely, [dimethyl 2,2′-((3-oxocyclohex-1-ene-1,2-diyl)bis(azanediyl))dibenzoate] for the turn on fluorescence sensing of Cu^{2+} ions. They concluded that the ICT process of enaminone moiety is responsible for quenching of fluorescence in this molecule.

Barbara and coworkers [82–84] studied experimentally the solvent effect on betaine dyes. Maroncelli and coworkers [79–81] experimentally studied the role of solvent on the ICT processes in several ICT-based molecules, including Coumarin 153 and 9-(4-biphenyl)-10-methylacridinium ($BPAc^+$), crystal violet lactone (CVL), and bianthryl derivatives. Huppert and

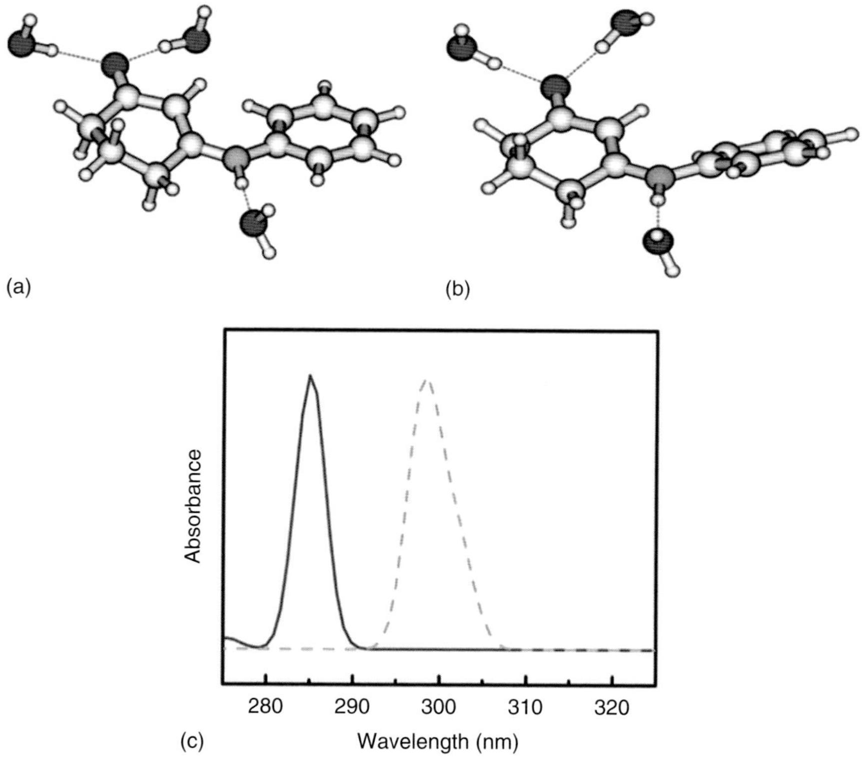

Figure 4.7 Theoretically computed structures of (a) 1 : 3 PACO-H_2O cluster in the gas phase, (b) 1 : 3 PACO-H_2O cluster macrosolvated in water, and (c) the corresponding simulated absorption spectra (solid and dotted line, respectively). (Misra *et al.* 2009 [94]. Reproduced with permission of American Chemical Society.)

coworkers [85–87] studied the effect of solvation on 6-(4-methylphenyl) amino-2-naphthalenesulfon-*N*,*N*-dimethylamide (TNSDMA), an electron transfer probe. The effect of solvents on D–A-based electron transfer probes, octadecylrhodamine B (ODRB) and *N*,*N*-dimethylaniline (DMA) have been reported by Fayer and coworkers [78]. The same group has reported studies of solvent effect on the CT process of rhodamine 3B. Bhattacharyya and coworkers [96–99] studied solvent effect on DMABN using theoretical tools. Guchhait and coworkers [100–103] studied the solvent effect on ethyl ester of *N*,*N*′-dimethylaminonaphthyl-(acrylic)-acid, 2-methoxy-4-(*N*,*N*-dimethylamino)benzaldehyde, two well-known ICT probes. There are several other examples of studies on the effects of a medium on the CT process. Interested readers may go through a review [104] on effect of solvents (especially hydrogen bonding) on the CT process.

We have noted earlier that several ICT molecules show dual emission in polar solvents from the LE and ICT species of those molecules, while a single emission peak is observed in nonpolar solvents from the LE species. The ratio of quantum yield of fluorescence of the ICT and the LE species is therefore an important parameter for studies of solvent effect on the ICT process. An ICT process can be illustrated using Scheme 4.1 [105].

In Scheme 4.1, the absorption of radiation ($h\nu$) leads the molecule to the LE state from which it leads to the ICT state. k_a and k_d are the rate constants of forward and backward ICT reaction, respectively. k_f (LE) and k_f (ICT) are the radiative rate constants, while τ_0 (LE) and τ_0 (ICT) are the fluorescent lifetimes of the LE and ICT species, respectively. The ratio of quantum yield of fluorescence of the ICT (ϕ_{ICT}) and the LE (ϕ_{LE}) species can be expressed as follows:

$$\frac{\phi(\mathrm{ICT})}{\phi(\mathrm{LE})} = \frac{k_f(\mathrm{ICT})}{k_f(\mathrm{LE})} \frac{k_a}{k_d + {}^1/_{\tau_0}(\mathrm{ICT})} \tag{4.22}$$

Zachariasse *et al.* [105] reported that except for very large reaction enthalpies, k_a and k_d are much larger than ${}^1/_{\tau_0}(\mathrm{ICT})$ at room temperature. Then one can rewrite the above equation as

$$\frac{\phi(\mathrm{ICT})}{\phi(\mathrm{LE})} = \frac{k_f\,(\mathrm{ICT})}{k_f\,(\mathrm{LE})} \frac{k_a}{k_d} \tag{4.23}$$

It has been reported that for DMABN, k_a to k_d ratio strongly depends on the solvents and usually increases with increase in solvent polarity. The aforesaid ratio is 0.40 in toluene ($\varepsilon = 2.37$) and increases to 2.3 in diethyl ether ($\varepsilon = 4.24$) at

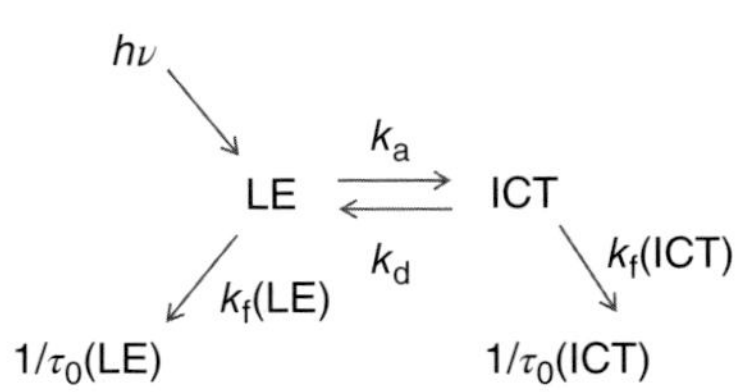

Scheme 4.1 Schematic representation of the processes involved during the ICT process (see text for details).

298 K. The k_a to k_d ratio of DMABN is reported to be 511 in acetone ($\varepsilon = 36.7$) at the same temperature [106]. On the other hand, the ratio of k_f (LE) to k_f (ICT) is much less dependent on the polarity. For example, $k_{f(ICT)}/k_{f(LE)}$ is 0.33 in toluene and 0.12 in acetonitrile. These results clearly indicate that the values of ϕ_{ICT}/ϕ_{LE} is dominantly dictated by the values of k_a/k_d. Fujiwara *et al.* [107] reported the ϕ_{ICT} to ϕ_{LE} of 4-fluoro-*N*,*N*-dimethylaniline (FDMA) to be about 2.0 in both nonpolar (*n*-hexane) and polar (acetonitrile) solvents. Results by Zachariasse *et al.* [105] cast doubts over this independence of ϕ_{ICT} to ϕ_{LE} ratio to solvent polarity and they did not find any signature of ICT in this molecule (they call it DMA4F). They assign the result to the high energy difference between the two lowest singlet states of this molecule ($\Delta(S_1,S_2)$), which they reported to be 8300 cm^{-1} in *n*-hexane. The value of $\Delta(S_1,S_2)$ of DMA4F is even larger than that of *N*,*N*-dimethylaniline (DMA). The authors also claimed that the formation of TICT in FDMA (or DMA4F) in TDDFT calculation by Fujiwara and coworkers could be due to artifact of the computational method.

Recently, Bohnwagner and Dreuw [108] investigated the deactivation pathways of FDMA in the gas phase as well as in solution using high-level quantum chemical methods like TDDFT, coupled cluster (CC2) theory, and the algebraic diagrammatic construction scheme (ADC(2), ADC(3)). They simulated the effect of solvent on the absorption and emission properties of this molecule using two distinct solvation models, namely, C-PCM in a perturbative state-specific correction scheme (ptSS-PCM) and state-specific equilibrium C-PCM (SS-PCM). The simulated absorption and emission energies were found to be reasonably matching with the experimental results. Their calculations revealed that the S_2 state was an optically allowed bright $\pi\pi^*$ state, while the S_1 state which was also of $\pi\pi^*$ character but having much weaker oscillator strengths than the former. The experimentally observed intense blueshifted absorption was assigned to excitation to the S_2 state, while the weaker redshifted absorption was attributed to the excitation to the S_1 state from the ground state. The authors estimated the probability of ISC by computing the energy gap and spin orbit coupling constants between the lowest lying singlet and triplets state by considering three crucial geometries, namely, the S_0 equilibrium geometry, the most stable S_1 geometry, and approximate S_1/S_0 conical intersection (CI) geometry. They found that the computed spin–orbit couplings were small, irrespective of the geometry. The authors predicted two pathways of deactivation of radiation after initial excitation to the S_2 state, although the probability of radiative and non-radiative transitions remained inconclusive. Firstly, the S_2 state of FDMA might relax to a planar S_1 state minimum via a well-accessible S_2/S_1 CI from which the fluorescence could take place. Secondly, the S_2 state might decay through nonradiative pathway to the ground state through a low-lying S_1/S_0 CI along the prefulvene-like vibrational mode of the benzene ring. The proposed deactivation mechanism of FDMA is depicted in Figure 4.8. The authors conclude that although FDMA is structurally related to DMABN and PBN, the molecule does not show dual fluorescence, irrespective of the polarity of the solvents. They also inferred that the twisted S_1 state or TICT state predicted earlier was due to computational artifact of the TDDFT method and the TICT state vanishes when the amount of nonlocal Hartree–Fock exchange in the functional is increased.

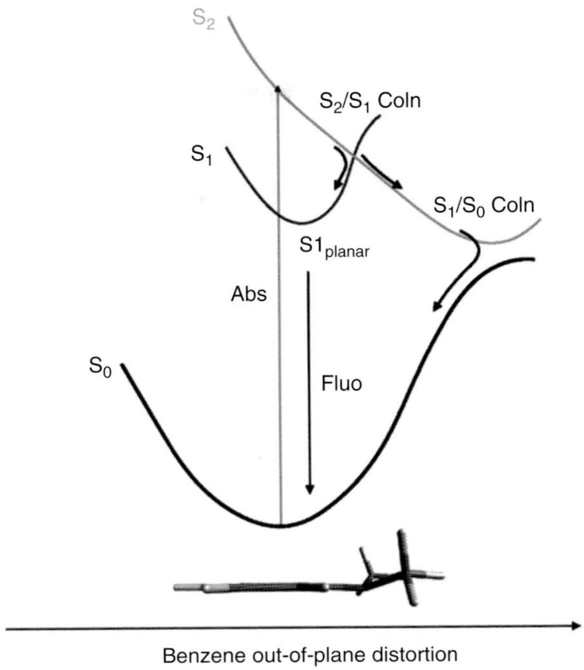

Figure 4.8 Schematic representation of the deactivation mechanism of FDMA as proposed by Bohnwagner and Dreuw. After photoexcitation to the bright S_2 state, the excited molecule may either relax to an S_1 state minimum $S1_{planar}$ and fluoresce or it may decay to the ground state non-radiatively via internal conversion. (Bohnwagner and Dreuw 2017 [108]. Reproduced with permission of American Chemical Society.)

As mentioned several times in this book, the properties of the medium play a crucial role in shaping the photophysics of an ICT probe. Sen and coworkers [109] reported the ultrafast excited-state dynamics of thioflavin-T in chloroform and methanol using time-resolved absorption and fluorescence spectroscopy in conjunction with quantum chemical calculations. To rationalize the 30 times higher fluorescence quantum yield of thioflavin-T in chloroform than that in methanol, the authors proposed formation of an additional ICT pathway (they called it ICT-2) along with the usual ICT state (they termed it ICT-1). Excitations lead the molecule to the LE state from which it could deexcite through either of these pathways. The authors noted that the ICT-1 pathway was extremely fast along the torsional coordinate of the dimethylanilino moiety, relative to the benzothiazole ring. They found that although the LE to ICT-1 pathway was most likely barrierless in nature, an inherent activation barrier was present between the LE and the ICT-2 states. The aforesaid activation barrier was found to be solvent dependent as its value was reported to be 0.59 and 1.58 kcal mol^{-1} in methanol and chloroform, respectively. Their TDDFT calculations and temperature-dependent studies also supported the existence of the activation barrier between the LE and ICT-2 states. The authors mentioned that the amplitude motion along dimethylamino torsional coordinate leads to the ICT-2 state. They reported the lifetime of the ICT-1 and ICT-2 states to be 0.8 and 2.5 ps in methanol and 9.1 and 50.6 ps in chloroform, respectively. They attributed the higher quantum yield of thioflavin-T in chloroform to higher activation barrier between the LE and ICT-2 states in this solvent. In methanol, the energy of the molecule at room temperature was reported to be sufficient to cross the energy barrier to reach the ICT-2 state, resulting in smaller

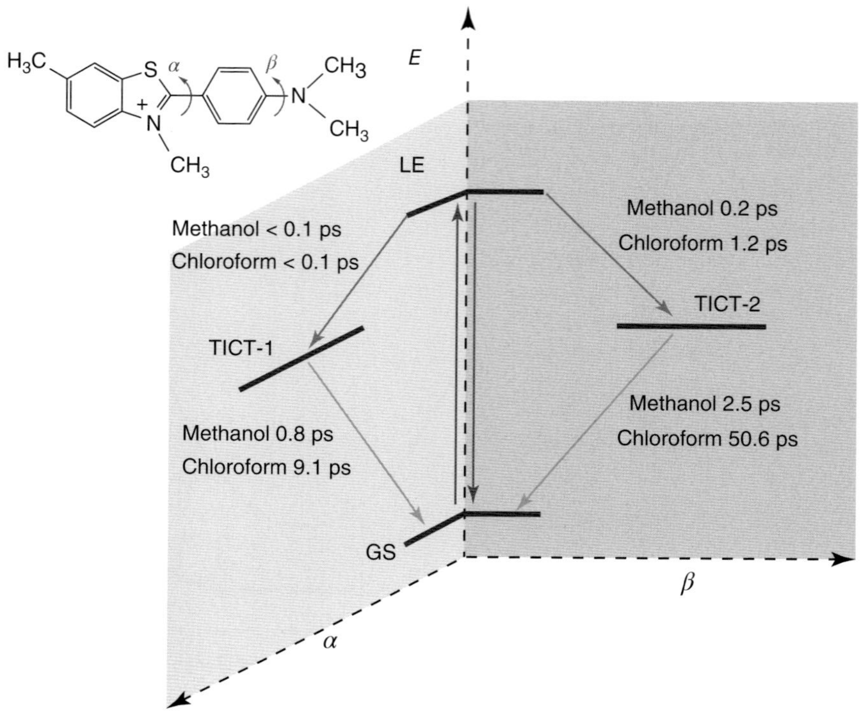

Figure 4.9 Schematic representation of the excited state relaxation pathways of thioflavin-T in chloroform and methanol. Although the locally excited molecule is likely to decay through the ICT-1 mechanism, the ICT-2 mechanism may be operational, depending on the solvent. (Mukherjee *et al.* 2016 [109]. Reproduced with permission of Elsevier.)

quantum yield of the probe in methanol. Using the aforesaid results the authors argued that the LE population of thioflavin-T mostly decayed through the ICT-1 pathway, although the ICT-2 pathway could be active in some solvents. The proposed relaxation pathways of thioflavin-T are shown in Figure 4.9. Some other examples of using femtosecond transient absorption (TA) and fluorescence up-conversion techniques to understand the excited-state photophysics of some other ICT molecules can be found in Chapter 3.

4.5 Effect of Hydrogen Bonding on ICT

The hydrogen bonding process in molecular systems is of fundamental importance in chemistry and biology. Therefore, investigations of hydrogen bonding through studies on structure and dynamical properties of molecular clusters or in bulk solvents provide deep insight into the microscopic as well as macroscopic behavior of molecules. The studies of molecular clusters with specific size provide an extra opportunity to study a particular behavior of a molecule in a tailored environment. Investigation of microclusters of a molecule with solvents facilitates investigations of the effects of bulk solvation on a molecule in a systematic way. Most of the studies on H-bonding reported so far have been devoted to the ground state of a molecule. Recently, with the advent of several state-of-the-art time-resolved spectroscopic techniques, studies of hydrogen bonding in the electronically excited states have been receiving increasing attention. TA spectroscopy with subpicosecond resolution, femtosecond time-resolved fluorescence spectroscopy, time-resolved vibrational spectroscopy, and so on are being widely used for studying the medium effect on the excited-state properties of a molecule. Some examples of these studies are discussed in the preceding section and also in Chapter 3. Complementary quantum chemical calculations have also been used widely to explore different facets of excited-state hydrogen bonding in recent years. Configuration interaction, configuration interaction singles (CIS), configuration interaction singles and doubles (CISD), TDDFT and, very recently, CASSCF methods are being used successfully for this purpose. Calculations at the TDDFT level of theory for this purpose has been a popular choice as optimization of the excited states of large organic molecules using multi-configuration self-consistent field (MCSCF) methods are costly. The advancement of computational facilities in terms of both hardware and software allowed the scientific community to use more and more sophisticated techniques for quantum chemical studies. Probably the most common way of calculating the stabilization energy (E_{SE}) due to the formation of microclusters, say AB_n, is to compute the energy difference as shown in Equation 4.24.

$$E_{SE} = -[E_{AB_n} - (E_A + nE_B)] \tag{4.24}$$

where E_A and E_B are the energy of the molecules A and B, respectively, while E_{AB_n} is the total ground-state energy of the molecular complex AB_n at equilibrium.

Molecules that undergo photoinduced ICT process are interesting for studying hydrogen bonding in excited states. It is now known that the electronic charge distribution in the excited states of an ICT molecule can be quite different from that in the ground state. Therefore, hydrogen bond donor and acceptor sites present in these molecules are expected to either be activated or deactivated in the excited state, affecting the interaction with the solvent molecule. This change in solute–solvent interaction results in either stabilization or destabilization of the ICT probe, relative to the ground state. The change in solute–solvent interactions often manifested in their spectroscopic properties. Although several studies have been devoted to study the effect of polarity and viscosity of the medium on the ICT process, the effect of hydrogen bonding between the ICT probe and solvent molecules on the aforesaid process is scarce. Krishnamoorthy and coworkers [104] reviewed the effect of hydrogen bonding on the ICT and proton transfer processes as well as its application in molecular recognition and sensing. It is generally believed that a charged donor forms hydrogen bond(s) with the solvent in the ground state. After photoexcitation, after formation of the ICT state, the donor acquires a positive charge and the hydrogen bond between the excited-state donor and the solvent molecule is expected to break. The ICT process enhances the electronic charge over the acceptor group that can help the acceptor group either to form a new hydrogen bond with the solvent or strengthen the existing one. DMABN, the molecule studied widely for ICT process in the excited state, is one of the well-studied probes for effect of hydrogen bonding in the ICT process. Coumarin dyes are another class of ICT molecules put under the scanner for the studies of hydrogen bonding that affects the formation of the ICT state. Nouchi *et al.* [110] presented one of the early studies on the effect of hydrogen bonding on the ICT process of DMABN. They studied the ICT process of DMABN in non-hydrogen-bonded polymer matrix. The authors claimed that formation of solute–solvent hydrogen bonding in the ground state is a precondition for ICT emission in the excited state. They also suggested that the twisted conformation of DMABN obtained in the ground state due to hydrogen bonding with solvents emits upon photoexcitation. This theory was later challenged by several groups and does not hold much ground as it was earlier reported by Pilloud *et al.* [111] that dual emission from the LE and ICT states of DMABN could be observed even using dry acetonitrile as solvent. They also reported that addition of water to this aprotic solvent quenches the intensity ratio of both the LE and ICT emission. Pal *et al.* [112] have experimentally studied the excited-state photophysics of coumarin-1 dye in several protic solvents. They authors reported that the quantum yield of fluorescence (ϕ_f) as well the excited-state lifetime (τ_f) of the dye decrease abruptly in polar protic solvents. The values of τ_f are temperature dependent in the high polarity solvents with solvent parameters (Δf) more than 0.28. From their experimental results, the authors conclude that formation of intermolecular hydrogen bonding between the probe molecules with solvents provides extra stability to the probe in addition to the stability obtained in the polar medium. Mitra *et al.* [113] used steady-state and time-resolved spectroscopic studies to investigate the ICT process in *trans*-ethyl *p*-(dimethylamino)cinnamate, abbreviated EDAC in solvent with varying polarities and hydrogen bonding abilities. The authors reported a large shift in the emission maxima due to increase in polarity of the medium as a signature of ICT process in EDAC. The markedly different emission properties in the hydrogen bonding solvents, compared to that in aprotic solvents led them to conclude that hydrogen bonding plays a crucial role in the ICT process of this molecule. From the decreased ICT emission in hydrogen bonding solvents, they also predict a competition between hydrogen bond formation between the donor and solvent molecules and the ICT process.

Datta and coworker [49] investigated the effect of solvent on the excited-state dynamics of 3-aminoquinoline (Figure 4.10). The authors found that a decrease in non-radiative decay rate with increasing polarity, with a sudden jump at a critical polarity. This result is counterintuitive as with increasing polarity in ICT molecules a large Stokes shift is expected, which should be accompanied by a decreasing quantum yield due to an increase in the non-radiative rate. To understand the effect of medium polarity and hydrogen bonding on the photophysics of this molecule quantitatively, the authors performed the Kamlet–Taft analysis on the changes in absorption and emission maxima. The authors also performed the same analysis on the non-radiative decay rates (k_{NR}) of this molecule in different solvents. They found that the contribution from the polarity of the medium was highest in shaping the absorption spectra of this probe, while the contribution from the hydrogen bonding parameter increased in shaping the emission profile. Their results also showed that the values of k_{NR} governed predominantly

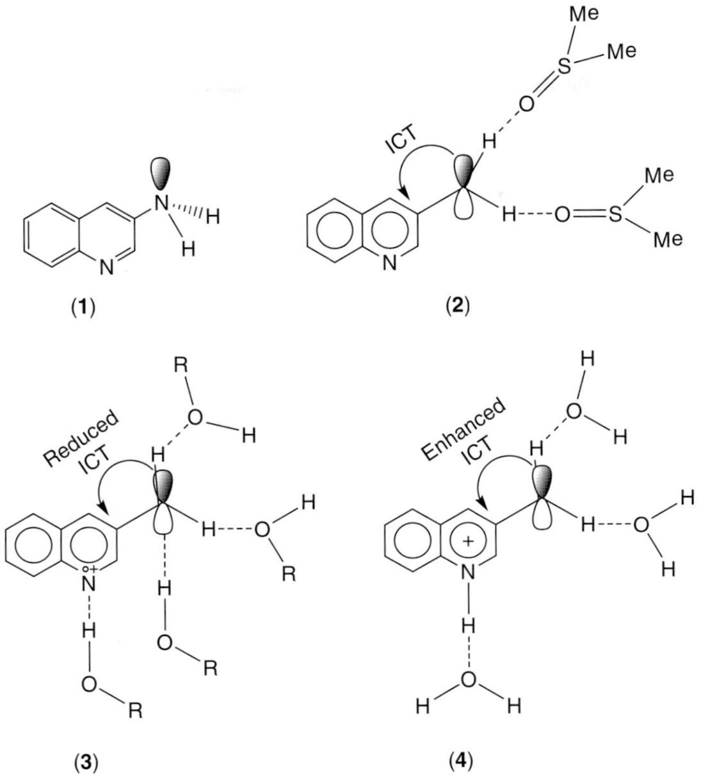

Figure 4.10 Schematic representation of the solvent effect on the ICT process in 3-aminoquinoline (see text for details). (Panda and Datta 2006 [49]. Reproduced with permission of American Institute of Physics.)

by the hydrogen bond accepting ability of the solvents. The authors concluded that the solvents interact with the hydrogen atoms of the amino group through their hydrogen bond accepting properties, thereby stabilizing the ICT state and hindering the flip-flop motion of the amino group. The hydrogen bond donor solvents stabilize the ICT state of this molecule by imparting partial positive charge on the ring nitrogen atom, although the authors found this process may hinder the ICT by blocking the lone pair of the amino nitrogen atom.

4.6 Resonance-Assisted Hydrogen Bonding (RAHB)

From the preceding discussion it is apparent that hydrogen bonding plays a crucial role in dictating the photophysics of several molecules. If the hydrogen bond donor and hydrogen bond acceptor groups are connected through a π-electron bridge, the hydrogen bonding and charge localization can affect each other. Gilli *et al.* [114] introduced the concept of resonance-assisted hydrogen bonding (RAHB) to account for this interplay between π-electron delocalization and hydrogen bonding. Misra *et al.* studied the RAHB process in β-enaminones in 1,4-dioxane-water mixtures [115]. The quantum chemical calculations were carried out at the DFT level to predict the hydrogen bonding lengths and angles in free PACO and PACO–nH_2O molecular clusters, where, $n = 1–3$. The results seem to be consistent with the trend suggested by the RAHB descriptor. The hydrogen bonding energies obtained also echoed the same pattern in the gas phase.

Gehlen *et al.* [116] reviewed the RAHB process in several ICT-based molecules, namely, derivatives of 1-aminopyrene, 9-amino-acridine and adenine derivatives that contain β-enaminone moiety. The authors noted that the electron-withdrawing nature of the substituents and the polarity of the medium affect the keto-amine/enol-imine equilibrium that leads to change in spectral properties of those molecules. The RAHB process was believed to be responsible for the change in fluorescence properties of these molecules. The authors also discussed the possibility of using some of the aforesaid RAHB dyes as molecular probes in radical copolymerization and as reporters in silver nanoparticle–dye interactions in solution.

4.7 Studies of ICT in Solvent Mixture and Confined Media

Sometimes it is difficult to decipher the desired property of a molecule as functions of continuously varying solvent properties using pure solvents only. To counter this problem, use of binary or ternary solvent mixtures has been growing very rapidly [85–88]. To understand the effect of hydrogen bonding on a solute, the mixture of a protic and aprotic solvent are popular choices for studying the photophysics of a probe. The mixture of a polar and nonpolar solvent may be used to probe the polarity-induced effect on a molecule. When a polar probe is dissolved in a binary mixture of a nonpolar and a polar solvent, the probe is expected to interact with these two solvents in different ways, which can give rise to what has been known as preferential solvation [117–120]. Firstly, the solvent can take part in hydrogen bonding with the solute to give rise a structure with specific shape and geometry. On the other hand, the solvent can also interact with the solute in a non-stoichiometric manner, that is, it can surround the polar probe through its opposite dipoles. This is known as dielectric enrichment [118]. An ICT probe is therefore an ideal probe for studies of solvation in solvent mixtures. The most important reason for this is that an ICT molecule is expected to possess higher dipole moment in the excited state than in the ground state (exceptions are there, e.g., betaine-30 has a lower dipole moment in the excited state than that in the ground state). Due to the polar nature of the ICT molecule the polar solvent will give much more stability to the excited state of the probe compared to its ground state. This is one of the reasons that the rate of formation ICT is generally increased in polar solvents. Therefore, the ICT molecules are widely used as solvatochromic probes in pure as well as binary solvent mixtures. (Interested readers may go through the studies by Huppert, Maroncelli, and Fayer mentioned in this chapter.)

From the discussion in the preceding section it is clear that the properties of the medium play an important role in the ICT process. Several studied are devoted to explore the ICT process in micelles, reverse micelles and ionic liquids. Sarkar *et al.* [121] used steady-state and time-resolved spectroscopic techniques to study the ICT process in Nile red in nanocavity of cyclodextrins. They reported that the non-radiative decay rate of this molecule decreases by about 2.5 times in β-cyclodextrin, while the same decreases slightly in γ-cyclodextrin, compared to that in pure water. The same group [122] has studied the rate of ICT of coumarin 490 in the water pool of aerosol OT (AOT)-heptane reverse micelle compared to normal water. They have reported a decrease in ICT about 3.5 times, although the solvation dynamics of the molecule decreases by several thousand times in the aforesaid reverse micelle, compared to that in pure water.

4.8 Studies of ICT in the Solid State

The preceding discussion reveals that the ICT process is mainly studied in the condensed phase. Only a few molecules show ICT in the solid state because of two factors – concentration quenching and structural restriction that may prevent them to acquire favorable structure needed for the ICT reaction. Very recently, Chujo *et al.* [123] reported emission from anthracene-*o*-carborane dyad (Figure 4.11) in the solid state that they assign to the TICT state. The dual emission observed from this molecule in the solution phase with emission maxima about 450 and 600 nm were assigned to the LE and ICT states, respectively. They have shown that the 600-nm peak gets redshifted with increase in solvent polarity, while the 450-nm peak is almost independent of the polarity of the medium. The Lippert–Mataga analysis reveals the slope of the approximate line of the emission band around 450 nm was almost zero, indicating a negligible change in the dipole moment between the ground state of this molecule to the LE state. The same analysis for the 600-nm peaks reveals a significant change in the dipole moment between the ground and the ICT state. The authors have found strong emission intensity of this molecule in the aggression and crystalline states, while that in the solution phase is very low. In the frozen matrix of 2-Me THF at 77 K, this molecule dominantly shows the emission from the LE state. The intensity of the TICT band increased with increase in temperature. This led the authors conclude that the rotation of the *o*-carborane moiety can be enhanced on heating. From the X-ray crystal structure of this molecule, they concluded that due to compact sphere structure of the *o*-carborane moiety, rotation of this group is possible after photoexcitation. Their quantum chemical calculations at the DFT [B3LYP/6-31G(d)] level of theory revealed the rotational barrier of this molecule in the gas phase was about 8 kcal mol^{-1} in the excited state. They also studied the energy barrier in the crystal packing in both ground and excited state employing quantum mechanics and molecular mechanics (QM/MM) method in which the central molecule was treated quantum mechanically (TDDFT method, to be more specific), while the surrounding molecules were modeled using the universal force field. Their QM/MM study revealed that the energy

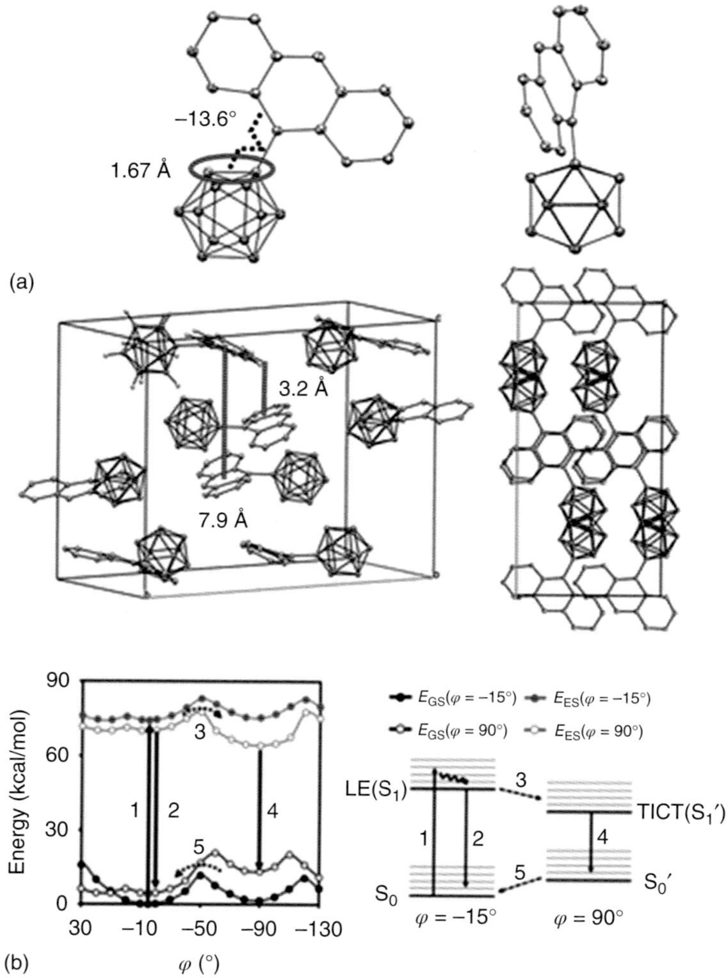

Figure 4.11 The molecular structure and packing diagram of anthracene-*o*-carborane dyad studied by Chujo *et al.* (a and b in upper panel). The lower panel shows the energy levels of ground and excited states of the dyad variable dihedral angle (ϕ) as obtained from the quantum chemical calculations. The proposed absorption and emission processes in the dyad are also shown in this panel. (Chujo *et al.* 2017 [123]. Reproduced with permission of John Wiley & Sons.)

barriers in the crystalline form were about 20 and 19 $kcal\,mol^{-1}$ in the ground and excited states, respectively. The authors claimed that although the energy barrier of rotation is increased from single molecule to crystal packing, the values are still very low for the rotation to occur for formation of the TICT state. From these studies they concluded that owing to the compact spherical shape of *o*-carborane moiety rotation of this group is possible even in crystalline state.

The same group has synthesized several multifunctional *o*-carborane–pyrene dyads and studied the emission properties in the solid state [124]. Their aim was to prepare multifunctional luminescence molecules that showed dual emission, aggregation-induced emission (AIE), as well as crystallization-induced emission (CIE) enhancement and luminescent color changes. They synthesized the pyrene-substituted *o*-carborane derivatives with extremely bright solid-state luminescence properties.

There are some other reports of ICT-based molecules showing solid-state emission properties. For example, Yuan *et al.* [125] have reported the solid-state emission from D–A molecules TPA3TPAN and DTPA4TPAN (Figure 4.12), based on triphenylamine (TPA) and 2,3,3-triphenylacrylonitrile (TPAN) units. The authors reported that although these molecules are non-fluorescent in common solvents due to aggregation-caused quenching (ACQ), they become highly emissive when they aggregate as nanoparticles, solid powders and thin films. They reported that a combination of ICT and AIE processes led the TPA3TPAN and DTPA4TPAN to give rise to a bright yellow emission with quantum efficiencies of 33.2% and 38.2%, respectively.

Wan *et al.* [126] reported the solid-state emission from triphenylamine-*o*-carborane dyads. The authors reported that triphenylamine-*o*-carborane dyads showed dual emission from the LE and ICT states in solution. As expected, the LE emission was solvent independent while the intensity and the wavelength of the ICT emission were determined by the properties of the media. The authors reported bright solid-state emission from their triphenylamine-*o*-carborane dyads through the AIE process.

From the preceding discussion it is apparent that the properties of the medium play a crucial role in shaping the photophysics of the ICT molecules. During the excited-state ICT process, the molecule is generally excited to the LE state, from which a new ICT species with lower energy and higher dipole moment is generated. The excited-state ICT process generally occurs in polar solvents as the polar medium exerts stability of the polar ICT state that can be seen through the redshift in the emission maxima. However, there are some exceptions as the ICT

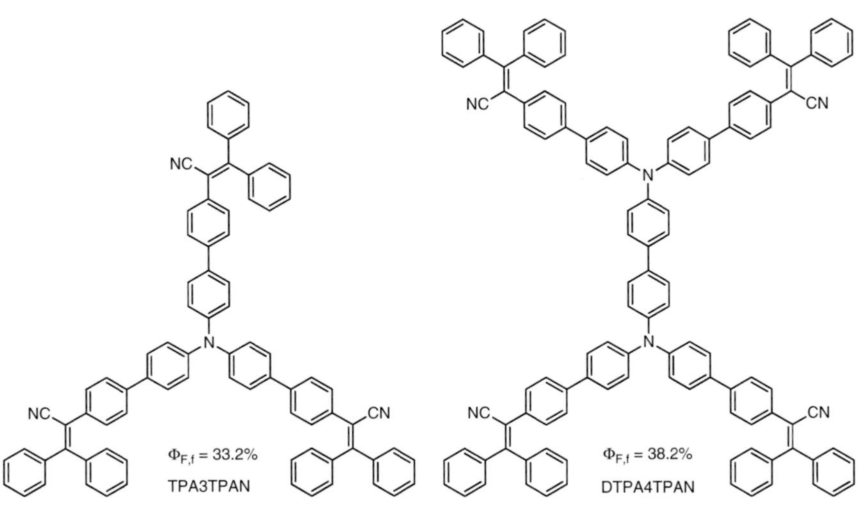

Figure 4.12 The chemical structures of TPA3TPAN and DTPA4TPAN. (Yuan *et al.* 2012 [125]. Reproduced with permission of American Chemical Society.)

process can lead to blueshifted emission maxima in molecules which are zwitterionic in the ground state. For these molecules, the dipole moment in the excited state is lower than that in the ground state. Although a majority of the studies of ICT are reported in the solution, a few studies in the gas phase can also be found in literature. Recently, several studies of ICT process in the solid state are reported with the aim of using the ICT-based materials for technological applications. As mentioned earlier, the polarity and the hydrogen bonding abilities of the medium play an important role in determining the rate of formation of the ICT state as well as its structure. It is therefore not surprising that the studies of ICT process in several media, including mixed solvents and confined media are being carried out to understand the underlying principle in detail. As ICT probes are being used as fluorescence sensors to detect molecules and ions in solution as well as in biological samples (see Chapter 6), more studies on solvent effect in the ICT process are expected to be reported in the near future.

References

1 Reichardt, C. (1998) *Solvents and Solvent Effects in Organic Chemistry*, Wiley-VCH Verlag GmbH & Co. KGaA, Weinheim.
2 Maronceli, M., MacInnis, J., and Fleming, G.R. (1989) *Science*, **243**, 1674.
3 Florusse, L.J., Peters, C.J., Schoonman, J., Hester, K.C., Koh, C.A., Dec, S.F., Marsh, K.N., and Sloan, E.D. (2004) *Science*, **306**, 469.
4 Cerny, J., Tong, X., Hobza, P., and Muller-Dethlefs, K. (2008) *J. Phys. Chem.*, **128**, 114319.
5 Sakota, K., Shimazaki, Y., and Sekiya, H. (2009) *J. Chem. Phys.*, **130**, 231105.
6 Larsen, R.W., Zielke, P., and Suhm, M.A. (2007) *J. Chem. Phys.*, **126**, 194307.
7 Tanabe, S., Ebata, T., Fujii, M., and Mikami, N. (1993) *Chem. Phys. Lett.*, **215**, 347.
8 Scharge, T., Luckhaus, D., and Suhm, M.A. (2008) *Chem. Phys.*, **346**, 167.
9 Barbu-Debus, K.L., Sen, A., Broquier, M., and Zehnaker, A. (2011) *Phys. Chem. Chem. Phys.*, **13**, 13985.
10 Raczyński, P., Dawid, A., Dendzik, Z., and Gburski, Z. (2005) *J. Mol. Struct.*, **750**, 18.
11 Hauchecorne, D., van der Veken, B.J., Moiana, A., and Herrebout, W.A. (2010) *Chem. Phys.*, **374**, 30.
12 Miller, D.J. and Lisy, J.M. (2008) *J. Am. Chem. Soc.*, **130**, 15381.
13 Li, X., Oomens, J., Eyler, J.R., Moore, D.T., and Iyengar, S.S. (2010) *J. Chem. Phys.*, **132**, 244301.
14 Pathak, A.K., Samanta, A.K., Maity, D.K., Mukherjee, T., and Ghosh, S.K. (2010) *J. Phys. Chem. Lett.*, **1**, 886.
15 Sobolewski, A.L. and Domcke, W. (2007) *J. Phys. Chem. A*, **111**, 11726.
16 Brutschy, B. (2000) *Chem. Rev.*, **100**, 3891.
17 Robertson, W.H., Diken, E.G., Price, E.A., Shin, J.W., and Johnson, M.A. (2003) *Science*, **299**, 1367.
18 de Abreu e Silva, E.S., Duarte, H.A., and Belchior, J.C. (2006) *Chem. Phys.*, **323**, 553.

19 Danilov, V.I., Van Mourik, T., and Poltev, V.I. (2006) *Chem. Phys. Lett.*, **429**, 255.
20 Sobolewski, A.L. and Domcke, W. (2008) *Chem. Phys. Lett.*, **457**, 404.
21 Knapp, C.J., Xu, Y., and Jäger, W. (2011) *J. Mol. Spectrosc.*, **268**, 130.
22 Buck, U. and Huisken, F. (2000) *Chem. Rev.*, **100**, 3863.
23 Miller, D.J. and Lisy, J.M. (2007) *J. Phys. Chem. A*, **111**, 12409.
24 Pathak, A.K., Mukherjee, T., and Maity, D.K. (2008) *J. Phys. Chem. A*, **112**, 744.
25 Sett, P., Mishra, T., Chowdhury, J., Ghosh, M., Chattopadhyay, S., Sarkar, S.K., and Mallick, P.K. (2008) *J. Chem. Phys.*, **128**, 144507.
26 Watanabe, T., Ebata, T., Tanabe, S., and Mikami, N. (1996) *J. Chem. Phys.*, **105**, 408.
27 Zakharov, M., Krauss, O., Nosenko, Y., Brutschy, B., and Drew, A. (2009) *J. Am. Chem. Soc.*, **131**, 461.
28 Vaden, T.D. and Lisy, J.M. (2005) *J. Phys. Chem. A*, **109**, 3880.
29 Striplin, D.R., Reece, S.Y., McCafferty, D.G., Wall, C.G., Friesen, D.A., Erickson, W., and Meyer, T.J. (2004) *J. Am. Chem. Soc.*, **126**, 5282.
30 Druzhinin, S.I., Kovalenko, S.A., Senyushkina, T.A., Demeter, A., and Zachariasse, K.A. (2010) *J. Phys. Chem. A*, **114**, 1621.
31 Weigel, A. and Ernsting, N.P. (2010) *J. Phys. Chem. B*, **114**, 7879.
32 Tarakeswar, P., Kim, K.S., Djafari, S., Buchhold, K., Reimann, B., Barth, H.-D., and Brutschy, B. (2001) *J. Chem. Phys.*, **114**, 4016.
33 Miyazaki, M., Fujii, A., Ebata, T., and Mikami, N. (2004) *Science*, **304**, 1134.
34 Dickinson, J.A., Hockridge, M.R., Kroemer, R.T., Robertson, E.G., Simons, P., McCombie, J., and Walker, M. (1998) *J. Am. Chem. Soc.*, **120**, 2622.
35 Pathak, A.K., Mukherjee, T., and Maity, D.K. (2006) *J. Chem. Phys.*, **124**, 024322.
36 Reiman, B., Buchhold, K., Vaupel, S., Brutschy, B., Havlas, Z., Spirko, V., and Hobza, P. (2001) *J. Phys. Chem. A*, **105**, 5560.
37 Han, K.-L. and Zhao, G.-J. (2011) *Acc. Chem. Res.* doi: 10.1021/ar200135h
38 Carmona, C., Galan, M., Angulo, G., Munoz, M.A., Guardado, P., and Balon, M. (2000) *Phys. Chem. Chem. Phys.*, **2**, 5076.
39 Tsuji, N., Ishiuchi, S., Sakai, M., Fuji, M., Ebata, T., Jouvet, C., and Dedonder-Lardeux, C. (2006) *Phys. Chem. Chem. Phys.*, **8**, 114.
40 Banno, M., Ohta, K., Yamaguchi, S., Iría, S., and Tominaga, K. (2009) *Acc. Chem. Res.*, **42**, 1259.
41 Cramer, C.J. and Truhler, D.G. (1999) *Chem. Rev.*, **99**, 2161.
42 Kirkwood, G. (1934) *J. Chem. Phys.*, **2**, 351.
43 Nakano, M. *et al* (2017) *Chem. Lett.*, **46**, 536.
44 Gutierrez, F., Trzcionka, J., Doloncle, R., Poteau, R., and Chouini-Lalanne, N. (2005) *New J. Chem.*, **29**, 570.
45 Gedeck, P. and Schneider, S. (1997) *J. Photochem. Photobiol., A*, **105**, 165.
46 Georgieva, I., Aquino, A.J.A., Plasser, F., Trendafilova, N., Kohn, A., and Lischka, H. (2015) *J. Phys. Chem. A*, **119**, 6232.
47 Lippert Von, E. (1957) *Z. Electrochem.*, **61**, 962.
48 Mataga, N., Kaifu, Y., and Koizumi, N. (1956) *Bull. Chem. Soc. Jpn.*, **29**, 465.
49 Panda, D. and Datta, A. (2006) *J. Chem. Phys.*, **125**, 054513.

50 Pereira, R.V. and Gehlen, M.H. (2006) *J. Phys. Chem. A*, **110**, 7539–7546.
51 Moyon, N.S., Chandra, A.K., and Mitra, S. (2010) *J. Phys. Chem. A*, **114**, 60–67.
52 Mukherjee, S., Sahu, K., Roy, D., Mondal, S.K., and Bhattacharyya, K. (2004) *Chem. Phys. Lett.*, **384**, 128–133.
53 Zhu *et al* (2016) *Sci. Rep.*, **6**, 24313.
54 Misra, R. and Kar, S. (2012) *Chem. Phys.*, **397**, 65.
55 Ramkumar, V. and Kannan, P. (2016) *J. Lumin.*, **169**, 204.
56 Liptay, W. (1965) *Z. Naturforsch.*, **20a**, 1441.
57 Sumalekshmy, S. and Gopidas, K.R. (2005) *Photochem. Photobiol. Sci.*, **4**, 539.
58 Sumalekshmy, S. and Gopidas, K.R. (2004) *J. Phys. Chem. B*, **108**, 3705.
59 Rettig, W. (1982) *J. Mol. Struct.*, **84**, 303.
60 Kamlet, M.J., Abboud, J.L.M., Abraham, M.H., and Taft, R.W. (1983) *J. Org. Chem.*, **48**, 2877.
61 Kamlet, M.J., Abboud, J.M., and Taft, R.W. (1981) *Prog. Phys. Org. Chem.*, **13**, 485.
62 Reichardt, C. (1994) *Chem. Rev.*, **94**, 2319.
63 Marcus, Y. (1991) *J. Solution Chem.*, **20**, 929.
64 Catalan, J. (2009) *J. Phys. Chem. B*, **113**, 5951.
65 Liu, X., Cole, J.M., and Low, K.S. (2013) *J. Phys. Chem. C*, **117**, 14731.
66 Renger, T. *et al* (2008) *Proc. Natl. Acad. Sci. U.S.A.*, **105**, 13235.
67 Hicks, J.M., Vandersall, M., Babarogic, Z., and Eisenthal, K.B. (1985) *Chem. Phys. Lett.*, **116**, 18.
68 Simons, J.D. and Su, S.G. (1990) *J. Phys. Chem.*, **94**, 3656.
69 Das, K., Sarkar, N., Nath, D., and Bhattacharyya, K. (1992) *Spectrochim. Acta*, **48A**, 1701.
70 Chang, T.L. and Cheung, H.C. (1990) *Chem. Phys. Lett.*, **173**, 343.
71 Haidekker, M.A., Brady, T.P., Lichlyter, D., and Theodorakis, E.A. (2005) *Bioorg. Chem.*, **33**, 415.
72 Wandelt, B., Tukewitsch, P., Stranix, B.R., and Darling, G.D. (1995) *J. Chem. Soc., Faraday Trans.*, **91**, 4199.
73 Ghosh, R. and Palit, D.K. (2014) *ChemPhysChem*, **15**, 4126.
74 Singh, P.K., Kumbhakar, M., Pal, H., and Nath, S. (2010) *J. Phys. Chem. B*, **114**, 5920.
75 Stsiapura, V.I., Kurhuzuzenkau, S.A., Kuzmitsky, V.A., Bouganov, O.V., and Tikhomirov, S.A. (2016) *J. Phys. Chem. A*, **120**, 5481.
76 Dixon, D.W. *et al* (1999) *Inorg. Chem.*, **38**, 5526.
77 Escudero, D. (2016) *Acc. Chem. Res.*, **49**, 1816.
78 Fenn, E.E., Moilanen, D.E., Levinger, N.E., and Fayer, M.D. (2009) *J. Am. Chem. Soc.*, **131**, 5530.
79 Reynolds, L., Gardecki, J.A., Frankland, S.J.V., Horng, M.L., and Maroncelli, M. (1996) *J. Phys. Chem.*, **100**, 10337.
80 Li, X., Liang, M., Chakraborty, A., Kondo, M., and Maroncelli, M. (2011) *J. Phys. Chem. B*, **115**, 6592.
81 Li, X. and Maroncelli, M. (2011) *J. Phys. Chem. A*, **115**, 3746.
82 Barbara, P.F. and Jarzeba, W. (1988) *Acc. Chem. Res.*, **21**, 195.
83 Barbara, P.F., Walker, G.C., and Smith, T.P. (1992) *Science*, **256**, 975.

84 Reid, P.J. and Barbara, P.F. (1995) *J. Phys. Chem.*, **99**, 3554.
85 Solntsev, K.M. and Huppert, D. (1999) *J. Phys. Chem. A*, **103**, 6984.
86 Molotsky, T. and Huppert, D. (2003) *J. Phys. Chem. A*, **107**, 2769.
87 Molotsky, T. and Huppert, D. (2003) *J. Phys. Chem. A*, **107**, 8449.
88 Agmon, N. (2002) *J. Phys. Chem. A*, **106**, 7256.
89 Eisenthal, K.B. *et al* (1987) *J. Phys. Chem.*, **91**, 6162.
90 Rettig, W. *et al* (1998) *J. Phys. Chem. A*, **102**, 7754.
91 Belau, L., Haas, Y., and Rettig, W. (2002) *Chem. Phys. Lett.*, **364**, 157.
92 Parusel, A.A. (2000) *Phys. Chem. Chem. Phys.*, **2**, 5545.
93 Bohnwagner, M.V., Burghardt, I., and Dreuw, A. (2016) *J. Phys. Chem. A*, **120**, 14.
94 Misra, R. *et al* (2009) *J. Phys. Chem. B*, **113**, 10779.
95 Yun, S.H. *et al* (2017) *Sens. Actuators, B*, **240**, 988.
96 Bhattacharyya, K. and Chowdhury, M. (1993) *Chem. Rev.*, **93**, 507.
97 Nag, A., Kundu, T., and Bhattacharyya, K. (1989) *Chem. Phys. Lett.*, **160**, 257.
98 Sarkar, N., Das, K., Nath, D.N., and Bhattacharyya, K. (1994) *Langmuir*, **10**, 326.
99 Majumdar, D., Sen, R., Bhattacharyya, K., and Bhattacharyya, S.P. (1991) *J. Phys. Chem.*, **95**, 4324.
100 Chakraborty, A., Kar, S., and Guchhait, N. (2006) *Chem. Phys.*, **324**, 733.
101 Singh, R.B., Mahanta, S., Kar, S., and Guchhait, N. (2007) *Chem. Phys.*, **342**, 33.
102 Samanta, A., Paul, B.K., Mahanta, S., Singh, R.B., Kar, S., and Guchhait, N. (2010) *J. Photochem. Photobiol., A*, **212**, 161.
103 Samanta, A., Paul, B.K., and Guchhait, N. (2012) *J. Lumin.*, **132**, 517.
104 Chipem, F.A.S., Mishra, A., and Krishnamoorthy, G. (2012) *Phys. Chem. Chem. Phys.*, **14**, 8775.
105 Zachariasse, K.A. *et al.* (2017) *J. Phys. Chem. A*, **121**, 1223.
106 Zachariasse, K.A. *et al.* (2006) *J. Phys. Chem. A*, **110**, 2955.
107 Fujiwara, T. *et al.* (2013) *Chem. Phys. Lett.*, **586**, –70.
108 Bohnwagner, M.V. and Dreuw, A. (2017) *J. Phys. Chem. A*, **121**, 5834.
109 Mukherjee, P., Rafiq, S., and Sen, P. (2016) *J. Photochem. Photobiol., A*, **328**, 136.
110 Nouchi, G. *et al* (1989) *Chem. Phys. Lett.*, **157**, 393.
111 Pilloud, D. *et al* (1987) *Chem. Phys. Lett.*, **137**, –130.
112 Pal, H. *et al* (2005) *Chem. Phys.*, **315**, 277.
113 Mitra, S. *et al* (2007) *J. Lumin.*, **127**, 508.
114 Gilli, G. *et al* (1989) *J. Am. Chem. Soc.*, **111**, 1023.
115 Misra, R. *et al* (2015) *J. Photochem. Photobiol., A*, **302**, 23.
116 Gehlen, M.H., Simas, E.R., Pereira, R.V., and Sabatini, C.A. (2012) in *Reviews in Fluorescence 2010* (ed. C.D. Geddes), Springer.
117 Raju, B.B. and Costa, S.M.B. (1999) *Phys. Chem. Chem. Phys.*, **1**, 3539.
118 Henseler, A., von Raumer, M., and Suppan, P. (1996) *J. Chem. Soc., Faraday Trans.*, **92** (3), 391.
119 Zurawski, W. and Scarlata, S. (1994) *Photochem. Photobiol.*, **60**, 343.
120 Khajehpour, M. and Kauffman, J.F. (2000) *J. Phys. Chem. A*, **104**, 7151.

121 Sarkar, N. *et al* (2004) *Chem. Phys. Lett.*, **388**, 150.
122 Sarkar, N. *et al* (2001) *Chem. Phys. Lett.*, **342**, 303.
123 Chujo, Y. *et al.* (2017) *Angew. Chem. Int. Ed.*, **56**, 254.
124 Nishino, K., Yamamoto, H., Tanaka, K., and Chujo, Y. (2016) *Org. Lett.*, **18**, 4064.
125 Yuan, W.Z., Gong, Y., Chem, S. *et al.* (2012) *Chem. Mater.*, **24**, 1518.
126 Wan, Y., Li, J., Peng, X. *et al* (2017) *RSC Adv.*, **7**, 35543.

5

Nonlinear Optical Response of ICT Molecules

5.1 Introduction

Molecules and materials which are capable of transforming various parameters of an optical beam are in great demand and constitute a class of matter called nonlinear optical (NLO) molecules and materials. The parameters of optical beam that we have referred to are frequency, phase, amplitude, polarization, and so on. Such transformations are technically important and have been extensively pursued in nonlinear optics. In NLO, we know that the dielectric polarization of the medium through which light is propagating responds nonlinearly to the electric field of light. Such nonlinear responses appear when the intensity of the electric field of light is very high, typically of the same order as that of the interatomic electric field. Electric fields of such magnitudes are easily realized in LASERs and it is no wonder then that the field of NLO has grown very significantly and rapidly ever since LASER was discovered [1, 2]. The most commonly encountered NLO response is the second harmonic generation (SHG), also known as *frequency doubling*. Here, in the SHG, two photons (frequency ω) disappear and a single photon with the frequency 2ω appears (as demanded by energy conservation). The nonlinear property of the medium responsible for the SHG is the second order susceptibility $\chi^{(2)}$, while for molecules it is called the first hyperpolarizability (β). Similarly, other nonlinear responses like the third harmonic generation (THG) or higher harmonic generation (HHG) have been observed in appropriate materials and at appropriate intensities of the electric field of light. Of special interest has been the NLO process leading to the sum-frequency generation (SFG) in which two photons of different frequencies (ω_1, ω_2) disappear, leading to the appearance of a single photon of frequency $\omega = \omega_1 + \omega_2$. Among higher order NLO responses, electro-optical Kerr effect, intensity-dependent refractive index, and self-focusing have attracted a lot of attention. Some other nonlinear responses having technological implications are four-wave mixing, cross-phase modulations, multiphoton absorption (two or more photons are simultaneously absorbed), and multiphoton ionization.

One of the reasons of the unending interest in the intramolecular charge-transfer (ICT) molecules has been their ability to respond nonlinearly to the electric field of light [3, 118]. Many such molecules are known to possess high first- and second-order hyperpolarizabilities and some of them display

Intramolecular Charge Transfer: Theory and Applications,
First Edition. Ramprasad Misra and S. P. Bhattacharyya.

two-photon absorption (TPA) spectra. If such ICT chromophores can be converted into materials with high NLO responses (e.g., high $\chi^{(2)}$ and $\chi^{(3)}$), we have at our disposal a definite route to fabricate materials with desired properties. In fact, organic π-systems endcapped with an electron donor (D) and an electron acceptor (A) provide a class of molecules (D–π–A), conventionally known as *organic push–pull systems* in which the donor–acceptor interaction (the so-called ICT) is responsible for generating distinct optoelectronic properties of the D–π–A chromophores (also called charge-transfer chromophores). The ICT or the donor–acceptor interaction leads to the formation of a "new" low-energy π molecular orbital (MO) extended over the entire D–π–A system. It is easy to excite electrons into the new low-energy MO by visible light, accounting for the fact that such ICT chromophores are usually colored. The ICT also polarizes the chromophore and leads to the generation of non-zero dipole moment. In the valence bond language, the charge-transferred configuration (D^+–π–A^-) usually contributes more to the excited state than to the ground state, which means that D–π–A molecules will not only possess a low-energy absorption band (in the visible region) but also display a large change in the dipole moment ($\Delta\mu \gg 0$) when electronically excited. It turns out that these two attributes ensure that the molecule displays a non-zero first hyperpolarizability (D–π–A system is assumed to be non-centrosymmetric). The simple picture provides easy designing clue for the synthesis of ICT chromophores. It is needless to mention here that the transition from molecular chromophore to materials requires much more elaborate criteria because the materials must be mechanically robust, optically transparent, and resistant to photodamage, and so on. However, the possibility of theoretically designing molecular chromophores, which may be transformed into NLO materials, has excited keen interest in the theoretical investigations on the ICT molecules in the context of their NLO responses. The technologically relevant NLO materials used initially were all of inorganic origin. They have been and are still being used in high-speed all-optical switching, in optical computing, including storage, in telecommunication, and so on. For example, KH_2PO_4, $LiNbO_3$, and so on, are still being used as materials for SHG. These materials are, however, not flexible, or durable. Organic ICT chromophores provide enough flexibility and controlled durability so that the search for new optoelectric materials is increasingly being directed to organic ICT molecules.

5.2 The NLO Response to Electric Field

We briefly review how different nonlinear responses of a medium originate from dielectric polarization. Assuming that the optical fields that the medium is exposed to are not too large, the dielectric polarization density $P(t)$ at a specific instant of time can be expanded in a Taylor's series in electric field $E(t)$ at that instant of time as

$$P(t) = \varepsilon_0 \left[\chi^{(1)}E(t) + \chi^{(2)}E^2(t) + \chi^{(3)}E^3(t) + \cdots\right] \tag{5.1}$$

For the sake of simplicity, we will be treating $E(t)$ as a scalar. In Equation 5.1, $\chi^{(n)}$ is the nth-order susceptibility of the medium and models the nth-order nonlinearity of the medium. We note here that $\chi^{(n)}$ is a tensor of tank $(n+1)$, which represents the symmetry of the medium as well as the parametric nature of the interactions. To examine the kind of nonlinear responses that can arise from the susceptibility of a particular order, let us restrict ourselves to the second-order nonlinearity only. In this case, the nonlinear polarization density P_{NL} is given by

$$P_{\mathrm{NL}} = \varepsilon_0 \chi^{(2)} E^2(t) \tag{5.2}$$

Let us further assume that $E(t)$ is a two-component field – that is,

$$E(t) = E_1 \cos(\omega_1 t) + E_2 \cos(\omega_2 t) \tag{5.3}$$

which means that the medium is forced to oscillate at frequencies ω_1 and ω_2. We can rewrite Equation 5.3 in complex form by taking

$$E(t) = \frac{1}{2}E_1\left(e^{-i\omega_1 t} + e^{i\omega_1 t}\right) + \frac{1}{2}E_2\left(e^{-i\omega_2 t} + e^{i\omega_2 t}\right) \tag{5.4}$$

With $E(t)$ thus defined,

$$\begin{aligned} P_{\mathrm{NL}} = {} & \frac{\varepsilon_0 \chi^{(2)}}{4}\left[|E_1|^2 e^{-i2\omega_1 t} + |E_2|^2 e^{-i2\omega_2 t} + 2E_1E_2 e^{-i(\omega_1+\omega_2)t}\right. \\ & \left. + 2E_1E_2^* e^{-i(\omega_1-\omega_2)t} + \{|E_1|^2 + |E_2|^2\} + A\right] \end{aligned} \tag{5.5}$$

where A is the complex conjugate terms.

In this expression, the first two terms represent SHG at frequencies $2\omega_1$ and $2\omega_2$, the third and fourth terms refer to SFG and difference frequency generation, while the terms in curly brackets represent what is known as *optical rectification* (zero frequency). Thus, we have indentified all the nonlinear processes that are linked to $\chi^{(2)}$. We can extend the analysis to higher-order nonlinearities in the same manner and identify, for example, the process linked to $\chi^{(3)}$.

The response (linear as well as nonlinear) of an isolated molecule to a homogeneous static electric field can be represented by any of following expansions (they provide identical results if the Hellmann–Feynman theorem is strictly obeyed).

$$E_i(F) = E_0 - \sum_i \mu_i^0 E_i - \sum_{ij} \alpha_{ij} E_i E_j - \frac{1}{2!}\sum_{ijk} \beta_{ijk} E_i E_j E_k - \frac{1}{3!}\sum_{ijkl} \gamma_{ijkl} E_i E_j E_k E_l - \cdots \tag{5.6}$$

$$\mu_i(F) = \mu_i^0 + \frac{1}{2}\sum_j \alpha_{ij} F_j + \frac{1}{6}\sum_{jk} \beta_{ijk} F_j F_k + \frac{1}{24}\sum_{jkl} \gamma_{ijkl} F_j F_k F_l + \cdots \tag{5.7}$$

E_0 is the field free energy of the molecule, while μ_i^0 is the ith component of the permanent dipole moment of the molecule. α represents dipole polarizability tensor, β the first hyperpolarizability, and γ the second hyperpolarizability tensors of the molecule, respectively. The polarizabilities and hyperpolarizabilities can be easily linked to the corresponding susceptibilities $\chi^{(n)}$. If the applied electric field is time varying, the NLO responses of the molecule will become frequency-dependent quantities $\alpha(\omega)$, $\beta(\omega)$, $\gamma(\omega)$, and so on.

5.3 Theoretical Calculation of NLO Response of ICT Molecules

Theoretical calculation of NLO response properties of ICT chromophores has received sustained attention. The microscopic calculations have been carried out within the framework of wave-function-based representation of the problem with the molecule both in static and time-varying electric fields. That means the problem has been tackled at the level of both time-dependent and time-independent Schrodinger equations. At the time-independent level, the many-electron problem in the presence of a static electric field has been solved by invoking the time-independent finite-field Hartree–Fock approximation. Essentially, the method calculates energy of the molecule in a number of well-designed electric fields. The field-dependent energies (or dipole moments) are fitted to the expressions in Equation 5.6 or 5.7 and the required response properties are obtained by differentiation and extrapolation to zero value of the field. It has been found that the Hartree–Fock level of calculations is inadequate to reproduce a good match with experimental data available in the literature. Most of the wave-function-based calculations are therefore further refined through MP2 or MP4 calculations of the energy of the molecule in the presence of the electric field. If frequency-dependent response properties are targeted, the method usually involves either the time-dependent Hartree–Fock (TDHF) approach or a CPHF approach at a particular geometry of the molecule. A great majority of the calculations that are done nowadays invoke the time-dependent density functional theory (TDDFT) as it is the most easily implemented technique for large molecules. One must use a large enough basis set so that calculated quantities become truly field-invariant physical quantities. Triple-split Sadlej basis sets [4, 5] are usually considered sufficient for the purpose of polarizability calculations. The choice of the functional is a bit tricky – the range-separated functionals have been claimed to work well in practical computations of α, β, or γ values of the molecules. The most commonly targeted NLO response is the first hyperpolarizability, β (second-order response). β is a tensor of rank 3 with 27 components. The Kleinman symmetry of $3\times3\times3$ matrix representing the β-tensor is exploited to reduce the number to 10. For comparison with experimental data, the components are condensed into a single value (β_{total}), where

$$\beta_{\text{total}} = \left[\beta_x^2 + \beta_y^2 + \beta_z^2\right]^{1/2} \tag{5.8}$$

where $\beta_i = \beta_{iii} + \sum_{i\neq j}(\beta_{ijj} + \beta_{jij} + \beta_{jji})(i = x, y, z)$. Using Kleinman symmetry, $\beta_x = (\beta_{xxx} + \beta_{xyy} + \beta_{xzz})$. Similarly, we have $\beta_y = (\beta_{yyy} + \beta_{yxx} + \beta_{yzz})$ and $\beta_z = (\beta_{zzz} + \beta_{zyy} + \beta_{zxx})$ or β_{vector} (β_{vec}) projected on dipole $\beta\cdot\mu$, μ being the dipole moment of the molecule. The values of β_{vec} at the static limit can be calculated from the tensorial components of first hyperpolarizability using the following equation:

$$\beta_{\text{vec}} = \frac{(\beta_x\mu_x + \beta_y\mu_y + \beta_z\mu_z)}{\mu} \tag{5.9}$$

The earliest attempt to model the first hyperpolarizability of an ICT chromophore can be traced to Ouder [5, 6]. In his famous two-state model, the frequency-dependent first hyperpolarizability, $\beta(\omega)$ turned out to be

$$\beta_\omega = \frac{3e^2}{2\hbar}\left[\frac{\omega_{12} f_0 \Delta\mu_{12}}{\left(\omega_{12}^2 - \omega^2\right)\left(\omega_{12}^2 - 4\omega^2\right)}\right] \tag{5.10}$$

where the ω_{12} is the energy of the ICT transition with oscillator strength f_0, $\Delta\mu_{12}$ is the change in dipole moment that takes place accompanying the $1 \rightarrow 2$ transition ($1 \equiv$ ground state, $2 \equiv$ ICT state), while ω is the frequency of the probing light. Clearly, the dispersion of $\beta(\omega)$ as indicated in Equation 5.10 is characterized by large values of β at $\omega = \omega_{12}$ and $\omega_{12}/2$. One can therefore anticipate and generate second-harmonic light from such ICT chromophores. It is also apparent that a substantial bathochromic shift in the ICT transition energy in a polar environment will lead to enhancement of the NLO response represented by the first hyperpolarizability β. If we put $\omega = 0$, Ouder's model can be used to calculate the static first hyperpolarizability approximately. Ouder's simple model has been much used and even been abused, but has retained its usefulness even today. It has spawned a number of investigations [7, 8]. For example, Sen *et al.* [7] used the model to investigate theoretically the kind of solvent modulation of the first hyperpolarizability of 4-*N*,*N*-dimethylaminobenzonitrile (DMABN) that could be expected. These authors predicted significant solvent polarity effect on $\beta(\omega)$ in the low-solvent polarity regime. Sen *et al.* in a separate study [8] reported detailed theoretical modeling of hyperpolarizabilities of some twisted intramolecular charge-transfer (TICT) molecules and their analogs at semiempirical level wherein they also proposed and evaluated a simple model for computing hyperpolarizabilities of ICT molecules and predicting correlation between β values and the electronic structure. In the early 1990s, the calculations were mostly restricted to semiempirical levels like CNDO/S, INDO/S, MNDO, and so on. From the mid-1990s there was rapid advancement in computing electronic structure and response properties both in terms of software and hardware capabilities. The researchers then began to increasingly use *ab initio* electronic structure methods to compute and analyze the linear and nonlinear response properties of atoms and molecules. It turned out that electron correlation effects must be taken into account in calculating the values of β and γ for achieving accuracy. Sim *et al.* [9] studied the importance of taking electronic correlation into account in the calculation of $\beta(-2\omega; \omega, \omega)$ in the case of prototypical ICT molecules like *para*-nitroaniline. Albert *et al.* [10] used the semiempirical theory to probe what kind of bridges in D–π–A chromophores would lead to enhancement of NLO response. Their work suggested that the NLO response of D–π–A systems was more sensitive to the electron excessivity of the "bridges" than to their aromaticities. Electron-excess heterocyclic bridges are thus expected to increase the NLO response much more than aromatic bridges for the same pair of donor and acceptor. Abotto *et al.* [11] used *ab initio* coupled perturbed Hartree–Fock (CPHF) level of theory and demonstrated that the structure design of an ICT chromophore can cooperatively interact with solvent-polarity-induced effects leading to the emergence of a strong NLO response in a novel family of highly polarizable azinium-(CH=CH-thienyl) dicyanomethanido chromophores. Thus, media effects need to be adequately grafted into the theoretical tools for calculating hyperpolarizabilities experimentally obtained from electric-field-induced second harmonic generation (EFISH) experiments in solution phase. Wang *et al.* [12] exploited both resonant and nonresonant hyper-Rayleigh scattering (HRS) experiments to suggest that the dispersion of β values of a number of ICT chromophores indicate that the vibronic structure of the chromophore in the excited state is an important attribute in shaping up the NLO response of the chromophore. The authors used a 2-vibronic-state theoretical model to explain the observed dispersion.

On the basis of theoretical studies [13], it turns out that three avenues are available for optimizing NLO response of D–B–A (B ≡ bridge) systems. The first option is to select a conjugated bridging unit and increase the number of electron donor and acceptor units on the two sides of the bridge. The second option is to modify the nature of the bridge and the third option is to increase the conjugation length of the bridge. An altogether different approach has been to distort the nature of conjugation itself [14]. One therefore asks the question: given a donor (D), an acceptor (A), and a bridging unit (B), what is the best structural arrangement that will produce an optimal NLO response? Theoretical and experimental studies seem to indicate that a polyene-bridged D–B–A system can exhibit wide variations in the NLO response due to distortions in the conjugated bridging unit. The conjugated bridge, for example, can be distorted from a polyene-like structure (**II**) in Figure 5.1 to cyanine-like structure (**II**) to a polymethine-like structure (**III**). The distortion type depends on the strengths of the donor (X) and acceptor (Y) and solvent polarity.

A valence bond configuration interaction (VB-CI) continuum solvation model was proposed by Chen *et al.* [14] to predict the effects of solvation in a polar medium, on the NLO response properties of organic ICT molecules like 1,1′-dicyano-6-(dibutylamine)-hexatriene (Figure 5.2). The simple theoretical model leads to analytical expressions for the absorption frequency (ω), polarizability (α), and hyperpolarizabilities (β, γ, etc.) and the bond length alteration factor, using only one solvent-dependent parameter (ε – the dielectric constant of the solvent).

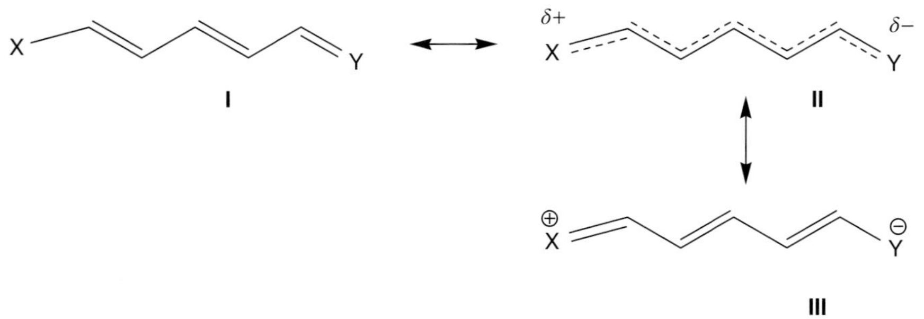

Figure 5.1 Resonance structures in DA systems undergoing long-range intramolecular charge transfer.

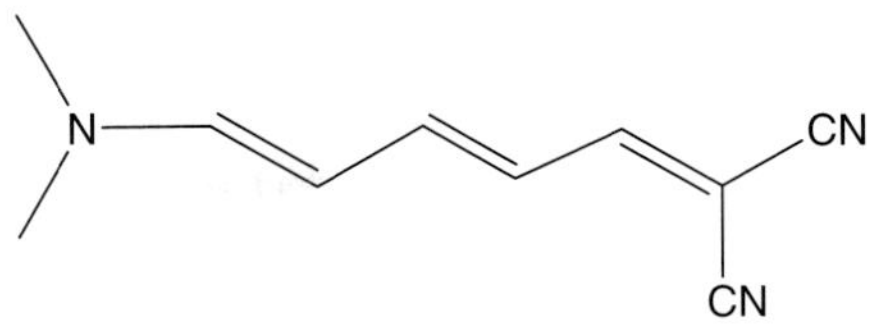

Figure 5.2 Chemical structure of 1,1′-dicyano-6-(dibutylamine)-hexatriene.

Thus, a simple variation on the theme of the two-state model could be very successfully utilized to understand different aspects of variation in structural motifs, enhancing the NLO response of ICT molecules [14]. Nandi *et al.* [15] reported theoretical estimates of solvent-modulated hyperpolarizabilities of *p*-nitroaniline, 5-dimethylamino-5′-nitro-2,2′-bithiophene (DNBT) and 1,1′-dicyano-6-amino-hexatriene (DCH), by employing the semiempirical finite-field self-consistent field (SCF) method with scaled self-consistent reaction-field method. These authors claimed that the theoretical $\beta \cdot \mu$ values compared well with β values predicted by EFISH experiments in solvents, while the theoretically estimated β_0 values compared successfully with β values obtained from HRS experiments.

Barzoukas *et al.* [16] advanced a two-form two-state description of donor–acceptor molecules, and identified important factors that characterize the molecule and the environment the molecule is in. In addition, they define a parameter that is proportional to the change of dipole moment of the DA molecule between the ground and excited states ($\Delta\mu$). The correlation between the hyperpolarizabilities computed by semiempirical methods and this parameter has been claimed to be remarkable. The method has been extended to account for the effects of solvent polarity on the hyperpolarizabilities. Jaquemin *et al.* [17] made use of long-range-corrected density functional theory (DFT) introduced by Ikura *et al.* [18] and Coulomb-attenuated model CAM-B3LYP of Yanni *et al.* [19] to compute static first hyperpolarizability of increasingly long polymethineimine oligomers. For medium and long oligomers, long-range corrected hyperpolarizabilities are slightly smaller than MP2 hyperpolarizabilities. CAM-B3LYP strongly improves the quality of hyperpolarizabilities predicted by B3LYP calculations. MP2-level static hyperpolarizabilities of azo-enaminone isomers have been reported by Fonesca *et al.* [20] who carefully analyzed the modulation of NLO response induced by different combinations of donors and acceptors. The hyperpolarizabilities were calculated analytically by CPHF method using different basis sets. The authors also reported MP2-level hyperpolarizability computed numerically using the finite-filed method. The authors have drawn attention to the geometry dependence of the computed first hyperpolarizability. It appears that the relative orientation of the donor group in the chromophore has an important bearing on the magnitude of the hyperpolarizability.

The dynamic linear and first hyperpolarizabilities of two geometrical isomers (*Z*, *E*) of donor–acceptor-substituted azo-enaminones (Figure 5.3) have been computed by Fonesca *et al.* [21] at the *ab initio* CPHF level. Static results at the MP2 level have also been reported. Correlated dynamic values were estimated by a multiplicative correction scheme. The authors predict that increasing donor strength enhances the diagonal components. Misra *et al.* [22] investigated linear

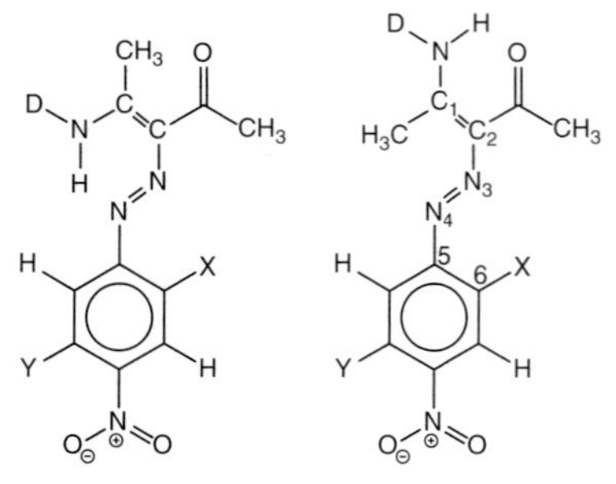

Figure 5.3 The chemical structure of the *Z* and *E* isomers of donor–acceptor-substituted azo-enaminones, studied by Fonesca *et al.* (Fonescca *et al.* 2017 [21]. Reproduced with permission of Elsevier.)

and NLO response properties of a number of β-enaminones by employing *ab initio* finite-field MP2 and TDHF methods with Sadlej pVTZ basis sets. These enaminones turn out to be a new class of ICT molecules with strong NLO response which can be modulated by introducing electron-releasing or electron-withdrawing substituents in the *para*-position of the phenyl ring.

Kang *et al.* [23] have recently synthesized and characterized a series of rather unconventional twisted π-electron system based electronic chromophores. These chromophores have large ring–ring dihedral twist angles (80–89°) in solid state which appear to persist unchanged even in the solution phase. High-level computations based on state-averaged complete active space SCF methods provide a new rationale for the experimentally observed exceptionally high (non-resonant $\mu{\cdot}\beta$ as high as $488\,000 \times 10^{-48}$ esu at 1907 nm) hyperpolarizabilities and explains the rather significant solvent modulation of NLO response of these molecules. The authors claim that their work suggests a new paradigm for molecular hyperpolarizabilities and electro-optics.

Zyss *et al.* [24] provide evidence in support of through-space charge-transfer in paracyclophane (pCP) molecules in which donor and acceptor moieties are made to interact through a sterically constrained π–π stack. The collective nonlinear polarization involving the full end-to-end molecular structure is confirmed by theoretical calculations using the collective electron oscillator approach. The calculations are indicative of a strong role of electron–hole-pair delocalization in the emergence of higher-order nonlinear response compared to the linear response properties. We may mention that strong through-space CT-dominated NLO molecules/materials are rare and the class of molecules reported by Zyss *et al.* [24] therefore deserves special mention.

DFT calculations of the first and second hyperpolarizabilities (β, γ) of $(ZnO)_n$ clusters have been recently reported by Li *et al.* [25]. The calculations were carried out using GGA/PBE, LDA/PZ, and B3LYP functionals and the results unambiguously point to strong size and shape dependence of the computed β and γ values. The authors provide plausible explanations for such dependence in terms of stability, electron delocalization volume, and chemical binding in ZnO clusters.

Recently, Misra [26] reported the linear and NLO response properties of some novel aryl-substituted boron-dipyrromethene (BODIPY) dyes (Figure 5.4). DFT calculations were performed using CAM-B3LYP functional and 6-31G(d,p), 6-31+G(d,p) and cc-pVTZ basis sets. MP2 calculations were performed on

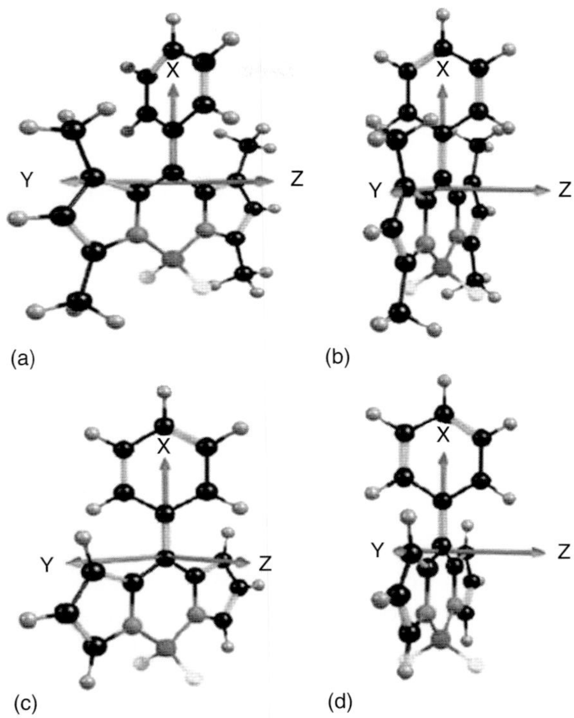

Figure 5.4 The front and side views of two representative aryl-substituted BODIPY dyes. The *x*, *y*, and *z* directions are shown for reference. (Misra 2017 [26]. Reproduced with permission of American Chemical Society.)

some of the systems for comparison. It was found that both ICT process and structural changes affect the NLO response in these molecules. The quantum chemical investigation suggests that the CT process in these molecules is mostly unidirectional. The total first hyperpolarizability (β_{total}) values of these molecules are also dominantly shaped by the response in the direction of CT. It is known that the ratio of β_{vec} and β_{total} also provides important information [27] about the direction of CT in the molecules as

$$\frac{\beta_{vec}}{\beta_{total}} = \cos\theta \tag{5.11}$$

where θ is the angle between the vector formed by β_{vec} components and the dipole moment vector.

The ratios of vector components of first hyperpolarizability (β_{vec}) to β_{total} of the probe molecules are close to unity, except for some molecules with acceptor group substitutions either alone or in conjunction with structural changes. Therefore, the values of $\cos\theta$ also support the unidirectional CT process in these molecules. It was found that the ratio of β_{vec} and β_{total} was not unity in some molecules with acceptor group substitution at the aryl ring. Using the MO pictures of these molecules, it was inferred that another CT pathway might be opened in some molecules with acceptor group substitution either alone or in conjunction with methyl group substitution in the BODIPY moiety.

It has been reported by several research groups that the electron bridge, that is, the π-conjugation network connecting the donor and acceptor units plays a crucial role in determining the extent of CT, affecting the NLO response

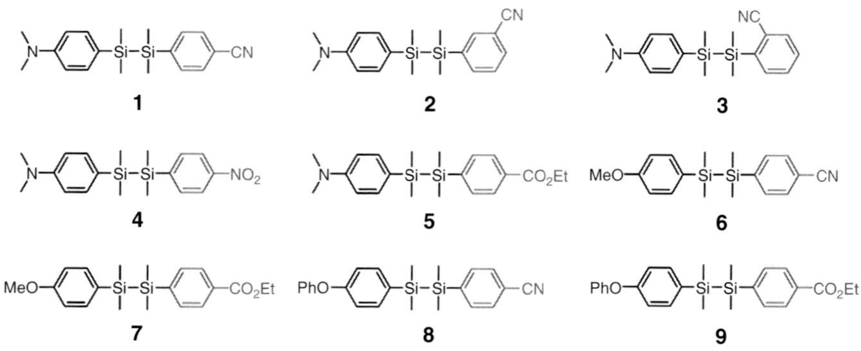

Figure 5.5 Chemical structures of the unsymmetrical 1,2-diaryldisilane derivatives containing electron-donating and electron-withdrawing groups studied by Shimada *et al.*. (Shimada *et al.* 2005 [28]. Reproduced with permission of American Chemical Society.)

of the molecule. Shimada *et al.* [28] synthesized a series of disilane-bridged donor–acceptor architectures that contain strong electron donor and acceptor groups (Figure 5.5) with the aim of studying their fluorescence and NLO response properties. The compounds showed purple to blue emission with emission maxima ranging between 360 and 420 nm and quantum yields up to 0.8 in the solid state. Their quantum chemical calculations also suggested that both the locally excited and ICT states contribute to the luminescence properties of those compounds. The compound with *p-N,N*-dimethylamino and *o*-cyano substitutions showed high optical SHG in the solid state.

Zhou *et al.* [29] reported the exceptionally large second-order NLO response in some donor–conjugated bridge–acceptor (D–B–A) systems, where graphene nanoribbons (GNRs) were used as the electron bridge (Figure 5.6). The authors have found the static first hyperpolarizability (β_0) up to 2.5×10^6 a.u. for the D(NH_2)-GNR-A(NO_2) system. The authors found that the size of the GNR plays a crucial role in shaping the NLO response properties of these compounds.

Nandi [30] studied the bridging effect on the second-order NLO response of a number of tetradehydrodinaptho[10]annulene derivatives using quantum chemical calculations. The author also investigated the effect of several DFT functionals on the predicted values of first hyperpolarizability of the molecules under investigation, which they found more or less consistent with each other. They have reported that the values of first hyperpolarizability obtained using CAM-B3LYP and M06-2X functionals matched with close margin. The authors, on the basis of their calculations, claimed that the first hyperpolarizability of the tetradehydrodinaptho[10]annulene derivatives can be tuned through modification of the acetylene linkages. The extents of CT as well as the static first hyperpolarizabilities (β_0) of these molecules were increased due to incorporation of different ring structures containing electronegative atoms. The authors found that use of furan and heteroazulene rings as bridging unit are most useful means to enhance the values of β_0. They invoked the two-state model to explain the results and found the model satisfactorily explained the variations of β_0 obtained due to use of different linkers.

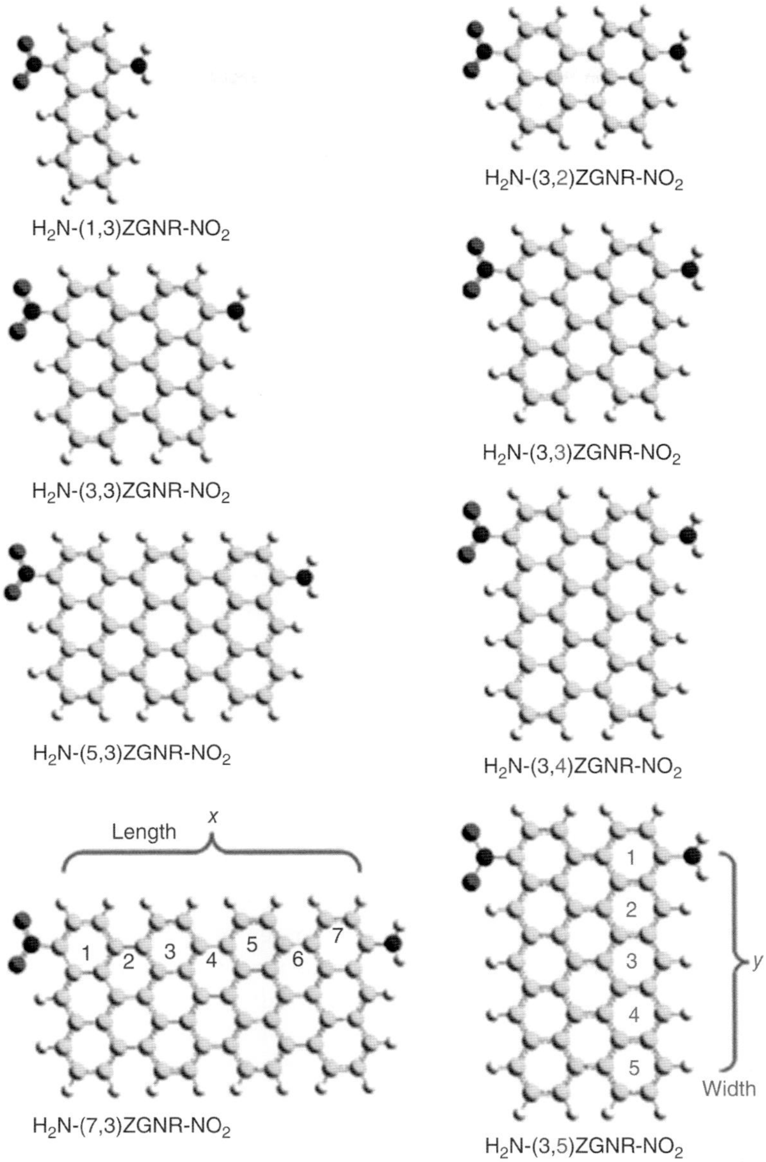

Figure 5.6 Structures of the NLO chromophores studied by Zhou *et al.* (Zhou *et al.* 2011 [29]. Reproduced with permission of John Wiley & Sons.)

Qiu and coworkers [31] have studied the second-order NLO response of some 4-(diethylamino) aryl group donor and strong tricyanovinyldihydrofuran (TCF) acceptor-based D–A–π–A and D–D–π–A systems and compared the performance with the respective D–π–A compounds. They synthesized compounds **A** and **C** (Figure 5.7) with either an additional acceptor (—CN) or donor group (thiophene) and they used compounds **B** and **D** for reference. They found that addition of cyano group as an additional acceptor (D–A–π–A) or thiophene as

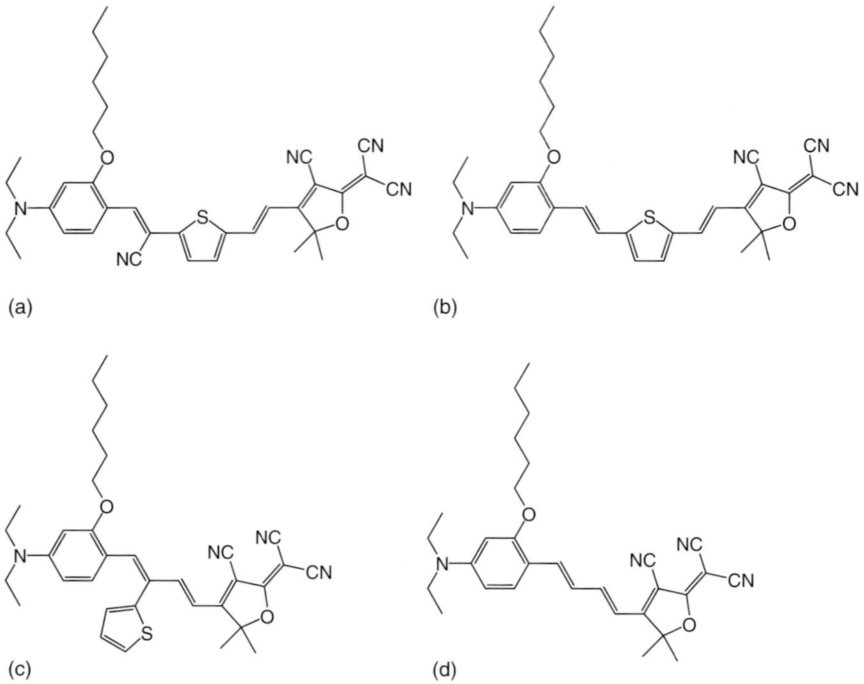

Figure 5.7 The chemical structures of the molecules studied by Qui and coworkers. (Qui and coworkers 2011 [31]. Reproduced with permission of Royal Society of Chemistry.)

an additional donor group (D–D–π–A) to the D–π–A systems showed similar or better NLO response and better optical transparency compared to their parent systems.

Xu *et al.* [32] explored the effect of replacement of heteroatoms in the donor and in the bridging unit on the second-order NLO response of a series of CT-based chromophores (Figure 5.8). Their crystal structure analysis and theoretical calculations predicted that incorporation of heteroatoms at suitable positions of those molecules not only enhanced the values of first hyperpolarizability but also diminished the intermolecular interactions that helped them convert the molecular hyperpolarizability into macro-electro-optic activity in electric field in polar films. The authors also found that the long chain of the donor or the bridging unit, acting as the isolation group, could reduce intermolecular electrostatic interactions that enhanced the macroscopic electro-optical activities of these molecules.

Ratner, Marks, and coworkers [33] studied the second-order NLO response properties of some multiply twisted electro-optic chromophores (Figure 5.9). The authors reported synthesis, structural, optical spectroscopic, and second-order NLO studies of a series of donor–acceptor poly-arylene chromophores, specifically, two-ring 2TTMC, dicyano(4-(3,5-dimethyl-1-(2-propylheptyl) pyridin-1-ium-4-yl)-3-methylphenyl)methanide; three-ring 3TTMC, dicyano (4′-(3,5-dimethyl-1-(2-propylheptyl)pyridin-1-ium-4-yl)-2,2′,3′,5′,6′-pentamethyl [1,1′-biphenyl]-4-yl)methanide; and four-ring 4TTMC, dicyano(4″-(3,

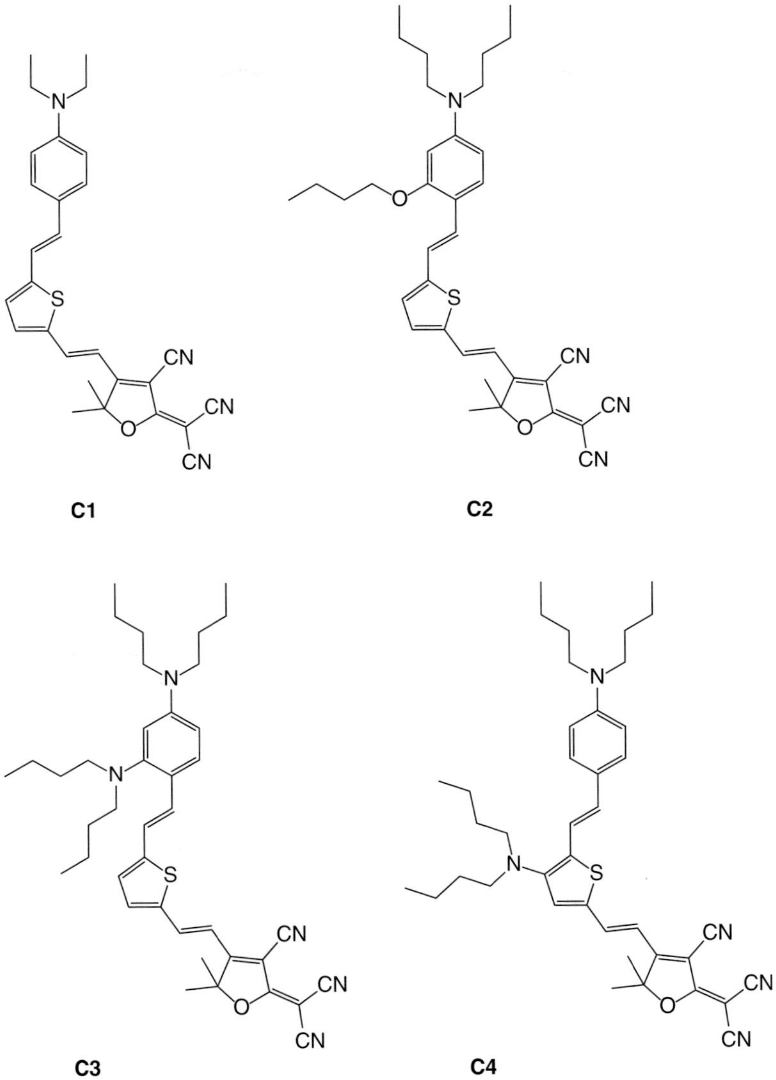

Figure 5.8 The chemical structures of the compounds studied by Xu *et al.* (Xu *et al.* 2015 [32]. Reproduced with permission of Royal Society of Chemistry.)

5-dimethyl-1-(2-propylheptyl)pyridin-1-ium-4-yl)-2,2′,3″,6,6′-pentamethyl[1,1′:4′,1″-terphenyl]-4-yl)methanide. The authors noted that the compounds have heretofore unachieved π-extension and substantial twisting from planarity. The authors carried out measurements of $\mu\beta_{vec}$, that is, the product of the chromophore dipole moment (μ) and the vector part of the molecular first-order hyperpolarizability β_{vec} tensor (see Equation 5.9) along the direction of μ, using the solution-phase DC EFISH method. The authors used single-crystal X-ray diffraction, DFT calculations, as well as B3LYP/INDO–SOS analysis to explain the high NLO response properties of these molecules. Their results suggested

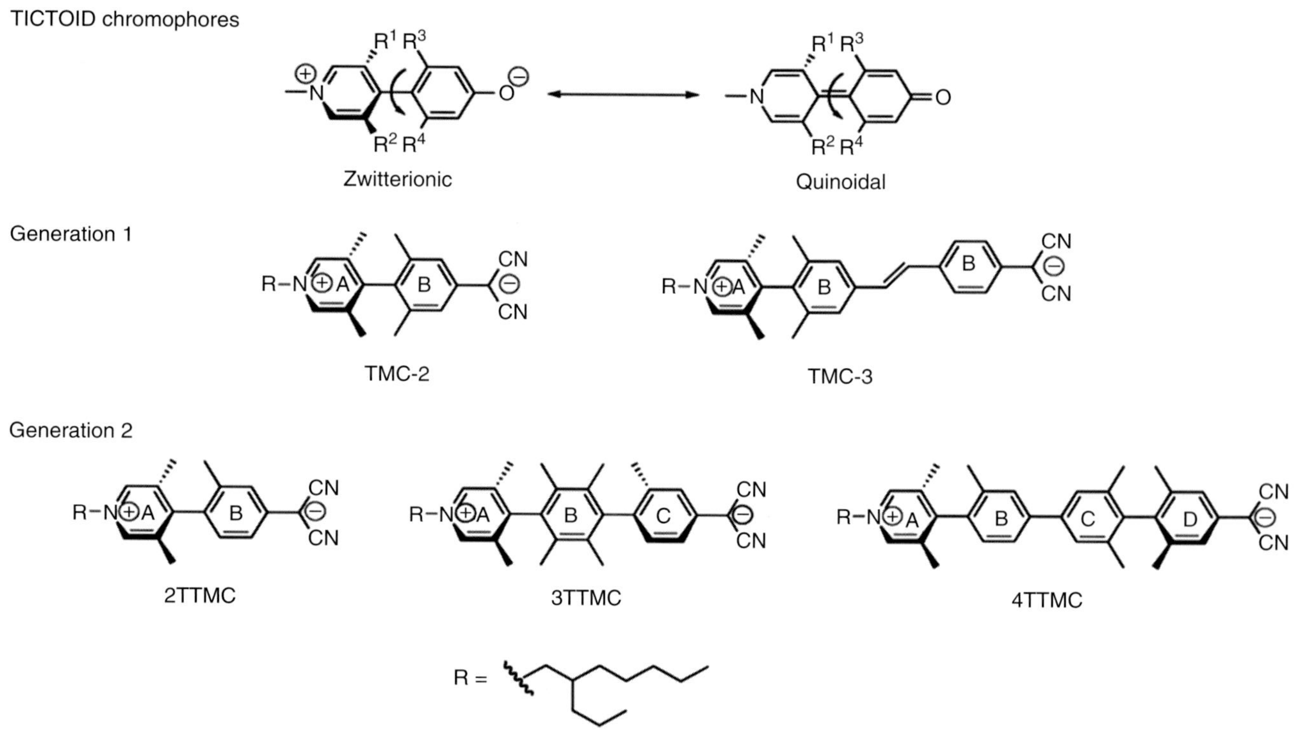

Figure 5.9 The ultralarge-response, twisted π-system electro-optic chromophores studied by Ratner, Marks, and coworkers. (Ratner, Marks, and coworkers 2015 [33]. Reproduced with permission of American Chemical Society.)

that increase in arylene catenation from two to three to four rings significantly increased the NLO response of the molecules studied. They also reported that for ring catenation of three or more, the large NLO response needs sterically enforced π-system twists near the donor and acceptor sites of the chromophores.

Misra *et al.* [34] studied the second-order NLO response properties of a series of 9,10-donor–acceptor-substituted anthracene derivatives. The variation of the first hyperpolarizability (β_{total}) among the molecules with the same donor but different acceptor and/or spacer indicated the influence of the ICT process in the NLO response properties of these molecules. The authors tried to find out if there were any correlations among calculated β_{total} with the donor strength, the strength of the acceptor, donor–acceptor interaction, and the separation between the donor and acceptor groups' acceptor separation using the genetic algorithm (GA) to search through the relevant parameter space. Their results indicated that the β_{total} of these molecules are dominantly determined by additive contributions from the donors and the acceptors.

We have mentioned earlier that several ICT molecules with donor (D) and acceptor (A) groups connected through a π-electron bridge (also called spacers) resulted in a high second-order NLO response. From the discussion presented in this section, it is apparent that the NLO response properties of several chromophores can be tuned through judicious choice of the donor and/or acceptor groups as well as by changing the nature of the connecting π-electron bridge. In addition to the reports described in this chapter, there are several other studies [35–47] that aimed to optimize the NLO response through the structural tuning.

Borini *et al.* [48] systematically investigated the structure hyperpolarizability correlation in *all-trans* polyacetylenes endcapped with different combinations of donor and acceptor groups. They computed the bond-length alternation (BLA) parameter, longitudinal polarizability, and first and second hyperpolarizabilities within the framework of density functional response theory with CAM-B3LYP functional. BLA, a parameter that measures the extent of delocalization, was found to correlate rather strongly with α and γ, and to a lesser extent with β. The observed trend (theoretical) confirms what is expected from conventional theoretical wisdom – that is, high NLO response can essentially depend on the choice of a backbone with high conjugation efficiency and appropriate choice of donor–acceptor pairs. The analysis revealed a simple power-law behavior of α and γ with respect to chain length. The β-chain-length dependence turned out to be more complex. The conjugation path-length dependence of NLO response was also investigated earlier by Whitaker *et al.* [40]. These authors examined the second-order NLO response of amine and phenol/ether derivatives of *p*-nitroaniline (PNA) and *p*-nitrophenol experimentally by means of EFISH measurements and theoretically by adopting semiempirical sum-over-states (SOSs) and finite-field self-consistent methods of calculations. Experimental data tends to suggest that *N*-phenyl substitution in 4-nitroaniline enhances the first hyperpolarizability much more than *N*-methyl substitution. In contrast, *O*-phenyl substitution produces much lower enhancement of the second-order NLO response compared to what is achieved by *O*-methyl substitution (Figure 5.10). The observed behavior is contrary to what is expected from conventional wisdom. Theoretical calculations explain the observed counterintuitive behavior

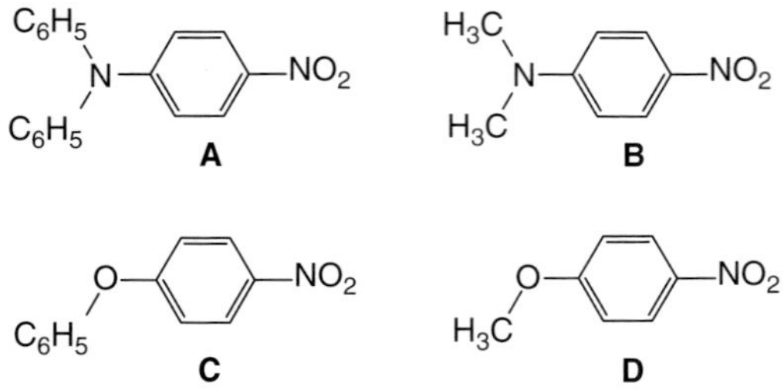

Figure 5.10 The chemical structure of some of the molecules studied by Whitaker *et al.*

in terms of a stronger participation of the N-phenyl moieties in the excited state and an extension of the conjugated path length brought in by the N-phenyl substitution. Thus, there may be a delicate interplay of inductive effect of a substituent and its ability to extend the π-conjugation network across which electronic polarization takes place in an external electric field.

Barlow *et al.* [49] investigated the metallocene-based second-order NLO dyes (Figure 5.11). The lowest energy transition in these compounds was assigned to a metal to acceptor transition as found from UV–vis and infrared (IR) spectroscopic studies, while the higher energy transition was found to be

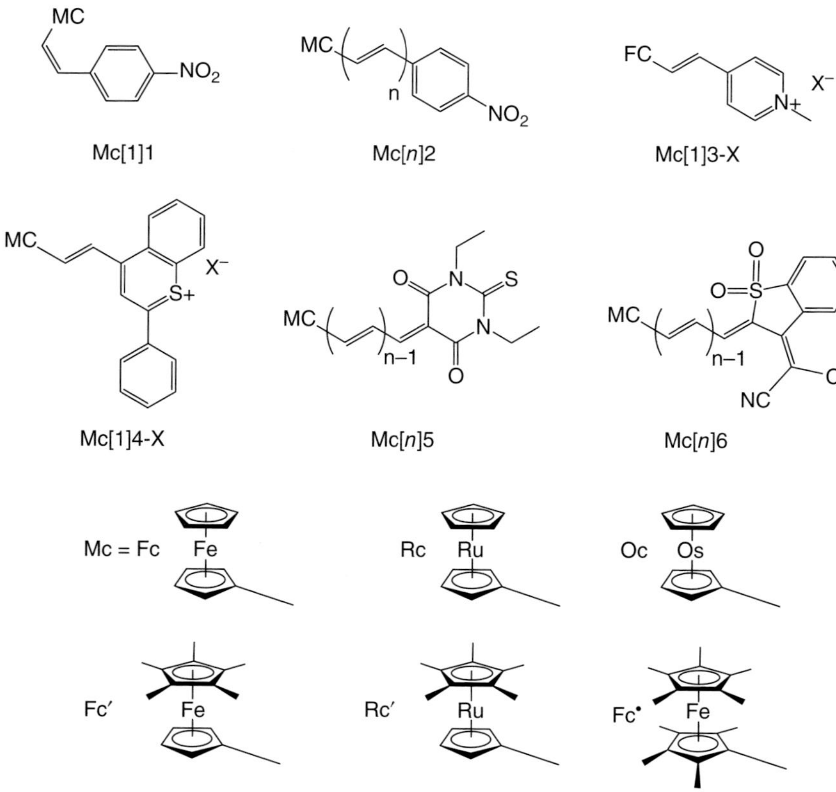

Figure 5.11 Structure of the metallocene-based nonlinear optical chromophores discussed by Barlow *et al.* (Barlow *et al.* 1999 [49]. Reproduced with permission of American Chemical Society.)

originating from a delocalized highest occupied molecular HOMO-3 orbital to an acceptor-based lowest unoccupied molecular orbital (LUMO). Stark spectroscopy confirmed that both the transitions trigger large changes in the dipole moment of the molecule and the theoretical expression for the first hyperpolarizability indicates that both the transitions contribute significantly to the first hyperpolarizability. The tuning of NLO response in these systems is therefore much more difficult.

Janjua *et al.* [50] quantum mechanically designed terpyridine-substituted hexamolybdates which are endowed with an enhanced second-order NLO response. They found that the theoretical designing suggested that $[Mo_6O_{18}(N_4C_{25}H_{14}(CF_3)_2(CN)_2)]^{2-}$ could be expected to have a large second-order NLO response. The electron-withdrawing ability of the electron acceptor groups like F, Cl, Br, I, CF_3, and CN at the end of the terpyridine ligand directs the CT to take place from the polyoxymetallate (POM) cluster to the terpyridine segment along the z-axis, which enhances the NLO response. The quantum chemical calculations were carried out using range-separated functionals like CAM-B3LYP and LC-B3LYP, which appear to be more effective when it comes to computing NLO response properties than the conventional methods.

It is now possible to use theoretical tools to design molecules of different classes with large second-order NLO response. The conversion of these molecules, organic or metallo-organic, into useful and effective NLO materials is not that straightforward. Several complicated conceptual and practical issues need to be tackled in this context. Marder and Perry [51] examined several conceptual issues related to the designing of new molecule-based NLO materials. One of these issues concerns whether a fundamental limit exists for the magnitude of second-order response (β) for a molecule of a given length (L); and if it exists, how does one go about designing a molecule to achieve that maximum second-order response. The second issue addressed by them is very important in practice and concerns how to pack the designed molecules in a noncentrosymmetric crystal so as to maximize a particular NLO response. The third issue examines how best one can improve the orientational stability of chromophores in poled polymers and if alternative alignment strategies could be developed. Marder *et al.* [52] also addressed the issue of tailoring molecular materials so that a new nonlinear phenomenon like photorefractivity emerges.

From a purely quantum mechanical point of view, the nonlinear polarization arises when an external electric field induces mixing among the ground and CT excited states of a molecule, causing masked shift of electron density to one end of the molecule. The length between the donor and the acceptor units in the molecule is therefore one factor that has been tailored to maximize or enhance the NLO response. It has been recently shown that

$$\beta \propto \left(\mu_{ee} - \mu_{gg}\right) \frac{\mu_{ge}^2}{E_{ge}^2} \tag{5.12}$$

where μ_{ee} is the dipole moment in the excited state and μ_{gg} is the dipole moment in the ground state. μ_{ge} is the transition dipole moment (TDM), while E_{ge} represents the energy of the g → e transition. The four-orbital generic model of

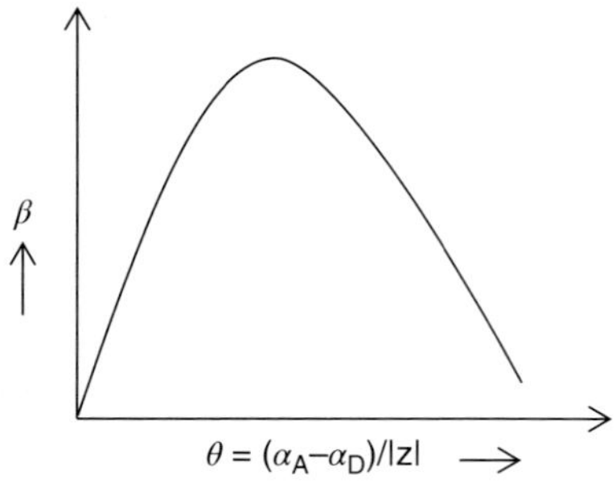

Figure 5.12 The variation of β with change in $\theta = (\alpha_A - \alpha_D)/|z|$.

Marder *et al.* predicts β to reach a peak value as a function of the difference in the Coulomb energy of the donor (D) and acceptor (A) end groups normalized by coupling strength $|z|$ within the orbitals of the bridge connecting the donor and the acceptor units, that is, $\theta = (\alpha_A - \alpha_D)/|z|$. The typical plot looks like what is shown in Figure 5.12.

Small values of $\theta = (\alpha_A - \alpha_D)/|z|$ indicate a regime marked by strong mixing of donor and acceptor orbitals, while large values of θ signify the decoupling of the D, A orbitals from the bridge. The authors observed that barring a few exceptions, molecules endcapped with sufficiently strong donors and acceptors have not been synthesized to reach the maximum on the β–θ curve. In fact, for a given type of bridge, finding the appropriate D–A combination is still an active area of research. The authors also analyzed β as a function of the BLA parameter and found that the peak in the β–θ curve is closer to the cyanine limit than to the bond-length alternated polyene limit. A related curve for γ was also developed and analyzed. The predicted behavior of γ was later confirmed experimentally [13]. It appears that broad designing clues can be better obtained from deftly constructed models rather than large-scale electronic structure and response calculations. The factors that govern or shape the values of α, β, and γ are now sufficiently well understood so that molecules with, say, the desired level of second-order response (β) can be designed successfully. There is still a wide gap in our understanding of the strategy to be adopted to obtain optimum NLO response from the corresponding molecule-based materials. It turns out that the majority of the molecules with a large NLO response absorbs in the visible region of the electromagnetic spectrum. This absorption reduces their usefulness as frequency doublers of light from diode or Nd:YAG lasers. New strategies are needed so that the NLO response of a molecule can be enhanced without sacrificing the "blue-transparency." The other issues concern crystal growth and processing of the organic materials technologically. Examples of some solid-state materials with ICT properties are discussed in Chapter 4. However, further research is needed to address the aforesaid issues to develop useful strategies for preparation of NLO-active materials for technological applications.

5.4 Studies of Two-Photon Absorption

TPA is theoretically defined as an NLO phenomenon that is related to the imaginary part of the third-order nonlinear susceptibility. As TPA is one of the simple and most studied NLO phenomena, several experimental and theoretical studies on this process have appeared in literature. In recent years, the studies of organic molecules with large TPA cross section have gained prominence due to their immense application potential in technology, including multiphoton fluorescence microscopic imaging, three-dimensional optical data storage, optical power limiting, and photodynamic therapy [53–55]. Several molecules, including ICT molecules, metalloporphyrins, molecular tweezers, and quantum dots were explored for this purpose [56–59]. In this section, we shall mainly discuss the TPA activities of some ICT-based molecules. It has been reported that the TPA activity of a molecule can be controlled through tuning of the strengths of the donor and acceptor groups, length of conjugation, the properties of the medium (polarity and hydrogen bonding ability), and self-aggregation, to mention a few [60–62]. During the TPA process (Figure 5.13), the system first absorbs a single photon of energy, $h\nu_1$, to get excited to some virtual state. Before relaxing back to the initial state, it absorbs another photon of energy, $h\nu_2$, almost instantaneously to reach the final state. In many cases, the energy of the two photons are the same, that is, $h\nu_1 = h\nu_2$. The two-photon activity of a system is generally represented in terms of TPA cross section (δ_{TP}). The values of δ_{TP} can be calculated either using the semiempirical SOS approach or the *ab initio* response theory.

He [63] reported and analyzed effects of ICT on the TPA behavior of multibranched triphenylamine derivatives. They found that the dominating factor is the extent of ICT upon photoexcitation rather than coupling between branches. It appears that it is more effective to change the electron-donating and electron-accepting abilities of the branches and the core rather than just increase the number of branches when it comes to enhancing the TPA ability of such multibranched molecules. Diaz *et al.* [64] demonstrated both theoretically and

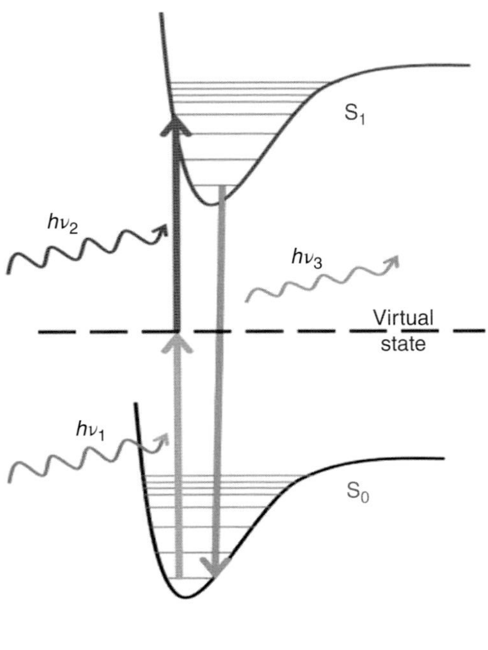

Figure 5.13 Schematic representation of the two-photon absorption process. (Alam *et al.* 2014 [62]. Reproduced with permission of American Chemical Society.)

experimentally that the TPA cross section and two-photon circular dichroism (TPCD) signals of hexahelicene (intrinsically chiral) molecules are strongly affected by the strength of ICT and the nature and extension of electron delocalization. TPA and TPCD spectra obtained by the double L-scan technique were theoretically simulated by invoking the modern analytical response theory using B3LYP, CAM-B3LYP functional, and aug-cc-pVDZ and 6-31++G(d,p) basis sets.

One important aspect of research in this connection has been the efforts to understand theoretically the structure – two-photon activity correlations – that is, what kind of structure manipulations would lead to the emergence of molecules with large TPA cross section. Kogej *et al.* [65], for example, investigated the structure – two-photon absorptivity relationships for donor–acceptor π-conjugated compounds. These relationships, established on the basis of electronic structure calculations, provide strategic clues for designing dyes for large TPA cross sections for fundamental photon wavelengths in the desired range of 0.6–1 μm region. Rumi *et al.* [66] examined structure–activity relationships in two-photon absorbing chromophores like bis donor diphenyl polyene and bis (styryl) benzene derivatives. They found that the TDMs for transition between the ground and first excited singlet state (μ_{ge}) and that between the two excited singlet states ($\mu_{ee'}$) play an important role in shaping up the TPA cross sections. It was established that increase in chain length resulted mainly in causing an increase in μ_{ge}, while addition of donor end groups or by moving from diphenylene-polyene to phenylene-vinylene type of bridges leads primarily to increase in $\mu_{ee'}$. These experimental findings matched well with what was predicted theoretically by the authors. These authors were thus able to provide a useful link between structural features of molecules of these classes and their dipole couplings and relevant state energies that could be exploited to design molecules with high-TPA cross sections.

Alam *et al.* [67] looked into the origin of large two-photon activity of 4,4′-dimethylamino nitrostilbene (DANS) molecules. They used quadratic response theory and a two-state model and found that neither the donor–acceptor pairs nor the π-conjugated bridge between the donor and acceptor moieties could individually account for the large two-photon activity of DANS. It is the cooperative interplay of the two factors that leads to a large overlap between the ground and virtual, and between virtual and CT states and shapes up the very large two-photon activity of the molecule concerned. The findings may be exploited as an important designing clue for TPA materials.

In fact, it is easy nowadays to design TPA active materials by controlling such factors as the length of conjugation, the dimensionality of the CT networks, strengths of donor and acceptor end groups, solvent polarity, self-aggregation, hydrogen bonding, and so on. An impediment to the otherwise well-understood designing process can, however, come from a rather unexpected source and a rather intriguing phenomenon that has come to be known as *channel interference*. It refers to the constructive or destructive interference between two different optical pathways for the two-photon transition in a given molecule (if such pathways exist). The constructive or destructive nature of the two alternative pathways is determined by the angle between the two TDM vectors. The phenomenon was first reported by Cronstrand *et al.* [68]. The explanation seemed to work well for two-dimensional CT systems. Alam and coworkers recently reviewed the phenomenon of channel interference [62] in the general context of three-dimensional molecules. They developed and examined a generalized version of the idea of channel control used successfully in the two-dimensional case, based on a series of TDDFT calculations of the relevant quantities. It turns out that the model can be applied to any molecule irrespective of dimensionality and the mechanism of CT (e.g., through-space or through-bond) and works equally well in gas phase or in solution. Based on the model, the authors provided chemical ways of controlling channel interference, albeit theoretically. It remains to be seen if the idea can lead to practical applications in the designing of efficient two-photon active materials.

Alam [69] studied the effect of relative positions of the donor and acceptor groups on channel interference in substituted biphenyls. The author used the acronym for the studied molecule as follows: PDOA (*para*-donor *ortho*-acceptor) for 4′-(dimethylamino)-2-biphenylcarbonitrile, PDMA (*para*-donor *meta*-acceptor) for 4′-(dimethylamino)-3-biphenylcarbonitrile, MDPA (*meta*-donor *para*-acceptor) for 3′-(dimethylamino)-4-biphenylcarbonitrile, and ODPA (*ortho*-donor *para*-acceptor) for 2′-(dimethylamino)-4-biphenylcarbonitrile (Figure 5.14). The gas-phase-optimized geometries of these molecules are shown in Figure 5.14. The author also tried to address whether the orientation of TDMs of these molecules could be predicted theoretically. The results obtained

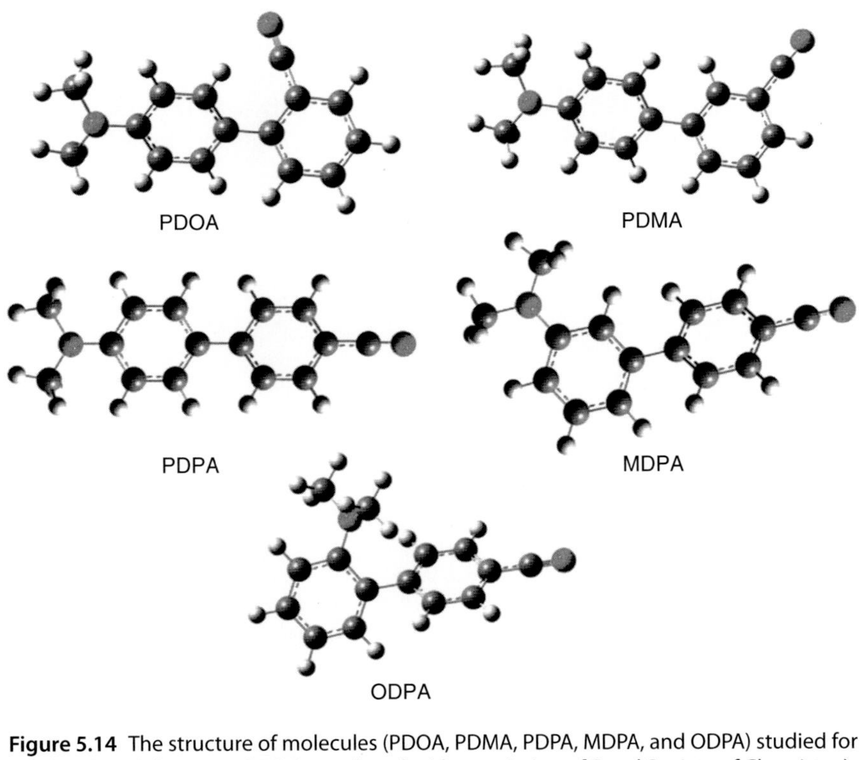

Figure 5.14 The structure of molecules (PDOA, PDMA, PDPA, MDPA, and ODPA) studied for TPA by Alam. (Alam 2015 [69]. Reproduced with permission of Royal Society of Chemistry.)

from the linear and quadratic response theory led the author to infer that substitution of the donor group at the *para*-position give rise to constructive interference, while incorporation of the donor in the *ortho*- or *meta*-position leads to destructive interference. The author explained the results in the light of relative orientations of the TDM vectors; and it was found that when the donor group is placed in the *para*-position to the acceptor, the vector oriented itself in such a way that the pairwise angles were both either greater than 90° or less than 90°. This kind of orientation led to the constructive interference in the molecule. The author also examined the effect of the medium on the two-photon activities of these molecules and found that solvents played major roles in shaping the values of δ_{TP}.

Jha *et al.* [70] reported the electronic structure, one-photon absorption and TPA properties of four commonly used reference fluorophores, namely, *p*-bis(*o*-methoxystyryl)benzene (abbreviated as bis-MSB), coumarin 307, fluorescein, and rhodamine B (Figure 5.15). The authors used TDDFT to calculate the one-photon absorption and TPA spectra of the aforesaid molecules and compared their TPA results with available experimental data. This led them to conclude that caution must be taken while comparing TPA cross sections obtained through different measurements/calculations. For example, while using calibration methods, one must be careful to check whether those methods are immune to systematic errors. They also cautioned that while comparing the TPA cross section data, the definition of it used must be the same in all cases. The authors cited the example of Z-scan technique, in which the TPA cross section is, in many cases, infected by stepwise contributions that are pulse length dependent.

Although molecules with excited-state ICT (generally with D–π–A architectures) are known to exhibit high TPA cross section [71], some molecules with

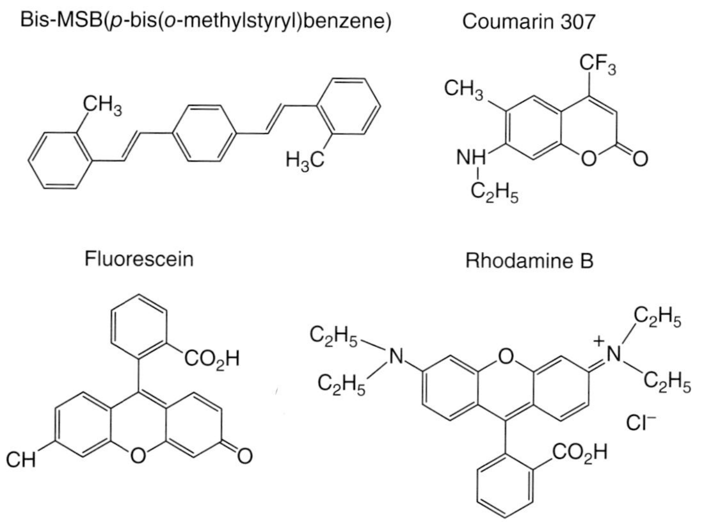

Figure 5.15 Chemical structure of the molecules investigated by Jha *et al.* (Jha *et al.* 2008 [70]. Reproduced with permission of John Wiley & Sons.)

the acceptor group placed between two donor groups (D–π–A–π–D) have been reported to show larger TPA cross section than with the donor group placed between two acceptors (A–π–D–π–A) [72, 73]. Ajayaghosh and coworkers [74] reported enhancement of TPA cross section of a bipyridine-centered D–π–A–π–D (donor–acceptor–donor)-type ratiometric fluorescent molecular probe due to binding of Zn^{2+} ions (Figure 5.16). The authors found an increase of TPA cross section by 13 times, while the fluorescence brightness was enhanced by 9 times at 620 nm due to binding of the Zn^{2+} ions to the probe molecule (they called it GBC). They demonstrated the ability of GBC in both *in vitro* and *in vivo* imaging of cells. They noted that the good cell permeability, high fluorescence quantum yield, large TPA cross section, and high two-photon fluorescence brightness of the probe molecule provided advantages over other TPA molecules for two-photon imaging of free zinc ions in live cells.

Odelius *et al.* [75] theoretically investigated the mechanism of Zn^{2+} sensing by the TPA dye reported by Ajayaghosh and coworkers [74]. The authors studied

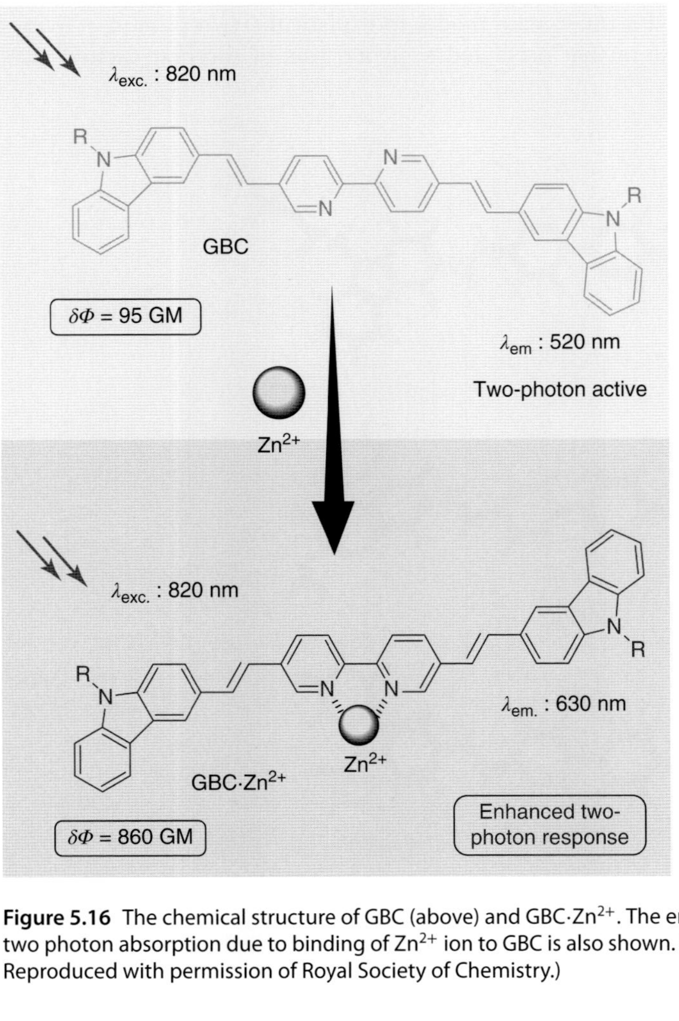

Figure 5.16 The chemical structure of GBC (above) and GBC·Zn^{2+}. The enhancement of the two photon absorption due to binding of Zn^{2+} ion to GBC is also shown. (Divya *et al.* 2014 [74]. Reproduced with permission of Royal Society of Chemistry.)

the linear and NLO properties of the bound and free probe molecule. They also considered the solute–solvent interactions implicitly and explicitly using the polarizable continuum model and quantum mechanics/molecular mechanics (QM/MM) model, respectively. The authors noted that due to the presence of four single bonds connecting a spacer with the carbazole group and the bipyridine moiety, each of their analyzed GBC species could adopt 10 stable conformers. As their results suggested that several possible conformers of GBC were accessible in solution at room temperature that contribute to the TPA process, the authors examined three molecular species, namely, *trans*-GBC, *cis*-GBC, and *cis*-GBC*Zn^{2+} (Figure 5.17), and their 10 stable conformers. They found that the one-photon absorption of both the free and bound forms of GBC was dominantly dictated by the $\pi \rightarrow \pi^*$ transition that leads the systems to the first excited state from the ground state. In contrast, their quantum chemical calculations revealed that the TPA cross section was negligible for the S_0 to S_1 transition for both the free and bound forms of GBC. The authors therefore surmised that the second excited state was involved in the strong TPA process. The authors invoked a three-state model to explain the two-photon activities of the aforesaid system that predicted a pivotal role of the first excited state

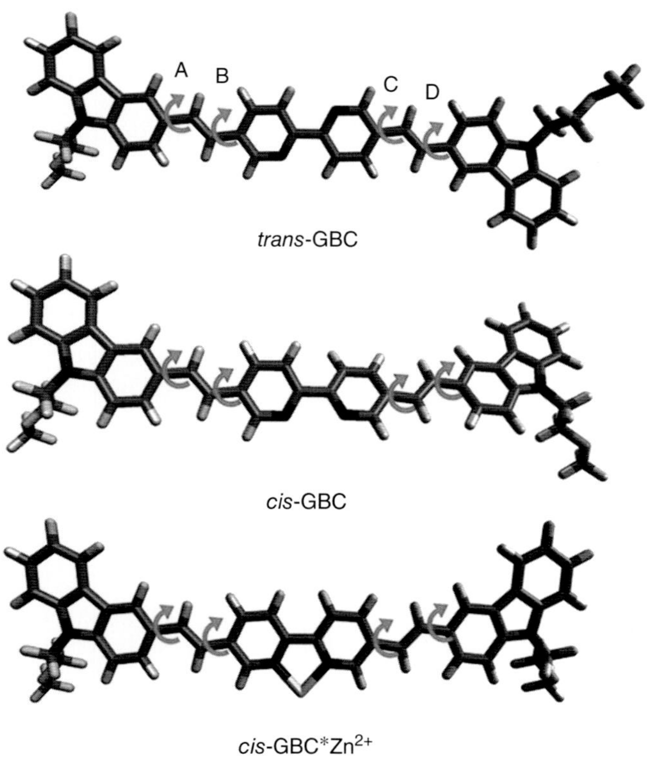

Figure 5.17 The most stable TTTT conformers of *trans*-GBC, *cis*-GBC, and *cis*-GBC*Zn^{2+}. The single bonds in the conjugation pathway are marked with arrows. (Bednarska *et al.* 2016 [75]. Reproduced with permission of American Chemical Society.)

that contributed to the TPA activity for the S_0 to S_2 transition through the SOS expression. Their studies also revealed that the increase in magnitude of transition moment from the first to second excited state was responsible for increase in two-photon activity by three times, which is a result of ICT due to binding of the zinc ion.

5.5 Third-Order NLO Response of ICT Molecules

So far our main focus has been on ICT-molecule-based materials with large β values. There is a growing demand for molecules and materials possessing large third-order optical nonlinearity. For molecules, it means that designing molecules with large second hyperpolarizability (γ) is a critically important problem. Such molecules can be processed into molecular materials for photonic applications like all-optical switching, data processing, protection of sensors, and so on. Garito *et al.* [76] proposed a mechanism that could bring in a spectacular increase in γ in conjugated linear π-molecular systems. They showed that introduction of polarization of the π-system that leads to a large difference between the dipole moments of the system in the ground and excited states could lead to the emergence of large γ values for the polarized polyenes. In fact, a parallel can be noticed in the semiconductor potential wells in which introduction of asymmetry has been found to give rise to rather spectacular second- and third-order nonlinear response in the mid- or far-IR portion of the spectrum.

Marder *et al.* [77] probed large molecular third-order optical nonlinearities in a series of polarized carotenoids in which there is an increasing ICT from polyenic chain to the acceptor moiety in the ground state and measured the γ values as a function of wavelength by THG. The authors found that the molecule with the largest CT registered a 35-fold enhancement of the γ_{max}, that is, γ measured at the peak of three-photon resonance, relative to the symmetric β-carotene molecule which itself is endowed with a very large third-order optical nonlinearity. The polarized molecules revealed, by Stark measurements, a large difference in the dipole moments ($\Delta\mu$) between the ground and excited states. Quantum chemical calculations at the INDO-MO-SCF level indicate that interaction among states with large $\Delta\mu$ values is an important determinant in the emergence of large third-order NLO response, validating the mechanism proposed by Garito *et al.*

Earlier, Naokioba *et al.* [78] synthesized a poly(arylene ethynylene) (PAE) type of π-conjugated polymers and evaluated the third-order susceptibility ($\chi^{(3)}$) of thin films of such materials. They observed a redshift in the electronic absorption band of PAE when heteroaromatic rings rich in π-electrons are introduced into the π-conjugation, indicating the role of CT interaction in shaping the redshift. The $\chi^{(3)}$ value of PAE was found to increase when CT interactions increased and attained a value of about 10^{-11} esu in the near-resonant region. The enhancement was attributed dominantly to increased CT interaction. It was claimed by the authors that introduction of CT interactions in the π-conjugation could be an effective route to design materials with large $\chi^{(3)}$ values. Agarwal *et al.* [79]

calculated off-resonant $\chi^{(3)}$ values for a one-dimensional conjugated polymer by adopting a tight-binding approximation. They obtained a simple relation between the linear and NLO parameter, for example, $\chi^{(3)} \propto N_d{}^6 \propto E_g{}^{-6}$, where N_d is an π-electron delocalization parameter and E_g is the optical band gap. N_d was seen to increase when the BLA parameter decreased. The theoretical relations were found to be obeyed by PAE materials in which $\chi^{(3)}$ became large as BLA tended to become zero.

A series of novel D–π–A molecular systems involving donor-substituted tetracyano ethylene (TCNE) and 7,7,8,8-tetracyano quinodimethane (TCNQ) adducts have been investigated by several groups [80, 81] by taking advantage of the "click-type" nearly quantitative [2 + 2] cycloaddition with appropriately activated alkynes and subsequent retro-electrocyclization. It turns out that the resulting nonplanar chromophores are naturally rather stable and can serve as ideal platforms for ICT processes. They have been explored as promising organic NLO materials. Wang *et al.* [82] recently designed and synthesized an oligo(phenylene-vinylene) (OPV)-bridged ICT compound, $(TCNQ)_2$-OPV_3 (Figure 5.18) and measured its third- and fifth-order nonlinear refraction indices (n_2 and n_4) by an NLO imaging technique with the phase object at the entry of a 4f coherent imaging system under NIR excitation. The authors point out several advantages of the method of measurement, namely, simple optical alignment,

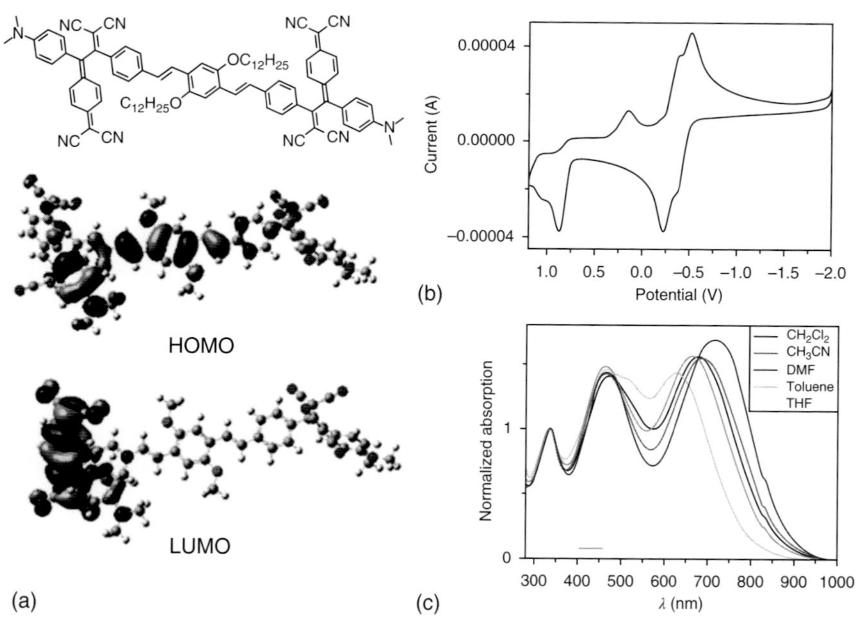

Figure 5.18 (a) Chemical structure of $(TCNQ)_2OPV_3$ (top), and LUMO and HOMO pictures (down) of the molecule, obtained from DFT [B3LYP/6-31G(d,p)] optimized geometry. The dodecyl substitutes were replaced by methyl groups for ease of calculation and hydrogen atoms were omitted for clarity. (b) Cyclic voltammogram of $(TCNQ)_2OPV_3$ measured in CH_2Cl_2 (+0.1 M nBu_4PF_6 at a scan rate of 0.2 V s^{-1}). (c) UV–vis absorption spectra (normalized) of the probe in different solvents. (Wang *et al.* 2017 [82]. Reproduced with permission of Royal Society of Chemistry.)

very high sensitivity along with insensitivity to statistical fluctuations of the laser beam, and the ability to measure the nonlinear absorption and refraction at the same time [83]. In addition, the method can extract information about both the magnitude and the sign of the nonlinear coefficients in one laser shot. No scanning (as in Z-scan method) is necessary. They reported the measured values of n_2 and n_4 as $-1.676 \times 10{-}19\,\mathrm{m^2\,W^{-1}}$ and $-1.273 \times 10{-}33\,\mathrm{m^4\,W^{-2}}$, respectively.

Geskin *et al.* [84] presented a quantum chemical analysis of the correlations among molecular structures and second- and third-order NLO response in ammonio/borato diphenyl polyene zwitterion and pointed out the important role of polarized aromatic groups. They obtained optimized molecular geometries at the MP2/6-31G level. The NLO responses were estimated via finite-field real-space as well as SOS methods at the INDO level of approximation. The real-space approach enabled them to identify the NLO active segments of the molecule directly, while the SOS calculations led to the identification of the virtual excitations and the CT pathways that are critical to the NLO response. Both the approaches highlighted the rather counterintuitive and remarkable result that the strongly polarized phenylene groups play the key role in the emergence of the large NLO response of these molecules.

Alain *et al.* [85] synthesized and investigated several series of push–pull diphenyl polyenes of varying chain lengths and donor and acceptor moieties. Their response properties were measured by electro-optical absorption experiments and third-order harmonic generation in solution. Each of the molecules exhibited a strong ICT band in the visible region accompanied by an increase in dipole moment ($\Delta\mu \gg 0$) upon excitation. The tuning of the responses was achieved by varying the D/A strengths, which led to a concomitant increase in $\Delta\mu$, β, and γ values. Giant $\Delta\mu$ values (up to 30 D) and enhanced nonresonant β and γ values ($\beta(0) = 500 \times 10^{-30}$ esu; $\gamma(0) = 8000 \times 10^{-36}$ esu) could be obtained without sacrificing the solubility, stability, and the transparency of the molecule.

Hasan *et al.* [86] reported a family of new compounds, which are covalently bonded hybrids of Anderson-type polyoxymetallates (CPOM) and porphyrin moieties. These molecules are all endowed with strong reverse saturable absorption and self-focusing effect at 522 nm with a pulse duration of 6 ns. They are therefore promising candidates for fabricating molecular device applications in photonics and opto-electronics. Quite counterintuitively, the hybrids wherein the POM is covalently linked to porphyrin by a shorter bridge has a stronger NLO response than the hybrid with linking through a longer bridge. Moreover, the hybrid with two porphyrins coupled to POM has a superior NLO response, as measured by the Z-scan method, compared to the hybrid containing a single porphyrin moiety. The authors point out that CT from the porphyrin to the POM unit has a key role in enhancing the NLO response, while the disposition of LUMO level of the POM moiety appears to be the most important factor that shapes the γ value.

Kato *et al.* [87] synthesized a new family of homoconjugated push–pull chromophores by [2 + 2] cycloadditions of 2,3-dichloro-5,6-dicyano-1,4-benzoquinone (DDQ) to donor-substituted alkynes and exploited their unique reactivity to generate a new class of donor–acceptor (D–A)-functionalized spiro compounds. The cycloadducts display strong ICT interactions, presumably due to effective homoconjugation arising from the rigid bicyclic framework. It appears that in the ferrocene-derived push–pull systems transannular CT interaction takes place and has a role in the emergence of remarkable third-order polarizabilities. TDDFT calculations at the TD-PBE1PBE/cc-pVTZ//PBE1PBE/6-31G(d) level predicted results that agree well with experimental transition energies. The longest wavelength absorption in all the chromophores are found to be derived from transition between the *N*,*N*-dialkylamino alkynes (DAA)-centered HOMO and the enedione-centered LUMO. The HOMO and LUMO orbitals in all the chromophores partially overlap on the cyclobutene fragment, identifying a main pathway for the observed homoconjugative CT interactions through the cyclobutane rings.

Teran *et al.* [88] recently reported the design and synthesis of a series of electron-excess thiophene containing donor–acceptor chromophores having condensed π-systems and inter-aryl twist angles regulated by steric factors. The idea was to exploit synergistic cooperation among multiple sources of nonlinearity leading to an enhanced third-order NLO response. Their suggested structures exploit two key mechanisms of nonlinearity. The first one is ICT, which is strongly enhanced by the presence of electron-excess thiophene rings that reduce the aromaticity at the chromophore. The second is a twisted chromophore geometry that produces a dense manifold of excited states (close lying) with dipole moment changes ($\Delta\mu$) between the ground and excited states that are roughly twice as much as $\Delta\mu$ values achievable in untwisted chromophores. Resonance forms of these tictoid thiophene-based chromophores are listed in Figure 5.19. These authors exploit nonlinear Z-scan measurements combined with spectroscopic and electrochemical data as well as quantum chemical calculations to elucidate the dependence of mechanisms of enhancement of nonlinear refractive index on molecular structure. It turns out that the real part of the second hyperpolarizability (Re(γ)] of twisted structures are far greater than the imaginary part (Im(γ)), the ratio achieving a value as high as 100 in the most strongly twisted structures. Such chromophores are expected to be promising candidates for fabricating materials for all-optical switching applications. Unlike the molecules and materials with large first hyperpolarizability, the designing clues for molecules and materials with large third-order nonlinear response are just beginning to emerge. One can anticipate exciting developments in the field of tailor-made NLO materials with large γ, which will find device applications in all-optical switching, telecommunication, and so on.

5.6 Experimental Studies on Nonlinear Optical Response

From the preceding discussion it is apparent that studies of NLO response properties have immense importance in contemporary science. It is therefore not surprising that several experimental studies in this direction have been reported in literature. HRS is probably the most popular technique to study

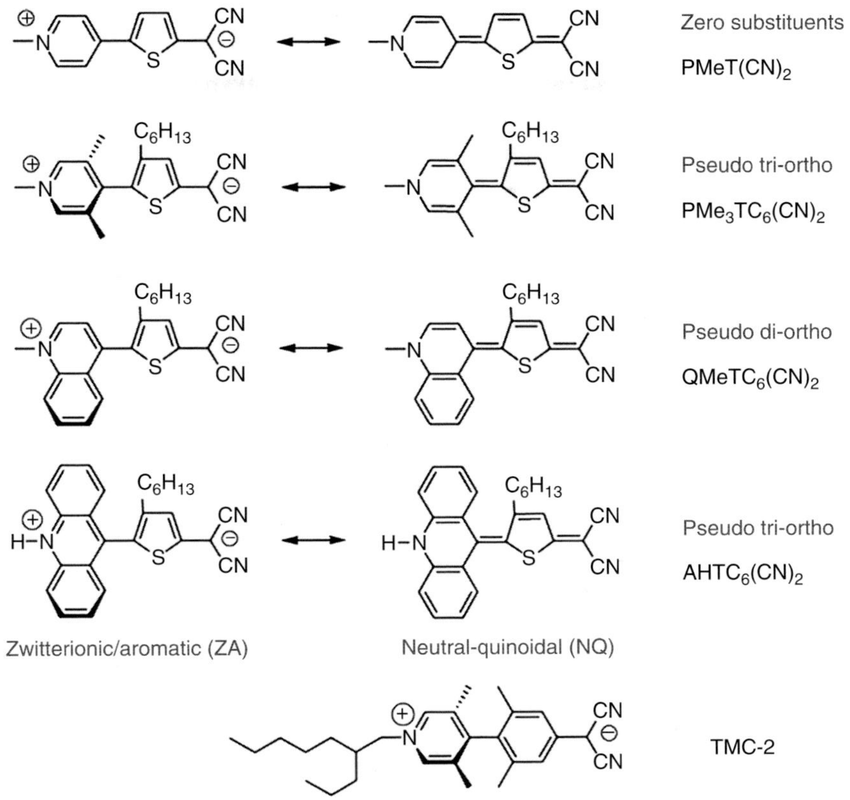

Figure 5.19 Resonance forms of tictoid thiophene-based chromophores considered by Teran *et al.* (Teran *et al.* 2016 [88]. Reproduced with permission of American Chemical Society.)

the second-order NLO response of a molecule. Several other experimental techniques, including EFISH, Z-scan, THG, degenerate four-wave mixing, optical Kerr gate, and Kurtz powder technique have been used to study the NLO response properties of organic molecules. The detailed discussions of the aforesaid techniques are out of the scope of this book and interested readers are referred to the following references for further details [89]. A relatively recent review summarized the use of some of these techniques for studies of metal acetylides and their derivatives [90].

Champagne and coworkers [91] reported a combined experimental and theoretical investigation on the second-order NLO properties of a series of oxazine derivatives. Their studies revealed that these molecules could have potential for use in NLO switches. In these molecules, the cleavage of the C—O bond at the junction of the indole and oxazine cycles was triggered by change in pH. This acido-triggered reduction leads to an open form (b^+) from a closed form "a" as shown in Figure 5.20. Earlier to this study, as the authors mentioned, oxazines have been reported to undergo three types of opening pathways through the cleavage of the C—O bond at the junction of the two heterocycles, namely, photochemical, redox, and acidic addition [92, 93]. These openings led to the

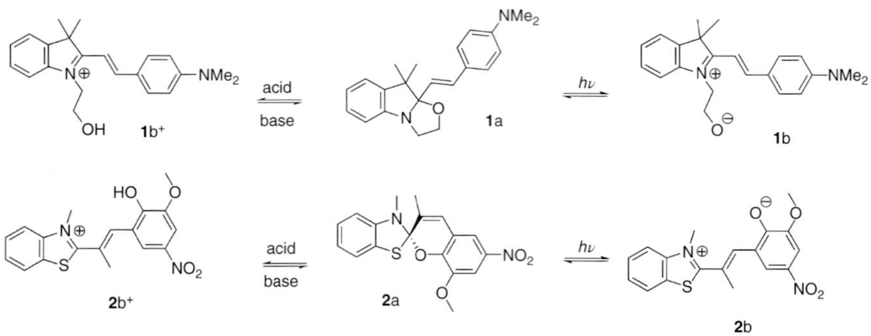

Figure 5.20 Chemical structure of the indolinooxazolidine (**1**a) and spiropyran (**2**a) NLO switches. The equilibria with their open forms, triggered by acid/base addition (**1**b$^+$/**2**b$^+$) or by irradiation (**1**b/**2**b), are also shown. (Beaujean *et al.* 2016 [91]. Reproduced with permission of American Chemical Society.)

formation of a phenolate if the process of C—O bond cleavage goes through photochemical and redox processes while it produces phenol chromophoric moiety when acid triggered and allows the π-conjugation between the R_1 group and the indoleninium unit. It was also reported that both the b and b$^+$ open forms showed significant changes in their absorption spectrum, compared to their closed-form counterparts.

It is worth mentioning that the two important parameters one can obtain from the HRS experiments are the β_{HRS} and the depolarization ratio (DR). The "full" HRS first hyperpolarizability can be defined as [94]

$$\beta_{HRS} = \sqrt{\langle \beta^2_{HRS} \rangle} = \sqrt{\langle \beta^2_{ZZZ} \rangle + \langle \beta^2_{ZXX} \rangle} \tag{5.13}$$

The DR is defined as

$$DR = \frac{\langle \beta^2_{ZZZ} \rangle}{\langle \beta^2_{ZXX} \rangle} \tag{5.14}$$

The full forms of $\langle \beta^2_{ZZZ} \rangle$ and $\langle \beta^2_{ZXX} \rangle$ without assuming Kleinman's symmetry conditions can be found in Ref. [94].

The DR gives information on the geometry of the NLO chromophore, that is, the part of the molecule responsible for the observed NLO response.

The first hyperpolarizability can also be decomposed into the sum of dipolar ($J = 1$) and octupolar ($J = 3$) tensorial components. Assuming Kleinman's conditions, one can write [94]

$$\langle \beta^2_{ZZZ} \rangle = \frac{9}{45}|\beta_{J=1}|^2 + \frac{6}{105}|\beta_{J=3}|^2 \tag{5.15}$$

$$\langle \beta^2_{ZXX} \rangle = \frac{1}{45}|\beta_{J=1}|^2 + \frac{4}{105}|\beta_{J=3}|^2 \tag{5.16}$$

The nonlinear anisotropy parameter (ρ), which is defined as the ratio of octupolar and dipolar contributions to the hyperpolarizability, encompasses the relative contributions of the octupolar and dipolar components and can be expressed as [91]

$$\rho = \frac{|\beta_{J=3}|}{|\beta_{J=1}|} \tag{5.17}$$

The term ρ is related to the DR as

$$\mathrm{DR} = 9\frac{\left(1 + \frac{2}{7}\rho^2\right)}{\left(1 + \frac{12}{7}\rho^2\right)} \tag{5.18}$$

Therefore, the term β_{HRS} can be expressed as follows [91]:

$$\beta_{HRS} = \sqrt{\frac{10}{45}|\beta_{J=1}|^2 + \frac{10}{105}|\beta_{J=3}|^2} = |\beta_{J=1}|\sqrt{\frac{2}{3}}\left(\frac{1}{3} + \frac{1}{7}\rho^2\right) \tag{5.19}$$

Using quantum chemical calculations, Champagne and coworkers [91] studied the structural, electronic, thermodynamic, linear, and NLO properties of the oxazine derivatives. From these calculations, they selected four compounds, namely, 3, 4, 11, and 12 that have intermediate to large NLO response (β_{HRS}) in their open forms; and β_{HRS} contrast ratios are clearly distinguishable for experimental investigations using HRS as well as UV–vis absorption spectroscopy. The authors characterized the open and closed forms of these compounds, as the photogenerated forms had quite a short lifetime, in the order of microseconds. The values of β_{HRS} for compounds **3** and **4** in chloroform, obtained through excitation at 1064 nm were 1910 . and 2050 a.u. in their closed (a) forms and 4820 and 4480 a.u. in their open (b^+) forms, respectively. The β_{HRS} for compounds **11** and **12** in similar conditions were 5210 and 1970 a.u. in their closed forms and 52 820 and 54530 a.u. in their open forms, respectively. The results clearly indicated the high open/closed β_{HRS} ratio, which made these compounds suitable for use as NLO switches. The authors also investigated the effect of donor/acceptor substitution on the electronic structure and properties of the investigated compounds. Their investigations revealed that substitution of donor groups led to larger β_{HRS} as well as β_{HRS} contrast ratios, compared to acceptor substitutions. The donor substitution also proved to be advantageous than the acceptor substitutions and the substitution of the former led to smaller excitation energies, larger opening-induced CT, reduction of the BLA, and smaller Gibbs energies of the opening reaction. The authors also carried out theoretical calculations on the zwitterionic open forms (b) revealing that the UV–vis spectra of both the open forms (b and b^+) were similar although their β_{HRS} values were significantly different and dictated by the nature of the substituent. The authors explained the variation in β_{HRS} of b and b^+ forms in terms of the presence of nitrophenol moiety in the b^+ form that presented a smaller β_{HRS} than its deprotonated counterpart nitrophenolate present in the b form.

Buckley *et al.* [95] reported the second-order NLO response of 14 dipolar cations that contain either methoxy or tertiary amino electron donor groups covalently attached to helquat acceptors (Figure 5.21). The authors synthesized

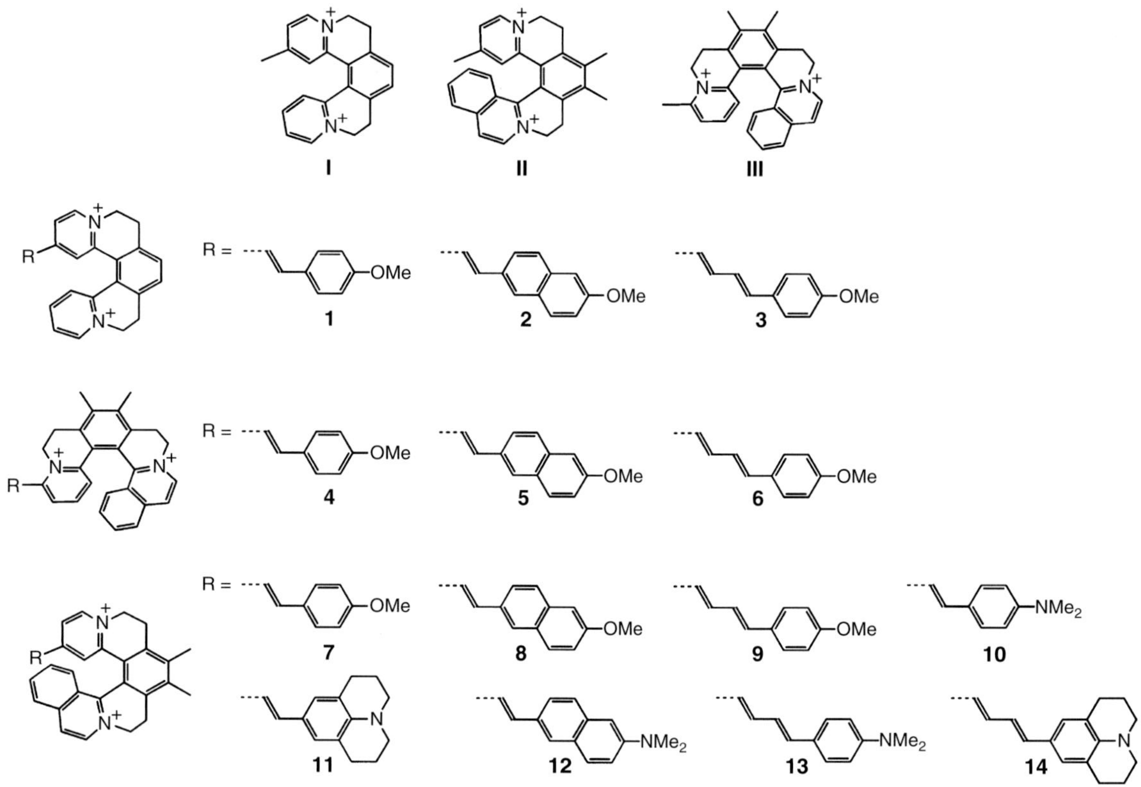

Figure 5.21 Chemical structure of the NLO chromophores studied by Buckley *et al*. (Buckley *et al*. 2017 [95]. Reproduced with permission of American Chemical Society.)

these derivatives and characterized them using nuclear magnetic resonance (NMR) as well as UV–vis absorption spectroscopy as their TfO^- salts. They assigned the intense, low-energy absorption with maxima around 400–600 nm to the ICT states of these chromophores. They determined the molecular quadratic NLO response of these salts using the HRS technique with an 800-nm laser. The NLO response was also measured indirectly using Stark (electroabsorption) spectroscopy for the low-energy absorption bands. The authors noted that the helquat motif could be viewed as a hybrid of the diquat and helicene structures. Therefore, the helical chirality of helquat in conjunction with its dye-like behavior could be interesting for NLO studies. Earlier, they studied the NLO response properties of helquat derivatives containing either tertiary amino or ferrocenyl electron donors [96]. In the helquat derivatives with methoxy or tertiary amino groups as electron donors, the authors showed that the linear and NLO properties of these chromophores could be widely tuned. The DFT and TDDFT calculations showed the chromophores behave as a two-state systems, that is, they show only one low-energy electronic transition that dominantly dictate their NLO response. The authors carried out femtosecond HRS measurements of TFO^- salts of the chromophores using the 800-nm laser in acetonitrile and also estimated their static first hyperpolarizabilities using the two-state model (β_0(H)). The β_0(H) values of the methoxy derivatives generally increase with the extension of π-conjugation network. This enhancement of β_0(H) was seen to be consistent with redshift of the low-energy ICT band of these derivatives. The values of β_0(H) were reported to be relatively modest and quite smaller than those with dimethylamino substituents. The authors attributed these results to the weaker electron-donating strength of the methoxy group, compared to the dimethylamino group. The authors carried out Stark (electroabsorption) spectroscopic studies in PrCN glass matrix formed at 77 K to derive the values of the change in dipole moment between the excited and ground state ($\Delta\mu_{12}$). They used the following equation to estimate the static first hyperpolarizability (β_0) from the Stark spectroscopic measurements. This equation is valid if the hyperpolarizability tensor has only nonzero elements along the ICT direction.

$$\beta_0 = \frac{3\Delta\mu_{12}(\Delta\mu_{12})^2}{(E_{max})^2} \tag{5.20}$$

In this expression, E_{max} is the energy of the ICT maximum, expressed in wavenumbers. The μ_{12} is the TDM that can be obtained by using

$$|\mu_{12}| = \left(\frac{f_{os}}{1.08 \times 10^{-5} E_{max}}\right) \tag{5.21}$$

where f_{os} is the oscillator strength of the transition.

The authors found that within each series the values of first hyperpolarizability obtained through Equation 5.21 (β_0(S)) for [**1–3**] [TfO_2], [**4–6**] [TfO_2], and [**7–9**] [TfO_2] increase steadily. They also found that the values of first hyperpolarizability obtained from quantum chemical experiment largely agreed with those obtained from HRS experiment and Stark spectroscopic studies.

Navarrete and coworkers [97] studied the electronic, vibrational, optical, and NLO properties of a series of π-conjugated push–pull chromophores built around

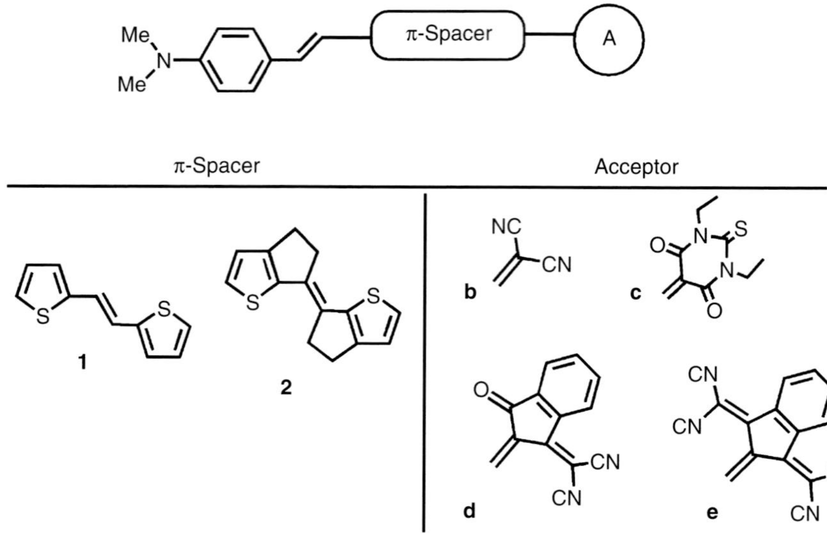

Figure 5.22 Chemical structures of the NLO chromophores studied by Navarrete and coworkers. The D/A substituents are connected to the end α,ω positions of the π-conjugated oligothiophene spacer group. The abbreviated notations for the chromophores used were according to their spacer and acceptor group combinations. (Navarrete and coworkers 2008 [97]. Reproduced with permission of American Chemical Society.)

dithienylethylene (DTE)-based spacers (Figure 5.22). The authors used UV–vis, IR, and Raman spectroscopic techniques as well as EFISH measurements in conjunction with quantum chemical calculations to explore the effect of the change in acceptor and spacers on NLO response properties of those molecules.

The authors tried to establish a structure–property relationship of a series of chromophores that had common donor groups but varied in the strength of the acceptor and π-conjugating spacers. They have either used DTE spacers or their covalently bridged (i.e., rigidified) analogs as spacers. The authors judiciously used the Raman spectroscopic data to compare the strength of ICT in the molecules studied. They noted that according to the effective conjugation coordinate (ECC) theory, the particular scattering is enhanced selectively with the collective C=C/C—C stretching vibrations of the π-conjugated backbone of a molecule due to existence of an effective electron–phonon coupling (or delocalization of electrons) in the π-conjugated molecules resulting from their quasi-one-dimensional structure [98]. As mentioned in Chapter 3, Raman spectroscopic techniques were reported to be useful in studying the structure–property relationship of several π-conjugated molecules [97, 99, 100]. The EFISH measurements by Navarrete and coworkers revealed that NLO chromophores that contained strong electron-acceptor groups connected via rigidified DTE spacers showed high values of $\mu\beta_0$. The authors noted the uncertainty value of $\mu\beta$ to be less than 15% for the compounds studied except for **1d** and **2d**. Due to the low solubility of compounds **1d** and **2d**, the uncertainty in the EFISH measurements could be higher. With these uncertainties, they reported the values of $\mu\beta$ to be the highest for compound **2d** ($13\,000 \times 10^{-48}$ esu), while it was the lowest for compound **4b** (1425×10^{-48} esu) in dicholoromethane. The $\mu\beta$ values reported in chloroform are quite close to those observed in dichloromethane. They observed intense UV–vis absorption bands with peaks in the range of 550–700 nm. Their quantum chemical studies revealed that the absorption in these molecules originated due to HOMO to LUMO transition. A redshift in the absorption maxima of these compounds due to increase in solvent polarity indicated the ICT character of the excited state. The authors assigned the enhancement of the $\mu\beta$ due to increase in acceptor strength and to the decreased excitation energy due to increase in the electron-withdrawing strength of the acceptor group. Their TDDFT calculations resulted in a decrease in excitation energy of 0.35 eV on moving from **1b** to **1e**, while the same was 0.24 on moving from **2b** to **2d**. The authors assigned the resemblance between the IR and Raman spectra to the occurrence of ICT. From the spectroscopic and NLO measurements, the authors surmised that the polarization of the ground electronic state was higher for the systems with covalently bridged spacers and strong electron acceptors.

Wu *et al.* [101] examined two pyrene derivatives for lossless broadband nonlinear refraction applications. These authors also reported earlier that properly designed pyrene derivatives could access ultrafast reverse saturable absorption in a wide spectral region spanning from 500 to 1000 nm [102]. Due to advancement of materials science, the researchers have shifted focus from designing materials with optical nonlinearity under specific wavelengths to broadband NLO materials. Wu *et al.* [101] synthesized and investigated the sign and magnitude of the nonlinear absorption and nonlinear refraction of (*E*)-1-(4-(dimethylamino)-3(pyren-1-yl)prop-2-en-1-one) (acronymed **DAPP-1**) and (*E*)-3-(4-(dimethylamino)phenyl-1-(pyren-1-yl)prop-2-en-1-one) (acronymed **DAPP-2**) (Figure 5.23) using the Z-scan experiment. The authors also studied the UV–vis absorption and emission properties, transient optical nonlinearity of these materials, and carried out quantum chemical calculations to interpret the results.

They carried out femtosecond Z-scan experiments to study the optical nonlinearity of **DAPP-1** and **DAPP-2**. The authors tuned the input wavelengths (λ) at 600, 650, 800, and 1030 nm to cover a wide spectral range. As the absorption

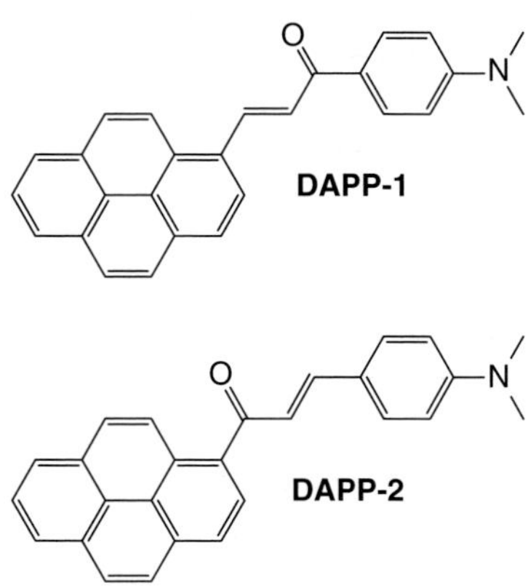

Figure 5.23 The chemical structure of **DAPP-1** and **DAPP-2**. (Wu *et al.* 2017 [101]. Reproduced with permission of Elsevier.

intensity of both the compounds reaches near zero at around 525 nm in the solvents used in their study, the Z-scan experiments were conducted at the off-resonance condition for all the aforesaid wavelengths. The authors found high linear transmittance of more than 90% including the reflection of quartz cuvette in each wavelength. They numerically fitted the measured Z-scan data using Sheik Bahae's theory [103] to evaluate the magnitude of intensity-dependent third-order optical nonlinearity. They extracted the third-order nonlinear absorptive coefficient (β) and refractive index (n_2) along with the real and imaginary part of the hyperpolarizability γ. They also calculated the figure of merit (F) of the compounds under different conditions using the relation

$$F = \frac{n_2}{\beta\lambda} \tag{5.22}$$

Their investigation revealed that the values of β of **DAPP-1** decreased with increased wavelength (i.e., decreased energy) of the incident photon. For example, the β of **DAPP-1** decreased from $1.5\times10^{-13}\,\mathrm{m\,W^{-1}}$ to $0.8\times10^{-13}\,\mathrm{m\,W^{-1}}$ as the wavelength of the incident radiation was changed from 600 to 800 nm. The β of **DAPP-1** further decreased to $0.2\times10^{-13}\,\mathrm{m\,W^{-1}}$ when the wavelength of the incident photon was 1030 nm. The other compound **DAPP-2** followed the same trend with an exception at 800 nm, where the value of β increased abruptly. As S_0 to S_1 transition plays an important role in the TPA of many non-centrosymmetric molecules [104], the authors argued that the abrupt increase in β could have originated from the two-photon resonant S_0 to S_1 transition. They found that **DAPP-1** displayed larger nonlinear absorption than **DAPP-2** and **DAPP-2** performed better as a nonlinear refractor although the molecular structures of these compounds were quite similar. The authors undertook quantum chemical calculations at the DFT level of theory to interpret the results. Their DFT calculations revealed that the optimized structure of **DAPP-1** could be roughly treated as a two-dimensional molecule. They also found that although the CT was dominated the HOMO to LUMO transition of this molecule, the transition showed a clear signature of π–π^* transition. As the extension of planar π-conjugation could significantly enhance the TPA, they assigned the approximate two-dimensional structure of **DAPP-1** to the larger nonlinear absorption of **DAPP-1**. Contrary to what was observed for **DAPP-1**, their DFT calculations predicted the out-of-plane rotation of the pyrene part of **DAPP-2**, while the rest of the molecule remained in the same plane. It had been reported earlier [105] that the twisted structure in many cases led to enhancement of the CT, resulting in larger dipole moment that in turn could lead to the enhancement of nonlinear refraction and decreased nonlinear absorption. The authors argued that the somewhat twisted geometry of the **DAPP-2** could be the reason for higher nonlinear refraction and lower nonlinear absorption, compared to that of **DAPP-1**.

Ricci *et al.* [106] studied the TPA process in two cationic chromophores, one dipolar and the other quadrupolar, and found that the TPA process in these molecules was enhanced with increasing excited state ICT process. They considered a dipolar probe of D–π–A^+ type and a quadrupolar probe of D–π–A^+–π–D type, in which the dibutylamino group is used as an electron

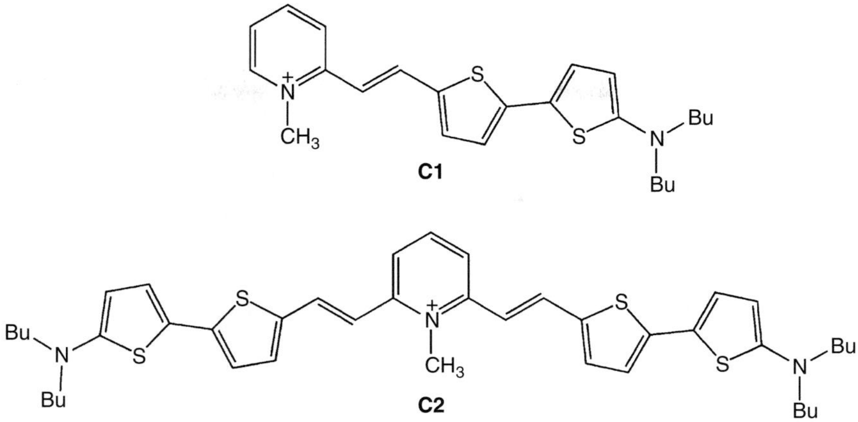

Figure 5.24 Molecular structure of **C1** and **C2**. (Ricci *et al.* 2017 [106]. Reproduced with permission of American Chemical Society.)

donor, methylpyridinium as electron acceptor, and bithiophenes as π-rich spacers. The compounds – they called them **C1** and **C2** (Figure 5.24) – showed absorption peaks near the red end of the visible spectrum. Those probes showed negative solvatochromism in their absorption spectra with increasing solvent polarity, which the authors rationalized in terms of changing electron density upon photoexcitation. For example, the absorption maxima of **C1** and **C2** are 627 and 654 nm in chloroform, respectively. The absorption maxima are blueshifted to 578 and 610 nm in ethanol and to 570 and 602 nm in chloroform, respectively. Their quantum chemical investigation predicted a large ground-state dipole moment for **C1**, which is reduced notably in the Franck–Condon geometry reached upon absorption of light. The authors found that the dipole moment of **C1** did not change much during the relaxation of the first excited state (S_1). The authors carried out femtosecond transient absorption and fluorescence up-conversion measurements that revealed that the excited-state dynamics of **C1** was mostly dominated by the solvent relaxation of the S_1 state. The lifetime of the S_1 state of **C1** as well as the quantum yield, albeit low, remained almost unaffected due to change in solvent polarity, which led the authors to state that the main photoinduced decay pathway of this molecule is internal conversion, irrespective of the properties of the medium used in their studies. In contrast to what was observed in **C1**, **C2** showed the signature of strong excited-state CT. The absorption spectra of **C2** were found to be redshifted compared to that of **C1**, although the emission spectra of these two molecules overlap. This indicated the localization of the excitation on only one branch of **C2**, from which fluorescence took place. This observation was supported by their theoretical calculation that predicted the symmetry of the **C2** to be broken in its relaxed S_1 geometry. The authors noted that the breaking of symmetry of **C2** in the S_1 state occurred in all the solvents used in their studies, irrespective of the polarity. Their ultrafast spectroscopic investigation revealed two notably different first excited singlet states in polar solvents, in which the ICT state was produced efficiently as a separate species. The formation of

ICT of **C2** was supported through the decrease in fluorescence quantum yield compared to its dipolar counterpart (**C1**). Their quantum chemical calculation predicted the TICT character of the relaxed first excited state of **C2**. It is worth mentioning that in spite of the difference in formation of the ICT state, the internal conversion was found to be the main decay pathway for both the molecules. The authors carried out femtosecond-resolved two-photon excited fluorescence (TPEF) measurements to estimate the TPA cross section of the aforesaid compounds [107]. As both the probes showed a one-photon absorption maximum at 650 nm, the authors used an excitation of 1300 nm, while they considered 800 nm to detect the maximum fluorescence peak. They used 2-(6-(4-dimethylaminophenyl)-2,4-neopentylene-1,3,5-hexatrienyl)-3-methylbenzothiazolium perchlorate (Styryl 9M) dissolved in ethanol as TPA reference to obtain the TPA cross section [108]. Their TPEF measurements found the TPA cross section to be 500 and 1400 GM for **C1** and **C2**, respectively. The authors argued that the higher TPA cross section of quadrupolar molecule **C2** was due to the efficient excited-state ICT process in this molecule, compared to its dipolar counterpart. Along with the possible application of **C2** in TPA-based devices, the authors also hoped that due to absorption at the red side of the visible spectrum, both **C1** and **C2** could be potential candidates for applications in organic photovoltaic solar cells as photosensitizers.

We may note here that due to wide applicability of TPA materials in up-conversion lasing, information technology, 3D-imaging, and so on, research on the ICT-based TPA materials is a very active area of research. Therefore, it is expected that several novel ICT-based organic molecules will be investigated for possible applications as TPA materials in the coming years.

5.7 Studies of NLO Molecular Switches

Studies of molecular switching have become popular due to a variety of reasons, including designing devices for molecular electronics applications and modeling of biological process [109, 110]. In 1999, Coe [111] reported the concept of switching of NLO response properties in some organic and metallo-organic molecular compounds. The NLO switching in a molecule could be achieved using photons, protonation of the molecule, redox effects, and magnetic behavior. Nakatani and coworkers [112] pointed out that the molecular NLO switching could be achieved if a molecule is accessible in two distinct states using external stimulus and the NLO response of those two states are completely different. Till date, several studies on NLO switching have been reported in literature. Some of the representative examples are discussed in this section. McCleverty *et al.* [113] reviewed the requirements of reversible NLO switching and cited several examples of second-order NLO response in solution. They noted that NLO switching could be obtained by modifying a part of an NLO-active molecule. The efficiency of the donor part in a donor–acceptor (D–A) molecule could be diminished through its protonation or oxidation. Similarly, protonation or reduction of the acceptor part could decrease its efficiency. Modification of the

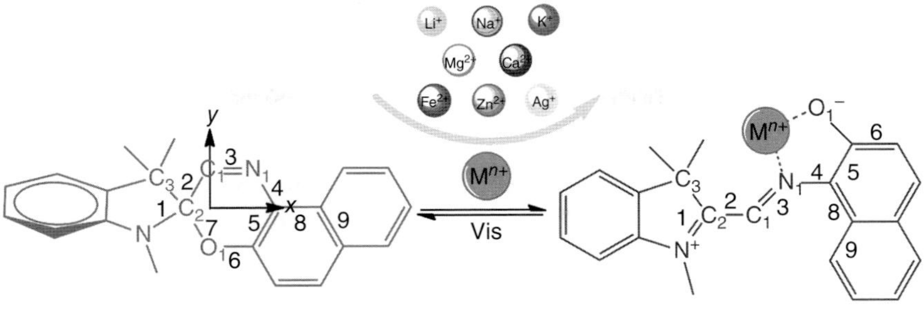

Figure 5.25 The structures of spirooxazine (left) and cation-complexed merocyanine forms. (Ye *et al.* 2017 [114]. Reproduced with permission of Royal Society of Chemistry.)

electron bridge connecting the donor and acceptor units could also be used to tune the NLO response of a D–A molecule. The authors reported the redox behavior of some metal-containing compounds and mentioned examples of NLO switching using the aforesaid tactics. Recently, Xie *et al.* [114] reported the studies of spirooxazine (Figure 5.25) as selective cation sensors using its NLO switching properties. Spirooxazine, a photochromic material, is known to respond to several external stimuli, like light, temperature, and metal ions to undergo reversible structural changes. Introduction of metal ion causes the closed form of spirooxazine to convert to open-form metallic merocyanine. This conversion from the closed form to open form is accompanied by changes in structure or π-electron distribution. The NLO response properties of these open and closed forms of spirooxazine are therefore notably different, making the molecule suitable for exploring the NLO switching and application in metal ion sensing. The authors carried out DFT calculations to optimize the structures and NLO response properties of the open and closed forms of spirooxazine as well as the metal ion complexes of merocyanine. Their calculations showed that spirooxazine transformed into metallic open-form merocyanine in the presence of metal ions (Li^+, Na^+, K^+, Ag^+, Mg^{2+}, Ca^{2+}, Fe^{2+}, and Zn^{2+}), which was accompanied by structural changes as well as formation of larger π-conjugation network. The resultant merocyanine derivatives showed 10- to 21-fold higher values of first hyperpolarizabilities, compared to that of spirooxazine. The authors used several functionals (CAM-B3LYP, BH and HLYP, M06-2X and ωB97XD) and five Pople's basis sets to access the effect of these parameters on the NLO response properties of these systems. They used the 6-31+G(d) basis set as well as the LanL2DZ basis set for metal ions to calculate the first hyperpolarizabilities of the spirooxazine and metal complexes of merocyanine forms in the gas phase as well in solution. They also found that β_x component dominantly dictated the values of β_{total} in the species studied, except for spirooxazine and Fe^{2+} complex of merocyanine in which β_y was dominant. The contribution from β_y and β_z components in shaping the values of β_{total} was small. The authors explained this result on the basis that the ICT occurs along the y-axis in spirooxazine and there is a small fraction of CT in the Fe^{2+} complex of merocyanine. In the rest of the complexes, the CT occurs along the x-axis. The quantum chemical studies performed by the authors revealed that introduction of various metal ions can enhance the first hyperpolarizability of spirooxazine that was converted to metallic open-form merocyanine with a different structure and larger π-conjugation network. The authors argued that this increase in NLO response was due to enhancement of CT from the indoline unit to the napthooxazine unit in the molecule.

Nakatani and coworkers [112] reported the NLO response properties of a bisboronate compound with five 6-membered ring heterobicycles and compared the response with the related monoboronate compound. Quantum chemical investigation at the DFT (B3PW91 functional and 6-31G* basis set) level of approximation showed that the diboronate derivative exhibits a centrosymmetric conformation in the ground state. So, the quadratic first hyperpolarizability of this compound vanishes due to its centrosymmetric nature. The authors found that application of an external electric field induced a large intramolecular rotation of the two "push–pull" subunits that reached 128°, giving rise to a β value of 49.2×10^{-30} cm^5 esu^{-1} at a field intensity of 10^{-3} a.u. The aforesaid result led the authors to study the NLO switching property of this compound in detail. The designing principle of the NLO-active molecule was to connect two "push–pull" NLO chromophores in which both of them are pinned at two ends of a chemical axis in such a way that they can rotate around it when subjected to an external electric field. They considered a well-known NLO-active molecule, DANS (molecule 1) and to prepare the ligand (molecule 2), a slight chemical modification was introduced for providing the linkage needed for pinning down the NLO chromophores on a chemical axis that is orthogonal to the CT direction. The ligand 2 was used to design the boronate compound **3** and bisboronate compound **4** (Figure 5.26).

The authors studied the NLO response of compounds **3** and **4** experimentally at room temperature using the EFISH technique. For this purpose, they used 1907 nm incident radiation that was obtained by Raman shifting of the 1064-nm fundamental beam of a nanosecond Nd:YAG pulse laser using a hydrogen cell (1 m long, 40 bar pressure). The EFISH experiments that provide $\mu \times \beta_{vec}$ showed that the value for compound **4** was 1.95 times higher than that of compound **3**. To rationalize this observation, the authors carried out quantum chemical studies on the NLO properties of compounds **3** and **4** by assuming different values of the torsion angle θ. The centrosymmetric ground state of compound **4** showed a β value of zero, as expected. The first hyperpolarizability increased with increasing external electric field and reached the value of 49.2×10^{-30} cm^5 esu^{-1} at an electric field of 10^{-3} a.u. At this condition, the $\mu \times \beta_{vec}$ value reached 776 D $\times 10^{-30}$ cm^5 esu^{-1}. A comparison of computed EFISH data of compounds **3** and **4** showed that the $\mu \times \beta_{vec}$ value of compound **4** is 1.95 times higher than that of compound **3** at a torsion angle of 73°. The authors found that the NLO properties of compound **3** were not field dependent, unlike the compound **4**. The increased NLO response of compound **4** was attributed to the dipole alignment within each molecular unit. From the aforesaid results, the authors concluded that although the ground state of the boronate chromophore made of two push–pull NLO subunits in quasi-free rotation around a chemical axis was centrosymmetric in nature with a first hyperpolarizability of zero, use of an external electric field could result in a significant increase in the β value.

1

2

3

4

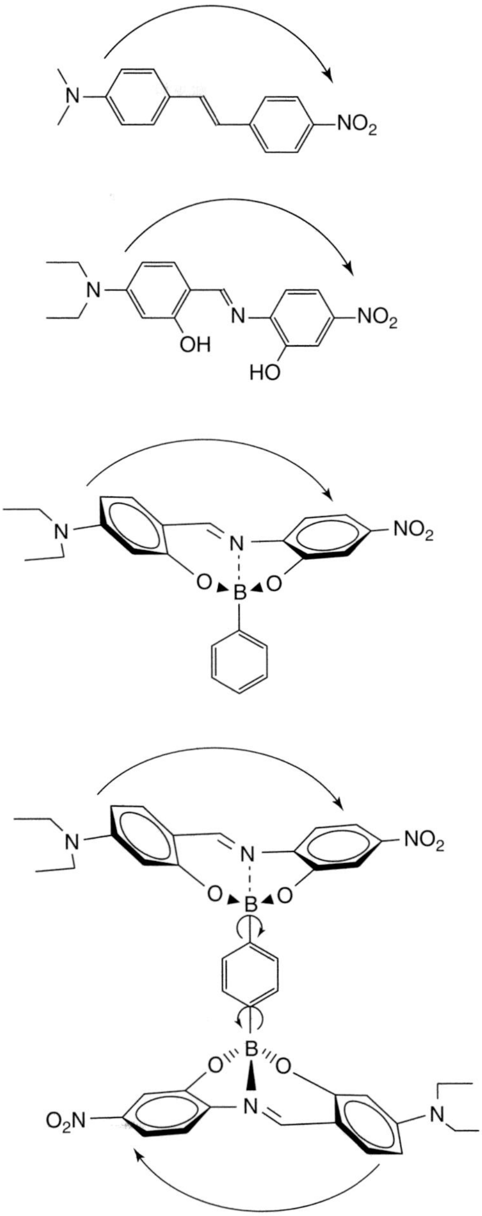

Figure 5.26 Chemical structures of DANS, the ligand, the boronate, and the bisboronate compounds studied by Nakatani *et al.* (Nakatani *et al.* 2006 [112]. Reproduced with permission of Royal Society of Chemistry.)

Su *et al.* [115] reported reversible switching and modulation of NLO response properties in dibenzoborole (DBB) derivatives with the aim of designing two-dimensional NLO molecular switch. DFT calculations revealed that the first hyperpolarizabilities of the four DBBs considered for the studies got affected either due to attachment with fluoride anion (F^-) and/or one-electron reduction. The authors noted that the three-coordinated DBBs turned into four-coordinated dibenzoborols (DBBF) due to attachment of F^- ions (Figure 5.27). They found

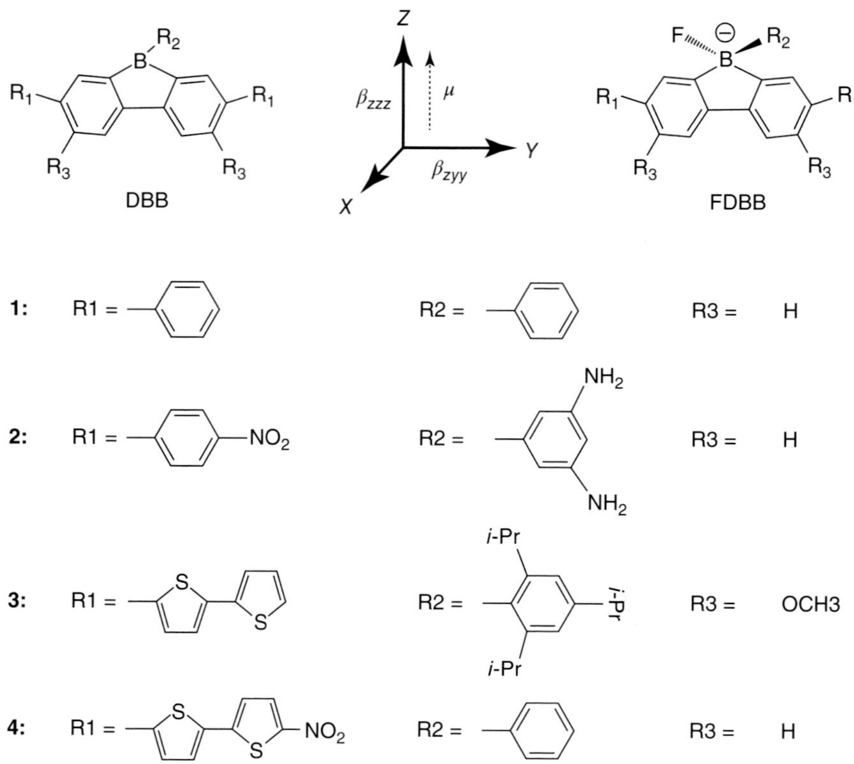

Figure 5.27 The chemical structures of DBB and DBBF derivatives studied by Su and coworkers. (Su and coworkers 2009 [115]. Reproduced with permission of American Chemical Society.)

that fluorinated derivatives of 5-(2,4,6-triisopropylphenyl)-2,8-dimethoxy-3,7-bithienyl-5*H*-dibenzo[*d*,*b*]borole (molecule 3) and 5-phenyl-3,7-bis-dinitrothenyl-5*H*-dibenzo[*d*,*b*]borole (molecule 4) showed about 12 times and 4 times higher values of β, respectively, compared to their non-fluorinated analogs. The values of β for fluorinated derivatives of molecules 3 and 4 are reported to be 64×10^{-30} and 272×10^{-30} esu, respectively. Similarly, the β values of one-electron reduced forms of molecules 3 and 4 were 47 times and 15 times higher than their neural counterparts, respectively. The authors found that the NLO switching was two-dimensional in character. They argued that the large off-diagonal tensors of β could be related to the CT transition, which were polarized perpendicular to the molecular dipolar axis. Their density of states (DOSs) and frontier MO analyses revealed that the binding of fluoride ion at the boron atom as well as due to the one-electron reduction process led to turning off the $p_\pi \rightarrow \pi^*$ conjugation of vacant p-orbital of the boron atom in LUMOs. The authors argued that the aforesaid process resulted in a higher extent of perpendicular CT and larger values of β. Using the aforesaid results, the authors inferred that attachment of F^- ion or reduction significantly affected the CT characters of the molecules studied. They also proposed that the molecules could be used as a new type of "On/Off" molecular switches.

There are some other reports on studies of NLO switches. For example, Xu *et al.* [116] reported the NLO switching processes in 4-diphenylamino-phenyl-substituted pyrazine derivatives. Liu *et al.* [117] studied the NLO switching properties of Ru(III/II) carboxylate complexes using quantum chemical calculations at the DFT level of theory. The Ru(II)COOH and Ru(III)COOH complexes can form Ru(II)COO^- and Ru(III)COO^- ions through the proton transfer process. The resulting COO^- group acts as a strong donor and the Ru(III) acts as a strong acceptor. The authors reported that the first hyperpolarizability (β) of Ru(III)COO^- was almost 36 and 48 times higher than that of Ru(II)COO^- and Ru(II)COOH, respectively. The β value of Ru(III)COO^- was reported to be almost 215 times higher than that of the Ru(III)COOH complex. Using their results, the authors proposed that the Ru(III/II) carboxylate complexes can be used as NLO switches.

In this chapter, we have discussed the theory of NLO response as well as experimental studies of NLO properties of ICT-based molecules. There are several reports of ICT-based molecules with high first hyperpolarizability, the measure of second-order NLO response while a few studies concerning the third-order NLO response can be found in literature. Studies of TPA, the imaginary part of the third-order nonlinear susceptibility, are also being reported to develop novel TPA-active materials. Experimental techniques, including Z-scan, HRS, and EFISH are being used either alone or in conjunction with high-level quantum chemical calculations to predict novel molecules for possible designing of new NLO/TPA active materials. Considering the importance of the field, more reports of NLO studies of ICT-based molecules can be anticipated in near future.

References

1 Bloembargen, N. (1965) *Nonlinear Optics*, World Scientific.
2 Boyd, R.B. (2008) *Nonlinear Optics*, 3rd edn, Elsevier.
3 Marks, T.J. *et al* (1994) *Chem. Rev.*, **94**, 195.
4 D.S. Chemla (Ed.), *Nonlinear Optical Properties of Organic Molecules and Crystals* (vol. 1), 2012, Elsevier.
5 Ouder, J.L. (1977) *J. Chem. Phys.*, **67**, 446.
6 Ouder, J.L. and Chemla, D.S. (1977) *J. Chem. Phys.*, **67**, 2664.
7 Sen, R. *et al* (1992) *Chem. Phys. Lett.*, **190**, 443.
8 Sen, R. *et al* (1993) *J. Phys. Chem.*, **97**, 7491.
9 Sim, F. *et al* (1993) *J. Phys. Chem.*, **97**, 1158.
10 Albert, I.D.L. *et al* (1997) *J. Am. Chem. Soc.*, **119**, 6575.
11 Abotto, A. *et al* (2003) *Chem. Eur. J.*, **9**, 1991.
12 Wang, C.H. *et al* (2001) *J. Appl. Phys.*, **89**, 4209.
13 Marder, S.R. *et al* (1993) *Science*, **261**, 186.
14 Chen, G. *et al* (1994) *J. Chem. Phys.*, **101**, 5680.
15 Nandi, P.K. *et al* (2001) *J. Mol. Struct. (THEOCHEM)*, **119**, 545.
16 Barzoukas, M. *et al* (1996) *J. Nonlinear Opt. Phys. Mater.*, **05**, 757.
17 Jaquemin, D. *et al* (2001) *J. Chem. Phys.*, **115**, 3540.
18 Iikura *et al* (2001) *J. Chem. Phys.*, **115**, 3540.

19 Yanni, T., Tew, D.P., and Handy, N.C. (2004) *Chem. Phys. Lett.*, **393**, 51.

20 Fonescca, T.L. *et al* (2005) *Chem. Phys. Lett.*, **413**, 356.

21 Fonescca, T.L. *et al* (2007) *Chem. Phys. Lett.*, **442**, 259.

22 Misra, R., Bhattacharyya, S.P., and Maity, D.K. (2008) *Chem. Phys. Lett.*, **458**, 54.

23 Kang, H. *et al* (2007) *J. Am. Chem. Soc.*, **129**, 3267.

24 Zyss, J. *et al* (2000) *J. Am. Chem. Soc.*, **122**, 11956.

25 Li, L. *et al* (2008) *Phys. Chem. Chem. Phys.*, **10**, 6829.

26 Misra, R. (2017) *J. Phys. Chem. C*, **121**, 5731.

27 Machado, D.F. *et al* (2016) *J. Phys. Chem. C*, **120**, 17660.

28 Shimada, M., Yamanoi, Y., Matsushita, T., Kondo, T., Nishibori, E., Hatakeyama, A., Sigimoto, K., and Nishihara, H. (2005) *J. Am. Chem. Soc.*, **137**, 1024.

29 Zhou, Z.J., Li, X.P., Ma, F., Liu, Z.B., Li, Z.R., Huang, X.R., and Sun, C.C. (2011) *Chem. Eur. J.*, **17**, 2414.

30 Roy, R.S. and Nandi, P.K. (2015) *RSC Adv.*, **5**, 103729.

31 Liu, F., Yang, Y., Cong, S., Wang, H., Zhang, M., Bo, S., Liu, J., Zhen, Z., Liu, X., and Qiu, L. (2014) *RSC Adv.*, **4**, 52991.

32 Xu, H., Yang, D., Liu, F., Fu, M., Bo, S., Liu, X., and Cao, Y. (2015) *Phys. Chem. Chem. Phys.*, **17**, 29679.

33 Shi, Y., Frattarelli, D., Watanabe, N., Facchetti, A., Cariati, E., Righetto, S., Tordin, E., Zuccaccia, C., Macchioni, A., Wegener, S.L., Stern, C.L., Ratner, M.A., and Marks, T.J. (2015) *J. Am. Chem. Soc.*, **137**, 12521.

34 Misra, R., Sharma, R., and Bhattacharyya, S.P. (2010) *J. Comput. Methods Sci. Eng.*, **10**, 149.

35 Chen, L.T., Tam, W., Marder, S.R., Stiegman, A.E., Rikken, G., and Spangler, C.W. (1991) *J. Phys. Chem.*, **95**, 10643.

36 Marder, S.R., Beratan, D.N., and Cheng, L.-T. (1991) *Science*, **252**, 103.

37 Brédas, J.L., Adant, C., Tackx, P., Persoons, A., and Pierce, B.M. (1994) *Chem. Rev.*, **94**, 243.

38 Kanis, D.R., Ratner, M.A., and Marks, T.J. (1994) *Chem. Rev.*, **94**, 195.

39 Dalton, L.R., Harper, A.W., Ghosn, R., Steier, W.H., Ziari, M., Fetterman, H., Shi, Y., Mustacich, R.V., Jen, A.K.Y., and Shea, K.J. (1995) *Chem. Mater.*, **7**, 1060.

40 Whitaker, C.M., Patterson, E.V., Kott, K.L., and McMahon, R.J. (1996) *J. Am. Chem. Soc.*, **118**, 9966.

41 Verbiest, T., Houbrechts, S., Kauranen, M., Clays, C., and Persoons, A. (1997) *J. Mater. Chem.*, **7**, 2175.

42 Papadopoulos, M.G., Leszczynski, J., and Sadlej, A.J. (eds) (2006) *Nonlinear Optical Properties of Matter: from Molecules to Condensed Phases*, Springer, Dordrecht.

43 Nandi, P.K., Panja, N., and Ghanty, T.K. (2008) *J. Phys. Chem. A*, **112**, 4844.

44 Nandi, P.K., Panja, N., and Ghanty, T.K. (2009) *J. Phys. Chem. A*, **113**, 2623.

45 Murugan, N.A., Kongsted, J., Rinkevicius, Z., and Ågren, H. (2010) *Proc. Natl. Acad. Sci. U.S.A.*, **107**, 16453.

46 Nandi, P.K., Panja, N., and Ghanty, T.K. (2010) *Theor. Chem. Acc.*, **126**, 323.

47 Johnson, L.E., Dalton, L.R., and Robinson, B.H. (2014) *Acc. Chem. Res.*, **47**, 3258.
48 Borini, S., Limacher, P.A., and Luthi, H.P. (2009) *J. Chem. Phys.*, **131**, 124105.
49 Barlow, S., Bunting, H.E., Ringham, C., Green, J.C., Bublitz, G.U., Boxer, S.G., Perry, J.W., and Marder, S.R. (1999) *J. Am. Chem. Soc.*, **121**, 3715.
50 Janjua, M.R.S.A., Guan, W., Yan, L., Su, Z.M., Karim, A., and Akbar, J. (2010) *Eur. J. Inorg. Chem.*, **10**, 3466.
51 Marder, S.R. and Perry, J.W. (1993) *Adv. Mater.*, **5**, 804.
52 Marder, S.R., Kippelen, B., Jen, A.K.Y., and Peyghambarian, N. (1997) *Nature*, **388**, 845.
53 Adegoke, O.O., Ince, M., Mishra, A., Green, A., Varnavski, O., Martínez-Díaz, M., Bauerle, P., Torres, T., and III Goodson, T. (2013) *J. Phys. Chem. C*, **117**, 20912.
54 Parthenopoulos, D.A. and Rentzepis, P.M. (1989) *Science*, **245**, 843.
55 Arnbjerg, J., Jiménez-Banzo, A., Paterson, M.J., Nonell, S., Borrell, J.I., Christiansen, O., and Ogilby, P.R. (2007) *J. Am. Chem. Soc.*, **129**, 5188.
56 Dichtel, W.R., Serin, J.M., Edder, C., Frechet, J.M.J., Matuszewski, M., Tan, L.S., Ohulchanskyy, T.Y., and Prasad, P.N. (2004) *J. Am. Chem. Soc.*, **126**, 5380.
57 Drobizhev, M., Karotki, A., Kruk, M., Krivokapic, A., Anderson, H.L., and Rebane, A. (2003) *Chem. Phys. Lett.*, **370**, 690.
58 He, J., Mi, J., Li, H., and Ji, W. (2005) *J. Phys. Chem. B*, **109**, 19184.
59 He, G.S., Tan, L.S., Zheng, Q., and Prasad, P.N. (2008) *Chem. Rev.*, **108**, 1245.
60 Ferrighi, L., Frediani, L., and Ruud, K. (2007) *J. Phys. Chem. B*, **111**, 8965.
61 Liu, K., Wang, Y., Tu, Y., Ågren, H., and Luo, Y. (2008) *J. Phys. Chem. B*, **112**, 4387.
62 Alam, M.M., Chattopadhyaya, M., Chakrabarti, S., and Ruud, K. (2014) *Acc. Chem. Res.*, **47**, 1604.
63 He, X., Xu, B., Liu, Y., Yang, Y., and Tian, W. (2012) *J. Appl. Phys.*, **111**, 053516.
64 Diaz, C., Vesga, Y., Echevarria, L., Stara, I.G., Stary, I., Anger, E., Shen, C., Moussa, M.E.S., Vanthuyne, M., Crassous, J., Rizzo, A., and Hernandez, F.E. (2015) *RSC Adv.*, **5**, 17429.
65 Kogej, T., Beljonne, D., Meyers, F., Perry, J.W., Marder, S.R., and Bredas, J.L. (1998) *Chem. Phys. Lett.*, **298**, 1.
66 Rumi, M., Ehrlich, J.E., Heikal, A.A., Perry, J.W., Barlow, S., Hu, Z., McCord-Maughon, D., Parker, T.C., Rockel, H., Thayumanavan, S., Marder, S.R., Beljonne, D., and Bredas, J.L. (2009) *J. Am. Chem. Soc.*, **122**, 9500.
67 Alam, M.M., Chattopadhyaya, M., and Chakrabarti, S. (2012) *J. Phys. Chem. A*, **116**, 11034.
68 Cronstrand, P., Luo, Y., and Årgen, H. (2002) *Chem. Phys. Lett.*, **352**, 262.
69 Alam, M.M. (2015) *Phys. Chem. Chem. Phys.*, **17**, 17571.
70 Jha, P.C., Wang, Y., and Årgen, H. (2008) *ChemPhysChem*, **9**, 111.
71 Bhaskar, A., Ramakrishna, G., Lu, Z., Twieg, R., Hales, J.M., Hagan, D.J., Stryland, E.V., and Goodson, T.J. III, (2006) *J. Am. Chem. Soc.*, **128**, 11840.
72 Zheng, Q., He, G.S., and Prasad, P.N. (2005) *J. Mater. Chem.*, **15**, 579.

73 Albota, M. *et al* (1998) *Science*, **281**, 1653.

74 Divya, K.P., Sreejith, S., Ashok kumar, P., Yuzhan, K., Peng, Q., Maji, S.K., Tong, Y., Yu, H., Zhao, Y., Ramamurthy, P., and Ajayaghosh, A. (2014) *Chem. Sci.*, **5**, 3469.

75 Bednarska, J., Zalesny, R., Murugan, N.A., Bartkowiak, W., Årgen, H., and Odelius, M. (2016) *J. Phys. Chem. B*, **120**, 9067.

76 Garito, A.F., Helfin, J.R. *et al* (1988) *Proc. Soc. Photo. Opt. Instrum. Eng.*, **971**, 2.

77 Marder, S.R. *et al* (1997) *Science*, **276**, 1233.

78 Ooba, N., Tomaru, S., Kurihara, T., Kaino, T., Yamada, W., Takaji, M., and Yamamoto, T. (1995) *Jpn. J. Appl. Phys.*, **34**, 3139.

79 Agarwal, G.P. *et al* (1978) *Phys. Rev. B: Condens. Matter*, **17**, 776.

80 Kivala, M. *et al* (2007) *Angew. Chem. Int. Ed.*, **46**, 6357.

81 Zhou, W. *et al* (2009) *Adv. Funct. Mater.*, **19**, 141.

82 Wang, C., Fan, C., Yuan, C. *et al* (2017) *RSC Adv.*, **7**, 4825.

83 Li, Y. *et al* (2008) *Opt. Express*, **16**, 6251.

84 Geskin, V.M., Lambert, C., and Bredas, J.L. (2003) *J. Am. Chem. Soc.*, **125**, 15651.

85 Alain, V., Redoglia, S., Blancard-Desce, M. *et al* (1999) *Chem. Phys.*, **245**, 51.

86 Hassan, S., Asif, H.M., Zhou, Y., Zhang, L. *et al* (2016) *J. Phys. Chem. C*, **120**, 27587.

87 Kato, S., Roberts Beels, M.T., La Porta, P. *et al* (2010) *Angew. Chem. Int. Ed.*, **49**, 6207.

88 Teran, N.B., He, G.S., Baev, A., Shi, Y., Swihart, M.T., Prasad, P.N., Marks, T.J., and Reynolds, J.R. (2016) *J. Am. Chem. Soc.*, **138**, 6975.

89 Sutherland, R.L. (1992) *Handbook of Nonlinear Optics*, Marcekl Dekker, New York.

90 Powell, C.E. and Humphrey, M.G. (2004) *Coord. Chem. Rev.*, **248**, 725.

91 Beaujean, P., Bondu, F., Plaquet, A., Garcia-Amoros, J., Cusido, J., Raymo, F.M., Castet, F., Rodriguez, V., and Champagne, B. (2016) *J. Am. Chem. Soc.*, **138**, 5052.

92 Zhu, S., Li, M., Tang, S., Zhang, Y.M., Yang, B., and Zhang, S.X.A. (2014) *Eur. J. Org. Chem.*, **2014**, 1227.

93 Deniz, E., Tomasulo, M., Cusido, J., Sortino, S., and Raymo, F. (2011) *Lamgmuir*, **27**, 11773.

94 de Wergifosse, M., Castet, F., and Champagne, B. (2015) *J. Chem. Phys.*, **142**, 194102.

95 Buckley, L.E.R., Coe, B.J., Rusanova, D., Joshi, V.D. *et al* (2017) *J. Phys. Chem. A*, **121**, 5842.

96 (a) Coe, B.J. *et al* (2016) *J. Org. Chem.*, **81**, 1912; (b) Buckley, L.E.R. *et al* (2017) *Dalton Trans.*, **46**, 1052.

97 Delgado, M.C.R., Casado, J., Hernandez, V., Navarrete, J.T.L., Orduna, J., Villacampa, B., Alicante, R., Raimundo, J.M., Blanchard, P., and Roncali, J. (2008) *J. Phys. Chem. C*, **112**, 3109.

98 (a) Zerbi, G., Castiglioni, C., and Del Zoppo, M. (1998) *Electronic Materials: the Oligomer Approach*, Wiley-VCH Verlag GmbH & Co. KGaA, Weinheim;

(b) Agosti, E., Rivola, M., Hernandez, V., Del Zoppo, M., and Zerbi, G. (1999) *Synth. Met.*, **100**, 101.

99 Casado, J., Miller, L.L., Mann, K.R., Pappenfus, T.M., and Lopez Navarrete, J.T. (2002) *J. Phys. Chem. B*, **106**, 3597.

100 Casado, J., Hernandez, V., Ruiz Delgado, M.C., Ponce Ortiz, R., Lopez Navarrete, J.T., Facchetti, A., and Marks, T.J. (2005) *J. Am. Chem. Soc.*, **127**, 13364.

101 Wu, X., Xiao, J., Sun, R., Jia, J., Yang, J., Shi, G., Wang, Y., Zhang, X., and Song, Y. (2017) *Dyes Pigm.*, **143**, 165.

102 Xiao, Z.G., Shi, Y.F., Sun, R., Ge, J.F., Li, Z.G., Fang, Y. *et al* (2016) *J. Mater. Chem.*, **21**, 4647.

103 Sheik-Bahae, M., Said, A.A., Wei, T.H., Hagan, D.J., and Van Stryland, E.W. (1990) *IEEE J. Quantum Electron.*, **26**, 760.

104 Pawlicki, M., Collins, H.A., Denning, R.G., and Anderson, H.L. (2009) *Angew. Chem. Int. Ed.*, **48**, 3244.

105 Shi, Y., Lou, A.J.T., He, G.S., Baev, A., Swihart, M.T., Prasad, P.N. *et al* (2015) *J. Am. Chem. Soc.*, **137**, 4622.

106 Ricci, F., Carlotti, B., Keller, B., Bonaccorso, C., Fortuna, C.F., Goodson, T. III,, Elisei, F., and Spalletti, A. (2017) *J. Phys. Chem. C*, **121**, 3987.

107 Xu, C. and Webb, W.W. (1996) *J. Opt. Soc. Am. B: Opt. Phys.*, **13**, 481.

108 Makarov, N.S., Drobizhev, M., and Rebane, A. (2008) *Opt. Express*, **16**, 4029.

109 Ward, M.D. (1995) *Chem. Soc. Rev.*, **24**, 121.

110 Lehn, J.M. (1995) *Supramolecular Chemistry – Concepts and Perspectives*, Wiley-VCH Verlag GmbH & Co. KGaA, Weinheim.

111 Coe, B.J. (1999) *Chem. Eur. J.*, **5**, 2464.

112 Lamere, J.F., Lacroix, P.G., Farfan, N., Rivera, J.M., Santillan, R., and Nakatani, K. (2006) *J. Mater. Chem.*, **16**, 2913.

113 Asselberghs, I., Clays, K., Persoons, A., Ward, M.D., and McCleverty, J. (2004) *J. Mater. Chem.*, **14**, 2831.

114 Ye, J.T., Wang, L., Wang, H.Q., Chen, Z.Z., Qiu, Y.Q., and Xie, H.M. (2017) *RSC Adv.*, **7**, 642.

115 Muhammad, S., Janjua, M.R.S.A., and Su, Z. (2009) *J. Phys. Chem. C*, **113**, 12551.

116 Xu, L., Zhu, H., Long, G. *et al* (2015) *J. Mater. Chem. C*, **3**, 9191.

117 Liu, Y., Yang, G.C. *et al* (2012) *Int. J. Quantum Chem.*, **112**, 779.

118 Champagne, B. (2009) *Polarizabilities and hyperpolarizabilities*, in *Chemical Modeling*, vol. 6 (ed. M. Springborg), Royal Society of Chemistry, p. 17.

6

Recent Technological Applications of ICT Molecules and Prospect of Designing New Molecules

6.1 Introduction

Intramolecular charge transfer (ICT) occurs between an electronic donor (D) and an electronic acceptor (A) in a(n) molecule/ion, connected through a π-electron bridge. A polyene-linker or an aromatic moiety often acts as a bridge in these molecules. Upon photoexcitation, several ICT molecules form a twisted intramolecular charge transfer (TICT) state, whose spectral properties are dependent on several factors, including polarity and viscosity of the medium as well as the temperature. As discussed in Chapter 4, several studies have been devoted to decipher the underlying principle that links the spectral properties of an ICT probe with the polarity and viscosity of the solvent. Of late, several ICT-based molecules are being used as sensors for polarity and viscosity of the medium, mostly in cases where conventional methods are either cumbersome or inefficient [1]. For example, molecular rotors, a class of ICT molecules, are being used to probe the microviscosity of several biological systems, including cytoplasmic fluid and blood plasma [1, 2]. Several other ICT molecules are being used to detect chemical species, including metal ions in solution [3]. As the scientific community is looking for cheap but efficient next-generation flat-panel displays to save energy that in turn will also help the environment, designing organic light-emitting diodes (OLEDs) has gained a momentum in recent years [3, 4]. Several reports can be found in literature studying the use of ICT-based molecules for potential application in OLEDs [5, 6]. ICT-based molecules are also being used in designing nonlinear optical (NLO) materials for several technological applications, including optical switching, optical computing, and telecommunication [7]. Due to the vast size of the literature studying NLO response of ICT molecules, we have ramified this part and discussed it in Chapter 5.

6.2 Application of ICT-Based Molecules

6.2.1 ICT Molecules as Fluorescence Sensors

It is now known that donor–acceptor-based ICT molecules undergo a rapid charge-transfer process from the initially formed locally excited (LE) state,

Intramolecular Charge Transfer: Theory and Applications,
First Edition. Ramprasad Misra and S. P. Bhattacharyya.

leading to formation of an ICT state. The new state with a higher dipole moment than that of the LE state often gets twisted, which is referred as *TICT* state. The LE to ICT conversion has been found to depend on several factors, including the polarity of the medium. In solvents with high polarity, the ICT process is expected to be faster as the polar environment will stabilize the more polar ICT state. The formation of the ICT state often gives rise to dual emission arising from the LE and ICT states. Several reports have mentioned that the twisting often leads to lower quantum yield as this structural change can enhance the non-radiative processes. As the LE to ICT photoreaction can be modulated using several external factors, like tuning the donor/acceptor strength, local polarity, and steric restrictions, several research groups have exploited these factors for designing ICT-based sensors for polarity, viscosity, as well as for sensing chemical species.

Sunahara *et al.* [8] designed a library of environmental polarity sensors, based on boron dipyrromethene (BODIPY) dyes (Figure 6.1). The authors systematically studied the fluorescence ON/OFF threshold in these molecules and also the role of charge transfer in this process. The emission from the unsubstituted BODIPY dye is generally insensitive to solvent polarity, but substitution with an aromatic group decreases the emission quantum yield. Steric restriction imposed by substitution of two methyl groups in the BODIPY moiety can enhance the quantum yield of the molecule. Use of this structural restriction, also called "pretwisting" is useful in controlling the charge transfer and hence the quantum yield of this molecule. Sunahara *et al.* [8] used various substitutions at the 8-position of the BODIPY moiety and found that the oxidation potential of the aromatic moiety became more positive, while the reduction potential of the BODIPY fluorophore became more negative with decreased solvent polarity. Therefore, the free energy change during the photoinduced charge-transfer process from the aromatic moiety became larger with decreasing polarity of the medium. The authors used these environment-sensitive molecules to probe the bovine serum albumin and also the living cells.

The pretwisting strategy was used to design polarity sensors based on push–pull biphenyl (4-(*N*,*N*-dimethylamino)-4′-formylbiphenyl) with a modulated dihedral angle of the aryl–aryl bond (Figure 6.2), achieved through use of a bridged structure or methyl groups [9]. The authors found that alteration in the torsional restriction between the *N*,*N*-dimethylaniline (donor) and benzaldehyde (acceptor) moieties affects the spectral properties in these molecules. The authors also reported that restricting the biphenyl chromophore to a twisted conformation increases the sensitivity of the probe toward solvent polarity. Shigeta *et al.* [10] reported studies of solvent-dependent ICT process in a D–A molecule with the aim of developing a solvent polarity sensor. The authors used a fluorenone-based molecule in which two diethylaminophenyl moieties substituted at the 2- and 7-positions act as electron-donating groups. They observed a switching between TICT emission and excimer emission by changing the solvent only, while no changes in the ground-state properties was observed. Their results suggested that the excimer formation was enhanced in nonpolar solvents, while with increasing solvent polarity the excimer quantum yield was decreased gradually with formation of the TICT peak.

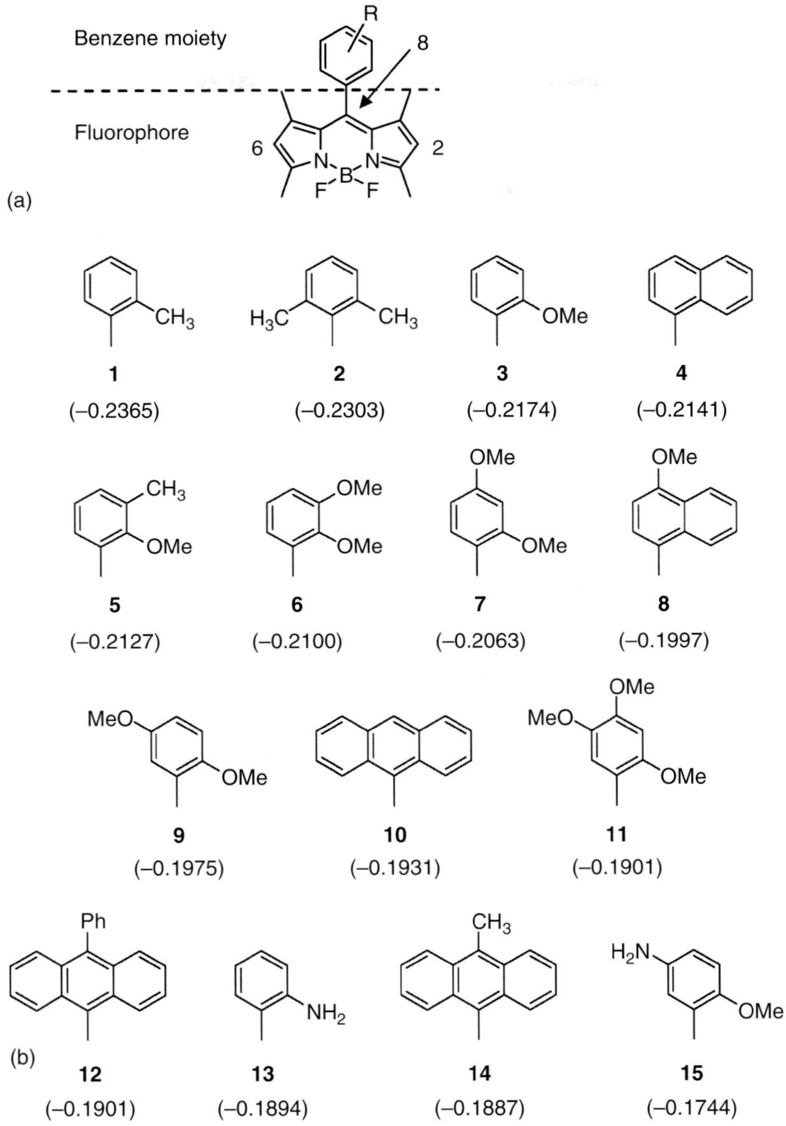

Figure 6.1 (a) The structure of the BODIPY fluorophores studied by Sunahara *et al.* (b) The structures of benzene moiety numbered **1–15** according to their HOMO energy. (Sunahara *et al.* 2007 [8]. Reproduced with permission of American Chemical Society.)

As discussed earlier, molecular rotors are used to probe viscosity of several biological systems. The molecular rotors are TICT molecules, where the donor and acceptor moieties are attached to each other, usually through a π-electron bridge. The donor and acceptor groups can rotate relative to each other following photoexcitation, leading to the TICT state. As change in viscosity may restrict the motion of the aforesaid group(s), molecular rotors are used widely as sensor for viscosity. It has been reported by several groups that many

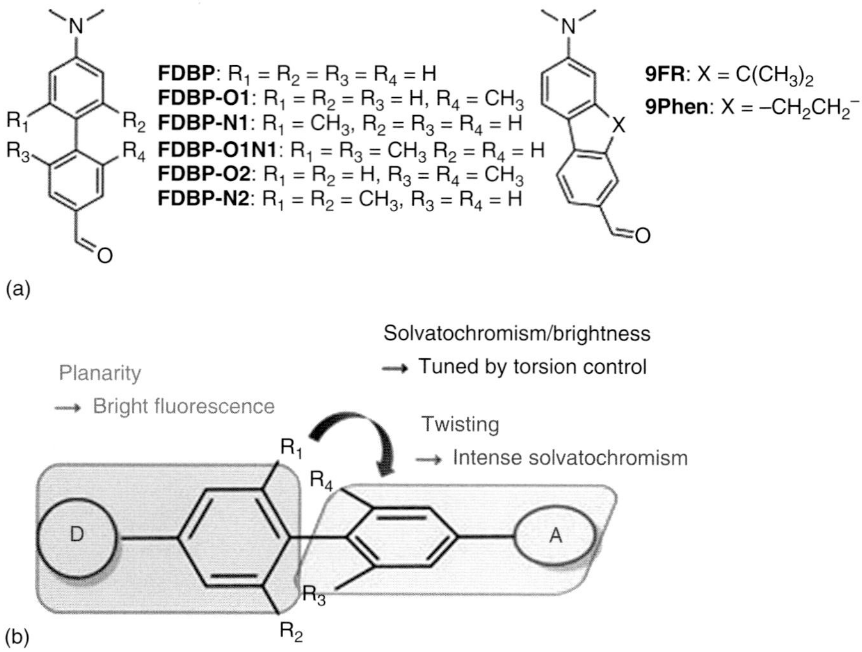

Figure 6.2 (a) Chemical structures of the push–pull biphenyl analogs studied by Sasaki *et al.* (b) The concept of "pretwisting" applied in their work. (Sasaki *et al.* 2014 [9]. Reproduced with permission of Elsevier.)

cellular and organismal functions are affected by the unusual change in the viscosity of their environment, which may lead to several diseases [1]. For example, health issues, including., diabetes, atherosclerosis, cell malignancy, and hypercholesterolemia, are reported to be linked to alterations of viscosity of the cell membrane. Alterations in cytoplasmic viscosity and viscosity changes in blood plasma are also believed to be reasons for the origin of several other diseases. As using conventional viscometers are cumbersome and they are inefficient to carry out real-time measurements of viscosity, suitable sensors for determining microscopic viscosity is the need of the hour. Molecular rotors are being used as microviscosity probes in the polymerization process, in probing the reorganization process in cytoskeleton, as well as in probing the viscosity of cell membrane. Several molecular rotors belonging to different chemical classes of molecules have been reported so far, for example, benzonitrile-based fluorophores, benzylidene malononitriles, stilbenes, and arylmethine-based molecules [11]. Structures of some of the molecular rotors are depicted in Figure 6.3. It is worthwhile to note here that there are several molecules, like DMABN, that show response to both solvent polarity and viscosity in their emission properties. Although it is extremely difficult to separate the effects of these solvent parameters completely, a clear understanding of the photophysics of a molecule is needed before using them as molecular rotors. Therefore, one must be cautious while using a molecule as viscosity sensor as influence from the polarity of the medium may lead to erroneous results.

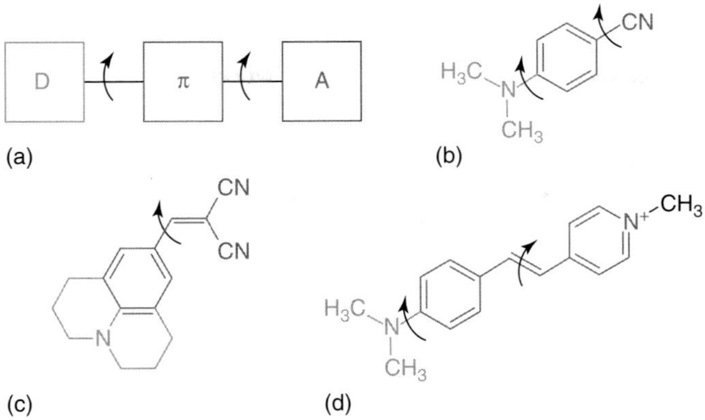

Figure 6.3 (a) The general structure of a molecular rotor for a D–π–A molecule. The chemical structure of some of the possible molecular rotors: (b) DMABN, (c) 9-(dicyanovinyl)julolidine (DCVJ), and (d) *p*-(dimethylamino)stilbazolium (*p*-DASPMI). (Copyright 2010 Haidekker and Theodorakis; licensed BioMed Central Ltd.)

Thioflavin-T is one of the most studied molecular rotors that has been used to detect amyloid fibrils.

Stsiapura *et al.* [12] studied the viscosity dependent properties of thioflavin-T (Figure 6.4) using both experimental and quantum chemical methods. They found that the minimum energy configuration of thioflavin-T has an torsional angle (φ) of 37° between the benzothiazole and aminobenzene rings, while the φ value of 90° corresponds to the minimum energy of the molecule in the first excited state. The authors concluded that incorporation of this molecule to amyloid fibrils restricts the relative twisting of the of the benzothiazole moiety compared to the aminobenzene ring, owing to enhanced rigidity of the medium with increasing concentration of the amyloid fibrils.

TICT-based molecules have been used to detect chemical species. For example, a BODIPY derivative (Figure 6.5) has been used as sensor for F^- ions [13]. Their results showed that this molecule binds F^- ions with high selectivity and stoichiometry, while it does not sense the other anions (Cl^-, Br^-, I^-, NO_3^-, OAc^-, HSO_4^-, $H_2PO_4^-$, CN^-) tested by the authors. Further studies also revealed formation of 1 : 1 clusters between the BODIPY dye and fluoride ion.

The N-oxide derivative of rhodamine B (called RhoNox-1) that shows weak emission in pure solvents while it gives rise to turn-on fluorescence in presence of Fe^{2+} ions [14] in living cells. The authors performed spectroscopic as well as electrochemical studies in conjunction with quantum chemical calculations at the density functional theory (DFT) level to understand the mechanism of the binding process. In physiological conditions, quenching of fluorescence of RhoNox-1 was observed, which the authors attributed to non-radiative deactivation of the excited state of tertiary amine N-oxide-substituted xanthene involving the TICT process and also partially to the photoinduced electron transfer (PET) from the N-oxide group. The molecule showed significant enhancement of fluorescence in Fe^{2+} loaded cells due to selective Fe^{2+}-mediated deoxygenation of the N-oxide group.

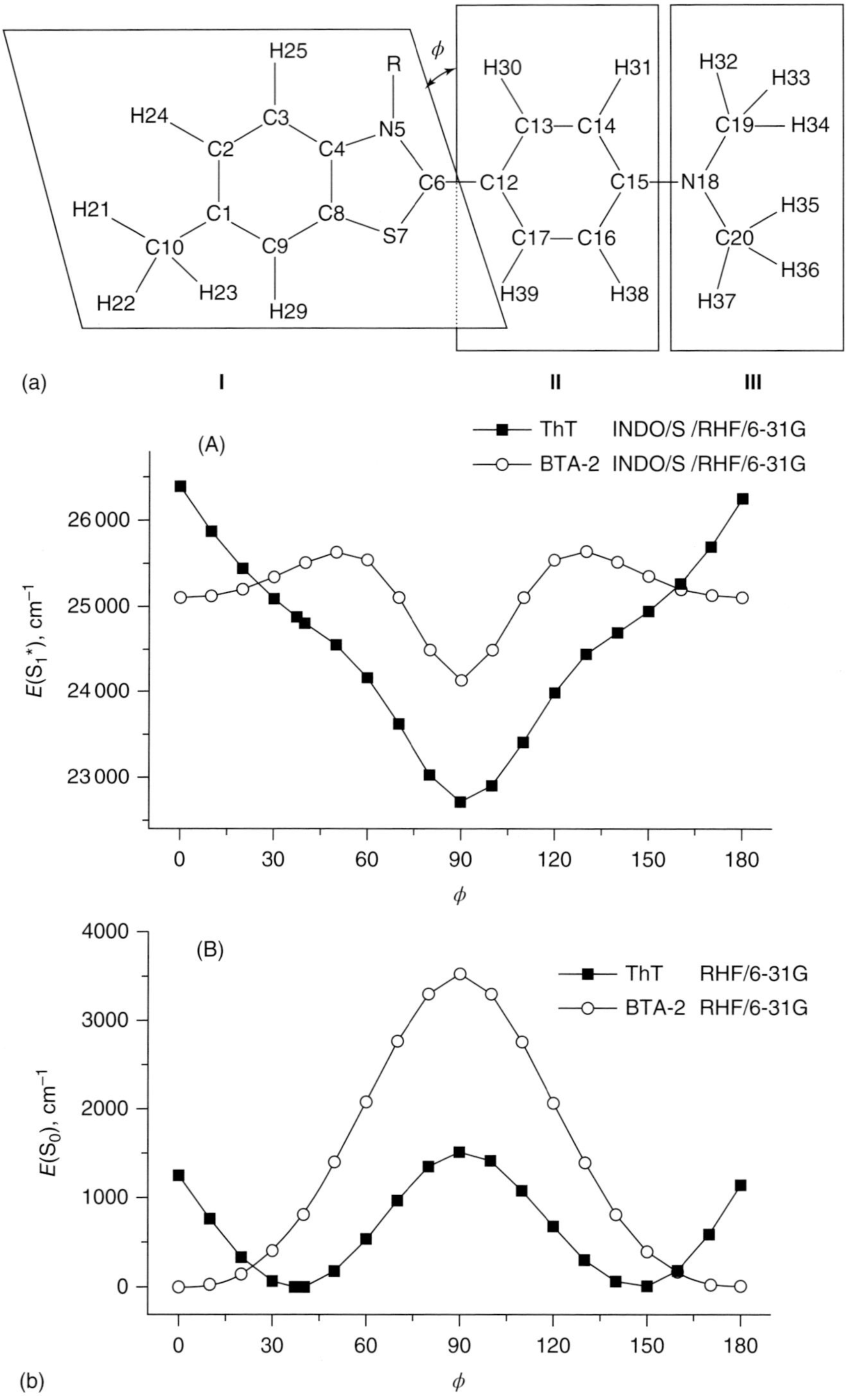

Figure 6.4 (a) The structure of thioflavin-T (R = —CH_3) and its derivative BTA-2 (R = —H). Three rings: benothiazole ring (**I**), benzene ring (**II**), and dimethylamino group (**III**) and the torsional angle ϕ between benothiazole and benzene rings are also shown. (b) The energies of ground state (S_0) and first excited state at Frank–Condon geometry (S_1) of these two molecules with change in φ. (Stsiapura *et al.* 2008 [12]. Reproduced with permission of American Chemical Society.)

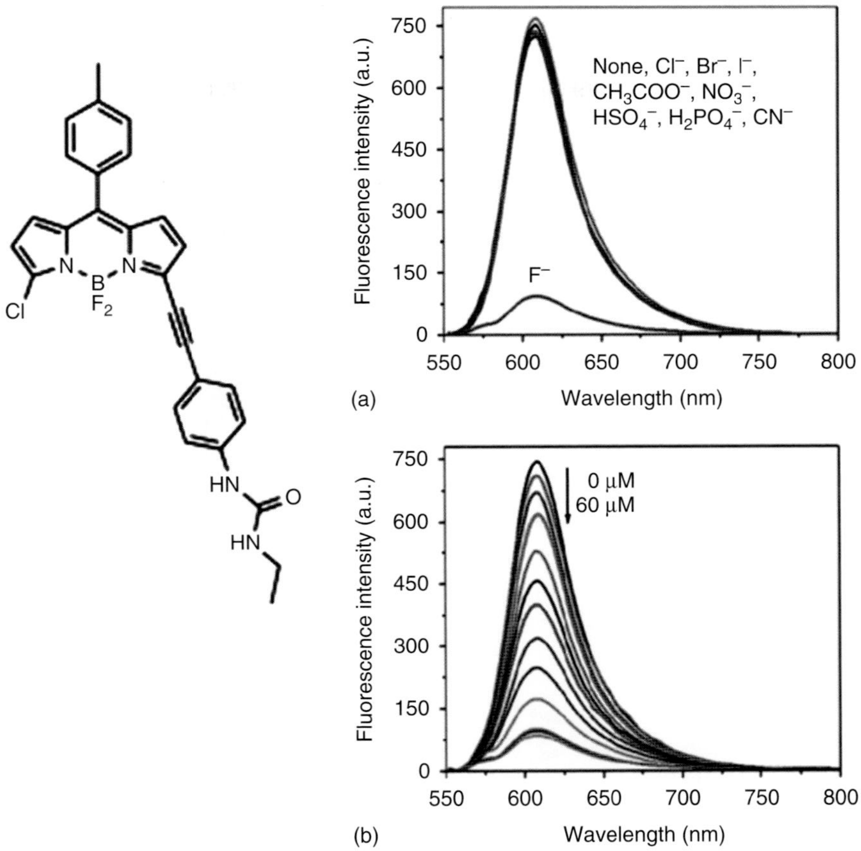

Figure 6.5 Chemical structure of the BODIPY derivative used by Liu *et al* (left). The right panel shows the change in emission intensity on addition of ions. (Liu *et al.* 2015 [13]. Reproduced with permission of Elsevier.)

Recently, Goel and coworkers [15] reported dimethylamino-based benzocrown ethers (Figure 6.6) as doubly twisted intramolecular charge-transfer (DT-ICT) molecules that can be used as acidic pH sensor. The authors claimed these sensors to be advantageous over the single-intensity-based pH sensors that rely on the fluorescence turn-on and turn-off response. The authors found high selectivity of the fluorescent rotor toward H^+ ions and its ratiometric fluorescence response manifested as changes in color from yellow to green to blue as they move from neutral pH (7.4) to acidic pH (1.5). Therefore, they concluded that this molecule can be used as a ratiometric acidic pH sensor.

Detection of the presence of metal ions in a sample quantitatively and selectively has immense importance in biological and environmental sciences. To know the presence of a specific ion, like fluoride, chloride, bisulfite, or a molecule like cystine, or hydrogen sulfide in a sample, several detection methods are being presently used. Of late, several ICT-based molecules are being used for detection of molecules and ions in solutions as well as in living cells. Many of these molecules were reported to be able to detect a species in

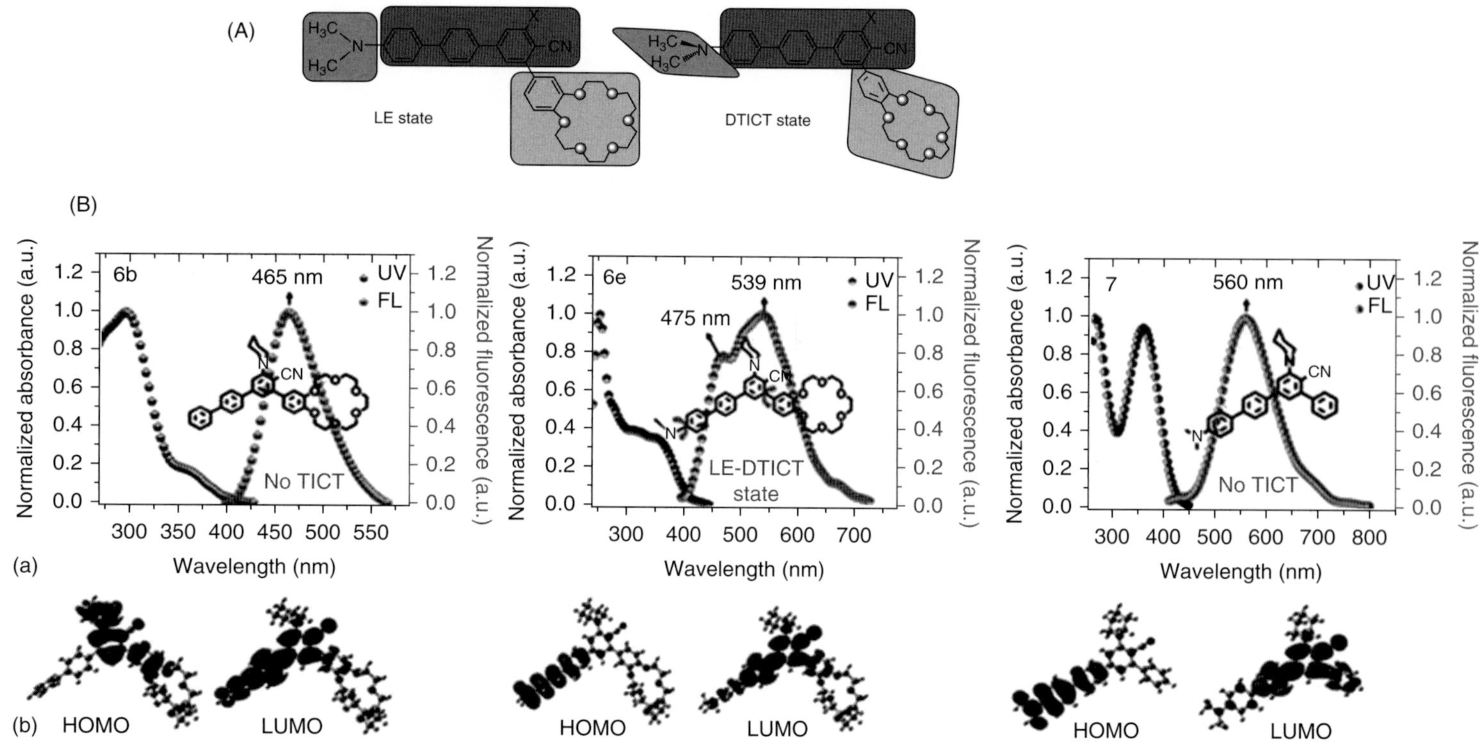

Figure 6.6 (A) The possible mechanism of double-twisted intramolecular charge-transfer (DT-ICT) process in one representative molecule studied by Goel *et al*. (B) Comparison of absorption, emission spectra, and molecular orbital pictures of three molecules. (Umar *et al*. 2016 [15]. Reproduced with permission of John Wiley & Sons.)

real time with high selectivity. In most of the cases, the sensing process relies on fluorescence "turn on" principle. That is, the TICT molecules are generally weak emitters; restriction of the twisting of the molecule that hinders the ICT process sometimes enhances their emission intensity. This fluoresce "turn on" process of TICT molecules is being utilized to detect molecules/ions with high selectivity. The fluorescence intensity or quantum yield of a fluorophore is dependent on several factors, including excitation wavelength, photomultiplier voltage, and excitation/detection slit width. To avoid these issues, ratiometric fluorescence sensors can be used. In these sensors, the ratio of fluorescence intensities at two different wavelengths is used to obtain information on sensing. Some of the representative examples of fluorescence sensing by TICT-based probes are discussed in this section.

Kumar and coworkers [16] designed a near-infrared (NIR) fluorescent TICT probe (Figure 6.7) for detection of human serum albumin (HSA). It is worth mentioning here that HSA is the most abundant transport protein in the circulatory system and presence of abnormal concentration of HSA in the human body is associated with several health conditions, for example, diabetes mellitus, kidney and liver diseases, and so on [16, 17]. Thus, quantitative detection of HSA in blood serum and other biological fluids has importance for diagnosis of the diseases associated with its abnormality. The authors have designed and synthesized the TICT molecule (they call it molecule "3"), whose fluorescence behavior is dependent on the occurrence of the charge-transfer process. That is, it shows fluorescence turn-on response due to inhibition of the TICT process. The spectroscopic studies in conjunction with theoretical calculations led the authors to conclude that the TICT probe interacts with the hydrophobic domain of HSA that inhibits TICT formation of the probe. The authors studied the ICT process of the molecule using steady-state emission spectroscopy. They found that the

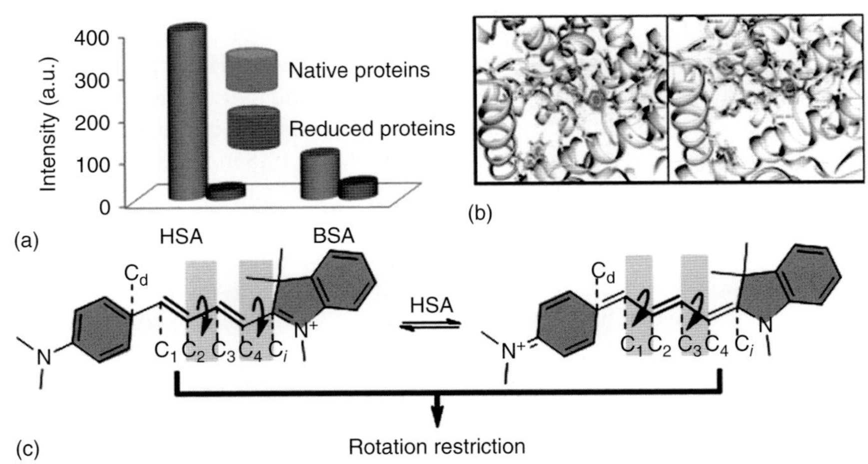

Figure 6.7 (a) The fluorescence bar diagram of the TICT-based compound studied by Kumar and coworkers. The emission intensity at 680 nm (λ_{ex} = 550 nm) due to addition of 15 and 60 μM of native and reduced HSA and BSA, respectively. (b) The docked images of the HSA-3 (left) and BSA-3 (right). (c) Proposed change in structure leading to fluorescence emission. (Reja *et al.* 2016 [16]. Reproduced with permission of Royal Society of Chemistry.)

emission maximum of the probe gets redshifted with increasing solvent polarity. The fluorescence quantum yield was found to decrease with increase in the polarity of the solvents. They assigned these behaviors to formation of the TICT state of the molecule. To find out whether restriction of TICT in this molecule leads to increase in emission intensity, the authors studied the emission behavior in water–glycerol mixtures with varying compositions. They found that the quantum yield of fluorescence increases with increase in percentage of glycerol in the mixture. In fact, they reported the quantum yield of fluorescence of the molecule increased about 12 times when going from 0% to 99% glycerol in the aforesaid solvent mixture. They attributed this observation to the restriction of rotational and vibrational modes in the viscous medium that in turn restricts the TICT process. The authors found that the fluorescence intensity at 680 nm of the probe in a solution of pH 7.4 was increased linearly with increase in concentration of HSA until it reaches 15 μM, after which a plateau was reached. The authors reported the detection limit of the probe for HSA to be 11 nM, which is believed to be sufficiently low for monitoring HSA levels in biological samples. They tested the selectivity of the probe for HSA by studying the emission behavior in the presence of several thiol-containing proteins and they have not observed any significant change in emission intensity due to their presence. The authors also reported the probe can be used to distinguish between HSA and bovine serum albumin (BSA), which many other sensors fail to detect selectively. The authors argued that the use of this TICT-based probe for quantitative detection of HSA has several advantages as the detection technique is based on reversible hydrophobic interactions rather than irreversible binding that could alter the native structure of the protein. They measured the HSA level in blood samples from healthy and hypertension patients and their results matched with standard detection processes, indicating the probe can be used in preclinical diagnosis.

Lin and coworkers [18] reported a TICT-based fluorescent probe (Figure 6.8) for selective and fast detection of hydrogen sulfide (H_2S). H_2S is known to be a noxious gas with an unpleasant smell of rotten eggs. Hydrogen sulfide is reported to be involved in a number of physiological and pathological processes. In addition to that, abnormal concentration of H_2S is associated with several diseases. Therefore, development of methods to detect H_2S in biological samples is necessary for diagnosis of H_2S-related health issues. Lin *et al.* designed and synthesized a TICT-based fluorescence probe (acronymed BH–HS) that can selectively detect H_2S over other reactive sulfur, nitrogen, and oxygen species with a fast response. The BH–HS contains a strong electron-withdrawing group named hemicyanine that has been used as recognition site for H_2S, and BODIPY as fluorescence reporter group. To confirm the TICT process in BH–HS, the authors have studied the effect of solvent polarity on the emission spectra of the probe. They found that the emission maxima of BH–HS get redshifted, while the emission intensity is decreased with increase in solvent polarity. In the TICT mechanism, it is believed that an intramolecular rotation is accompanied by the charge separation process in the excited state. A restriction of the intramolecular rotation inhibits the TICT process that in turn enhances the emission intensity of the molecule. The authors have studied the effect of viscosity on the fluorescence behavior of BH–HS in ethanol–glycerol mixtures with varying

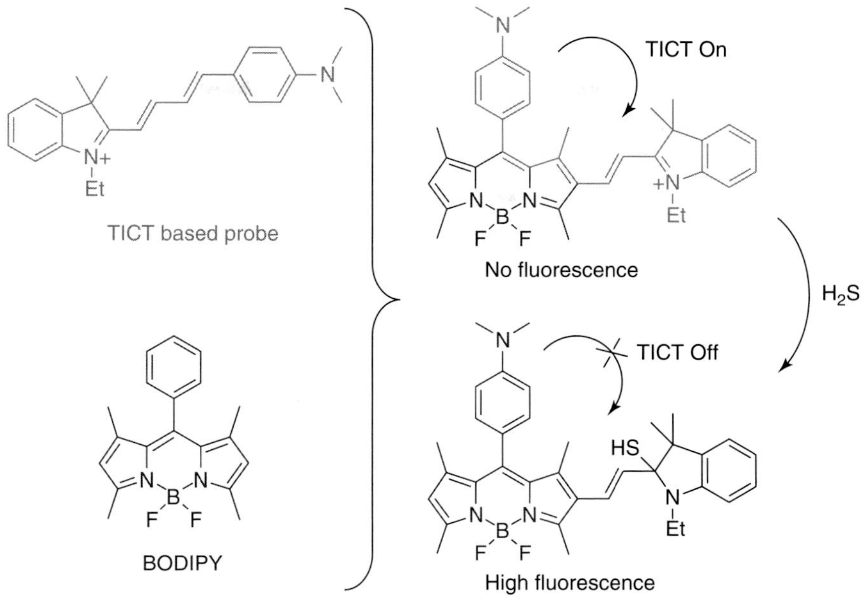

Figure 6.8 The chemical structure of BH–HS (upper right) and mechanism of fluorescence turn on through restriction of TICT process due to binding of H_2S. (Ren *et al.* 2016 [18]. Reproduced with permission of Royal Society of Chemistry.)

compositions. The increase in fluorescence intensity was about 12 times from pure ethanol to 50% ethanol in the ethanol–glycerol mixture. They assigned this to the TICT nature of the probe. DFT calculations indicated transition from the ground state to first excited state of BH–HS involves transitions from the highest occupied molecular orbital (HOMO) to both lowest unoccupied molecular orbital (LUMO) and LUMO + 1 orbitals. It was also reported that the HOMO of the molecule was mainly located on the dimethylaniline part, while LUMO and LUMO + 1 were mostly located on the BODIPY and hemicyanine units. As the HOMO of BH–HS does not overlap with the LUMO and LUMO + 1, the authors concluded that charge transfer in this molecule occurs from the dimethylaniline to BODIPY and hemicyanine part of the molecule. The binding of H_2S to BH–HS was probed by adding Na_2S, a commonly used H_2S source. The probe is almost non-fluorescent in a buffer solution of pH 7.4, while addition of Na_2S showed increase of emission intensity to about 57 times. The authors reported that the concentration of added Na_2S and emission ratios showed a linear relationship up to Na_2S concentration of 40 μM. The detection limit of H_2S by the TICT probe was reported to be 1.7 μM, which is quite lower than the physiological H_2S concentration in mammalian serum and in the brain. The pH of the medium does not have significant effect on the emission intensity of BH–HS. Maximum fluorescence response of the probe was observed in the range of pH 7.0–8.5 that implies that BH–HS can be used to detect H_2S in biological samples. On the other hand, increased sensitivity of the probe in the lower and higher pH ranges indicated that it could be used to detect H_2S in samples with wide range of pH values. The authors observed increase in emission intensity of BH–HS due to addition of Na_2S within seconds and a plateau was reached in about 2 min. They argued that the fast response of the probe enables it to detect H_2S in real time. To check the selectivity of the probe for H_2S, the authors added several other representative species like F^-, I^-, NO_2^-, $S_2O_3^{2-}$, SO_3^{2-}, and so on that they showed to have no notable fluorescence response. This indicated that the probe is highly selective for detection of H_2S. The authors also studied the ability of BH–HS to detect H_2S in living cells. They found the probe in low concentration did not show any notable cytotoxicity, thereby making it suitable for use in living cells. To detect the exogenous H_2S, they used HeLa cells that showed only weak fluorescence in presence of BH–HS alone but displayed strong fluorescence when treated with Na_2S for 30 min in the presence of BH–HS. The authors reported that the probe was able to detect endogenous H_2S that was biosynthesized inside the cells. They also found that BH–HS is membrane permeable, making the probe suitable for use in detecting H_2S in living cells.

Yu *et al.* [19] reported an ICT-based two-photon fluorescent probe (they call it DMPCA) for detection of bisulfite ion in aqueous medium (Figure 6.9).

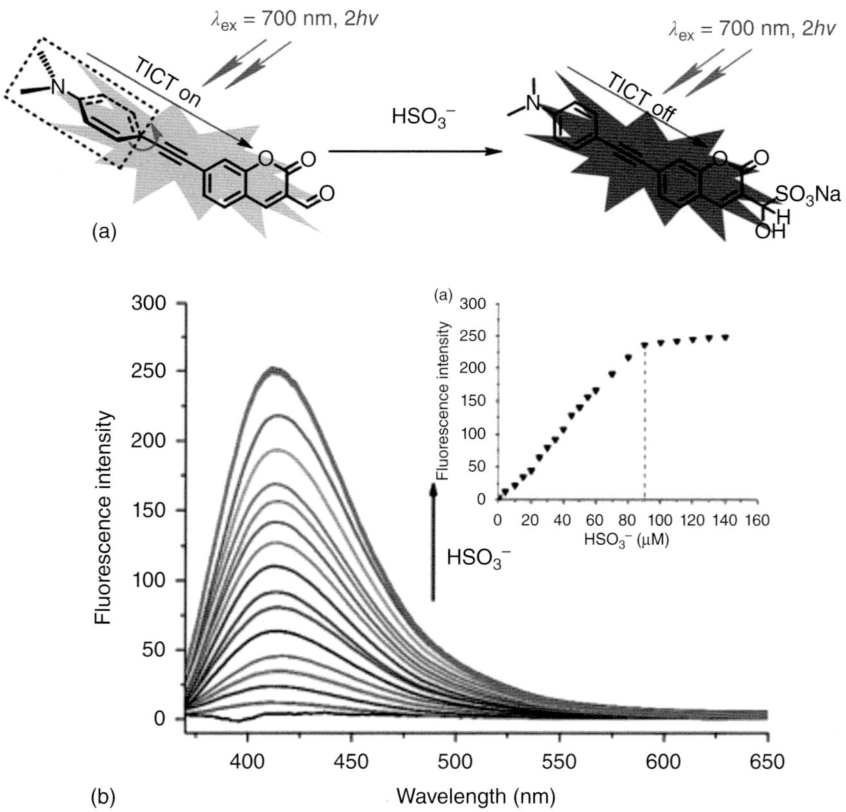

Figure 6.9 Panel (a) shows the restriction of TICT process in DCMPA due to binding of HSO_3^- ion. Panel (b) shows increase in emission intensity at 415 nm of DMPCA (10 μM) (λ_{ex} = 345 nm) on gradual addition of HSO_3^- (0–140 μM). The inset figure depicts the relationship between the fluorescence intensity (F_{415nm}) and the bisulfite concentrations. (Yu *et al.* 2016 [19]. Reproduced with permission of Elsevier.)

They found the emission intensity of the probe increases about 65 times due to addition of 10 equiv. of bisulfite anion to it. The design principle of this TICT-based two-photon fluorescence probe (Figure 6.9) lies in the following. The coumarin was used as the fluorescence group with *N*,*N*-dimethylamino as electron donor. The authors introduced the phenyl alkynyl group at the 7-position of the coumarin to achieve a good two-photon fluorescence. From the previous reports of reaction of aldehydes with bisulfite to form hydrogen sulfite adduct, they also anticipated that the aldehyde group of DMPCA will react with bisulfite ions that will affect the fluorescence spectra of the probe. To understand the effect of viscosity on the emission intensity of DMPCA, the authors studied the fluorescence spectra of the probe in water–polyethylene glycol medium. The increased emission intensity of the probe by about 30-fold with increasing viscosity of the medium was assigned to the TICT of the probe. Time-resolved fluorescence spectroscopy of the probe revealed the formation of two excited-state species of DMPCA with lifetimes of 0.65 and 2.74 ns. The authors assigned the slow component (2.74 ns species) to the TICT state of the molecule. The decay curve of the probe became monoexponential with lifetime of 1.08 ns due to addition of bisulfite ion. The authors argued that the formation of the TICT state was restricted due to addition of bisulfite anion that also enhances the emission intensity. The authors reported the fluorescence intensity of DMPCA at 415 nm got increased linearly with increasing concentration of the bisulfite ion up to 10 equiv. of the anion (the excitation wavelength was 345 nm). It was found that the fluorescence signal of DCMPA was stable in the pH range of 3.2–9.3. However, the fluorescence of the probe in presence of bisulfite ion was increased from pH 3.2 to pH 5, while it was decreased above pH 6. Therefore, the authors used a buffer solution of pH 5.0 to prepare the samples. The change in absorption maxima due to addition of bisulfite ion also indicated the formation of a new compound. The limit of detection of the anion by this probe was reported to be very low to use this probe for efficient detection of bisulfite anion. The cytotoxicity studies by the authors showed that the probe could be safely used for two-photon bioimaging at low concentrations. The authors found that the reaction between DCMPA and the bisulfite anion completed in about 2 min, indicating the molecule could be used to detect the aforesaid anion in real time. To check the selectivity of the TICT-based probe for bisulfite ion, the authors added several other anions (F^-, Cl^-, HS^-, CN^-, SCN^-, etc.) to the solution of DCMPA and found that no notable enhancement in the fluorescence intensity of the probe with addition of those anions. Mixing of these anions with bisulfite anion also did not affect the emission intensity of the probe much. The maximum two-photon absorption (TPA) cross-section of DCMPA with bisulfite anion was reported to be 725 GM at 700 nm. The authors argued that two-photon fluorescence probes that can be excited by TPA at the NIR region have several benefits over the one-photon microscopy that requires excitation with short-wavelength light. The excitation with short-wavelength lights has several limitations like shallow penetration depth, photo-damage, photo-bleaching, and auto-fluorescence. The authors also reported that DCMPA was able to detect bisulfite anion in the living cells.

Meng and coworkers [20, 21] have developed two ICT-based sensors for detection of cysteine and homocysteine in living cells. The cysteine and homocysteine molecules are known to play crucial roles in many physiological processes and they are associated with several health issues, including, slowed growth, hair depigmentation, lethargy, to mention a few. Both of these probes, they call them MQ and NQ, respectively (Figure 6.10) are based on two-photon fluorescence. Similar to the studies mentioned earlier, the TICT mechanism of NQ was studied using time-resolved fluorescence and viscosity-dependent emission studies. The authors also employed time-dependent density functional theory (TDDFT) calculations to examine the ICT mechanism in this molecule.

The cytotoxicity assays showed that NQ in low concentration was safe to use in detection of cysteine and homocysteine in live cells. The authors measured the TPA cross section of NQ that showed that the maximum TPA cross-section values of NQ bound to cysteine and homocysteine were 580 and 710 GM at 795 nm, respectively. The authors argued that the aldehyde group of NQ could form thiazolidine and thiazinane due to cyclization reaction with cysteine and homocysteine, respectively. Their cell culture and pH stability test results showed that NQ was cell permeable and the probe could be used for detection of cysteine/homocysteine under two-photon excitation in living cells.

Kitamura *et al.* [22] reported a ratiometric dual fluorescence probe based on TICT phenomenon for detection of an amino acid, namely, *N*-acetyl-D-phenylalanine (AcPhe). The authors designed the sensor based on a cholic acid derivative that also contained a TICT chromophore and the binding site for an amino acid. They introduced a 4-(*N*,*N*-dimethylamino)benzoate

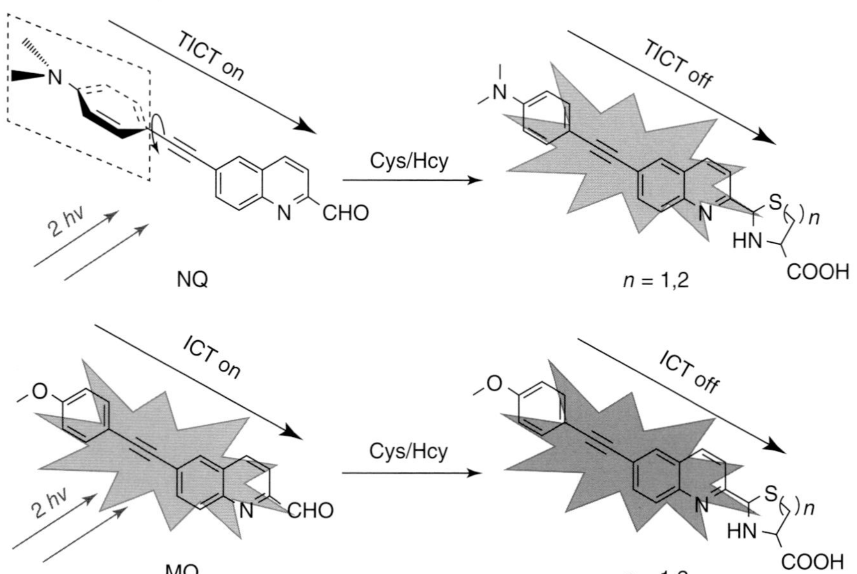

Figure 6.10 The chemical structure of TICT probes NQ and MQ, studied by Meng and coworkers. The authors showed that binding of cysteine (Cys) or homocysteine (Hcy) restricted the TICT process in these molecules, resulting into better fluorescence intensity. (Wei *et al.* 2016 [20]. Reproduced with permission of Elsevier.)

(DMAB) unit at the 12-position of the cholic acid as fluorescence sensor unit. The authors envisioned that binding of the amino acid to the sensor would change the TICT/LE fluorescence intensity ratio, which could give rise to quantitative information on the binding process. They showed that the ratio of fluorescence intensities of the sensor molecule at 440 nm to that of 350 nm decreased with increase in amino acid (AcPhe) concentration. The authors also found that the logarithm of ratio of TICT to LE quantum yields ($\ln(\varphi_{TICT}/\varphi_{LE})$) showed linear relationship to the inverse of temperature multiplied by 1000 ($1000\,T^{-1}$) with negative slopes both in presence and absence of AcPhe. They estimated the activation energies for the forward TICT formation processes to be 10.6 and 12.7 kJ mol^{-1} in absence and presence of AcPhe, respectively. From their studies the authors concluded that the TICT probe could be used for ratiometric detection of an amino acid.

Bhattacharya *et al.* [23] designed and synthesized some colorimetric ion-sensing probes based on anthra[1,2-*d*]imidazole-6,11-dione (Figure 6.11).

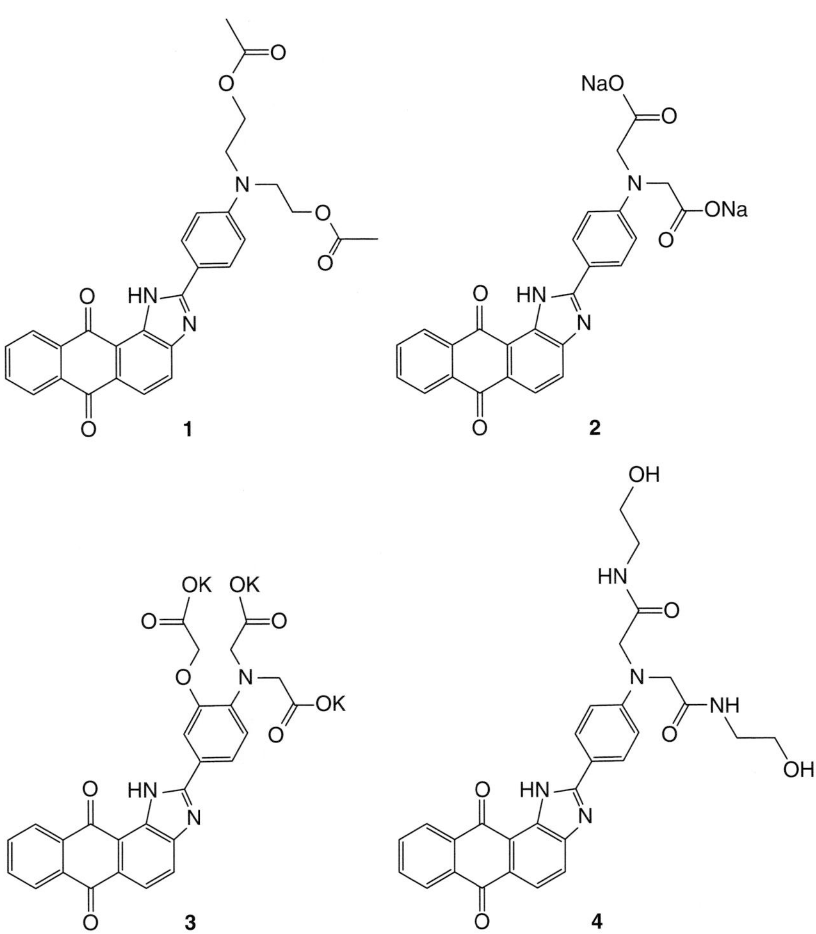

Figure 6.11 The chemical structure of the sensor molecules (**1–4**) studied by Bhattacharya *et al.* (Kumari *et al.* 2011 [23]. Reproduced with permission of American Chemical Society.)

All the four sensors were able to detect fluoride (F^-) as well as cyanide (CN^-) ions in organic media. They showed that the sensing mechanism involved the ICT process. One of the probes (molecule 2) was able to detect F^- and CN^- selectively as the change of color was from yellow to blue for F^- ion, while it turned into red due to addition of CN^- ion. It is worthwhile to mention that fluoride ion sensors could be useful in detection of uranium enrichment as well as detection of chemical warfare agent sarin [24]. On the other hand, cyanide is one of the most toxic ions known till date. Presence of a small amount of CN^- ion in mammalian cell could be dangerous as it inhibits the cellular respiration by interacting with the active site of cytochrome a_3 [25]. Development of sensors for fluoride and cyanide ions therefore has application in biological and environment sciences. The colorimetric sensor has an advantage over the other sensors as the signaling can be detected by the naked eye, while other sensors may need sophisticated instruments. The F^- and CN^- sensors described here are based on the ICT principle, in which a substituted nitrogen-linked aromatic unit functions as the donor, while the anthraquinone part in each molecule has an acceptor moiety. These sensors also work on the principle of "push–pull" effect of the donor and acceptor moiety, present in the molecule. If a negatively charged analyte binds to the electron-deficient acceptor in the system turns into "D–D" from "D–A" that reduces the charge-transfer character of the probe, weakening the signal [26]. Bhattacharya *et al.* [23] utilized the fluoride ion detection property of benzimidazole system that detects F^- ion through deprotonation of its NH proton. The authors found that each of the compounds studied showed colorimetric response upon addition of either F^- or CN^- ion, which was added in the form of their tetrabutyl ammonium salts in acetonitrile–DMSO (95 : 5) mixture. Addition of aforesaid ions to all the molecules showed the red.-shift of the ICT band. The molecule 2 showed a different behavior from that of the other probes as its color change due to addition of fluoride and cyanide ions was different. UV–vis absorption spectral studies showed about 120- and 81-nm bathochromic shifts of molecule 2 due to addition of F^- and CN^- ions, respectively. Other molecules also showed noticeable changes in the absorption maxima due to addition of the aforesaid ions. Out of the molecules studied, the authors found that the molecules 2 and 3 were able to detect fluoride ions even in the presence of cyanide ions. The shifts in absorption maxima of molecules 2 and 3 were also highest due to addition of the fluoride ion. The authors attributed this fact to the negatively charged donor atoms in these molecules. To prove the hypothesis, they compared the UV–vis absorption spectra of these two molecules with their neutral counterparts (ethyl ester derivatives). The absorption of the charge-transfer band of molecule 2 showed a peak at 468 nm vis-à-vis 450 nm for its ethyl ester derivative. The redshift of ester of molecule 2 was only about 40 nm, which is much less compared to molecule 2 that showed a redshift of 120 nm. The authors argued that the aforesaid spectral behavior was due to the presence of negative charge on the donor site. They also showed that the receptors interacted with the F^- and CN^- ions at 1 : 1 stoichiometry, indicating interaction of the added anions with benzimidazole proton, which is a more labile proton in the molecules studied. The authors also studied the possibility of using the aforesaid molecules as sensors in water. In aqueous medium, the probes failed to detect fluoride ions as F^- ion is highly solvated in water and loses its basicity due to hydration. However, all the molecules studied undergo changes in absorption spectra due to addition of CN^- ion in aqueous acetonitrile (9 : 1) medium. Addition of CN^- ion showed an increase in absorption intensity of molecule 2 at 504 nm and a decrease in intensity at 464 nm. The change of absorption intensity was completed with addition of 5 equiv. of the CN^- ion. The ratios of absorption intensities at 504 nm to that of 464 nm showed a linear relationship with the concentration of the CN^- ion, making the molecule ready for use as a ratiometric sensor for cyanide ion in aqueous media.

Recently, Bhat and Jha [27] reported their quantum chemical studies of selective binding of cyanide and fluoride ions to ammonium boranes and showed that the ICT process was involved in the sensing mechanism. The authors used DFT and TDDFT calculations to study the feasibility of binding of several anions to two isomeric boranes, namely, 4-[bis(2,4,6-trimethylphenyl)boranyl]-*N*,*N*,*N*-trimethylaniline or $[p\text{-}(Mes_2B)C_6H_4(NMe_3)]^+$ (where Mes represents mesitylene) and 2-[bis(2,4,6-trimethylphenyl)boranyl]-*N*,*N*,*N*-trimethylaniline or $[o\text{-}(Mes_2B)C_6H_4(NMe_3)]^+$, which they refer as 1 and 2 (Figure 6.12), respectively. They used CAM-B3LYP functional with 6-31G(d) basis set to optimize the ground-state structures of the aforesaid molecules as well as their adducts with the ions studied. The natural population analysis of these molecules indicated that the central boron atoms in both the molecules were the most active centers for nucleophilic addition of anions. The free energy changes (ΔG) due to addition of F^- and CN^- ions to 1 and 2 were negative, while the values were found to be positive due to addition of Cl^-, Br^-, NO_3^-, and HSO_4^- ions, indicating the molecules could bind fluoride and cyanide ions only among the studied ions. Their calculated binding energies also indicated that the binding affinity of the fluoride ion was higher than that of the cyanide ion. The other ions either did not bind to the molecules or showed insignificant binding energies. Their frontier molecular orbital analysis revealed that the first excited states (S_1) of both the molecules bore the signature of LE state of $\pi\sigma^*$ character. The same analysis resulted in the fourth excited state (S_4) of the cyano- and fluoride-bound forms of 1 (acronymed 1CN and 1F, respectively) and fifth excited state (S_5) of fluoride-bound 2 (acronymed 2F) of ICT character. The further analysis revealed that the synergistic effect of ICT and partial configuration changes resulted in quenching of fluorescence in 1CN, 1F, and 2F through internal conversion from S_4 (for 1CN and 1F) or S_5 (for 2F) to S_1.

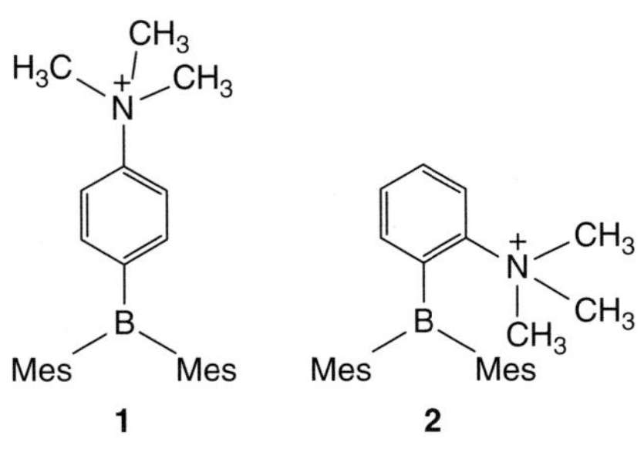

Figure 6.12 Chemical structure of $[p\text{-}(Mes_2B)C_6H_4(NMe_3)]^+$ (**1**) and $[o\text{-}(Mes_2B)C_6H_4(NMe_3)]^+$ (**2**). (Bhat and Jha 2017 [27]. Reproduced with permission of American Chemical Society.)

There are several other examples of utilizing the ICT process for detection of ions/molecules. Wei *et al.* [28] designed and synthesized a colorimetric and fluorescent chemosensor based on dicyanovinyl derivatives for selective detection of cyanide ion in aqueous medium. The probe molecule showed fluorescence "turn-off" response due to addition of CN^- ion. The solution of the probe molecule turned colorless from yellow, while the green fluorescence disappeared due to addition of CN^- ion. The authors also fabricated sensor strips based on the aforesaid molecule that served as efficient cyanide sensor. Thiagarajan *et al.* [29] reported a colorimetric and fluorescence chemosensor for detection of several anions, viz., F^-, AcO^-, $H_2PO_4^-$. They showed that the detection processes involved PET and ICT processes.

From the preceding discussion it is apparent that ICT-based fluorescent molecules can be used to detect several metal ions and molecules selectively in real time. The simplicity of implementation of these probes and compatibility with biological samples are other advantages of using these probes. Other than the ICT, there are several other phenomena like excited-state intramolecular proton transfer (ESIPT), electronic energy transfer (EET), fluorescence resonance energy transfer (FRET), and so on, that have been used for designing fluorescence chemosensors. Interested readers may go through a relatively recent review by Wang *et al.* that reviews some of the aforesaid topics [30].

6.2.2 In Organic Light-Emitting Diodes

Over the years, OLEDs have been used to generate inexpensive and efficient white light displays. OLEDs are believed to be the next-generation white-lighting source that will not only be inexpensive but also help reduce the energy consumption, which in turn will reduce the carbon dioxide emission to the environment. Studies on development of novel materials for OLED applications have gained momentum over the past three decades since the seminal work by Tang and VanSlyke [31]. To produce full-color displays and also to use in white-light-emitting devices, one needs OLEDs of three primary colors (red, green, and blue). Till date, red and green phosphorescent emitters based on organometallic platinum or iridium complexes are used as they match the requirements in terms of stability, efficiency, and color purity. However, efficient materials for blue color displays are still lacking. This drives the scientific community to develop novel materials for blue emitters that will match their red and green emitting counterparts in terms of operational stability, electroluminescence efficiency, and color index. Over the past few years, use of thermally activated delayed fluorescence (TADF) for designing OLEDs have gained momentum as they can harvest both triplet (T_1) and singlet (S_1) excitons for electroluminescence through efficient reverse intersystem crossing (RISC). Of late, TADF materials have caught the huge attention of the scientific community, owing to their high luminescence efficiency and their potentiality for designing OLEDs. Use of purely organic TADF materials can help get rid of precious-metal-based phosphorescent materials, which, needless to say, are expensive. TADF materials have the potentiality to harvest almost the entire populations of excitons from the singlet and triplet states, whereas these states individually can harvest 25% and 75% excitons, respectively. These materials can generate an additional exciton through efficient up-conversion from the lowest excited triplet state (T_1) to the lowest excited singlet state (S_1) through an RISC process. TADF materials are composed of an electron donor (D) and an acceptor (A) moieties. These D–A materials generally give rise to a small spatial orbital overlap between the HOMO and the LUMO to minimize the singlet–triplet splitting energy (ΔE_{ST}). This enhances the RISC process from the non-radiative T_1 state to radiative S_1 state, thereby enhancing the energy harvesting efficiency of the material. Several donor–acceptor (D–A) and donor–acceptor–donor (D–A–D) structured organic ICT materials have been used as chromophores for TADF materials [4, 5].

Bredas and coworkers [6] have studied the rates of up-conversion intersystem crossing (UISC), which we referred to as RISC, using quantum chemical methods at the DFT level of theory. The authors evaluated both the ΔE_{ST} and the spin–orbit coupling in a series of donor–acceptor molecules (Figure 6.13) in order to maximize the UISC rates that will facilitate TADF process in these molecules. Their results indicate that small ΔE_{ST} values and sizable spin–orbit coupling matrix elements are needed to be achieved to facilitate the UISC. The authors argued that a spatial separation of the HOMO and LUMO levels of the molecules, which is believed to be a dominant factor in obtaining small ΔE_{ST}, does not guarantee sufficient reduction of the singlet–triplet splitting energy. In fact, the author found that either a significant charge-transfer contribution to the triplet (T_1) state or a minimal energy difference between the LE and charge-transfer triplet states is required to obtain a small singlet–triplet splitting energy.

Adachi and coworkers [32] reported on blue TADF-based OLED molecules containing a 9,10-dihydroacridine or diphenylsulfone derivative (Figure 6.14). The authors claimed that these OLEDs have performance comparable to phosphorescence-based OLEDs and their device offered an external quantum efficiency of 19.5%. The authors identified a TADF-based blue-emitting compound MAD-DPS, in which diphenylsulfone (DPS) acts as an acceptor while the two units of 9,9-dimethyl-9,10-dihydroacridine (MAD) act as donor. They have carried out TDDFT calculations to study the excited-state properties of the molecules that provides insight into the interplay between the charge-transfer and possible locally excited triplet states. The authors also compared the blue emission of a phosphorescence based emitter, bis[(4,6-difluorophenyl)pyridinato-*N*,C^2](picolinato)iridium with that of MAD-DPS.

Recently, Yasuda *et al.* [33] reported twisted acridan–pyrimidine donor–acceptor-based delayed fluorescence molecules (Figure 6.15) for designing highly efficient blue electroluminescence materials. In these molecules the acridan/spiroacridan-based donor is connected to the pyrimidine-based acceptor through a phenylene π-electron bridge. The authors measured the electroluminescence of these materials doped in thin films and found to have very good (69–91%) quantum yields that are accompanied by TADF, originated from their small singlet–triplet splitting energy. The authors assigned the small ΔE_{ST} to the nearly orthogonal donor–acceptor structure of these molecules.

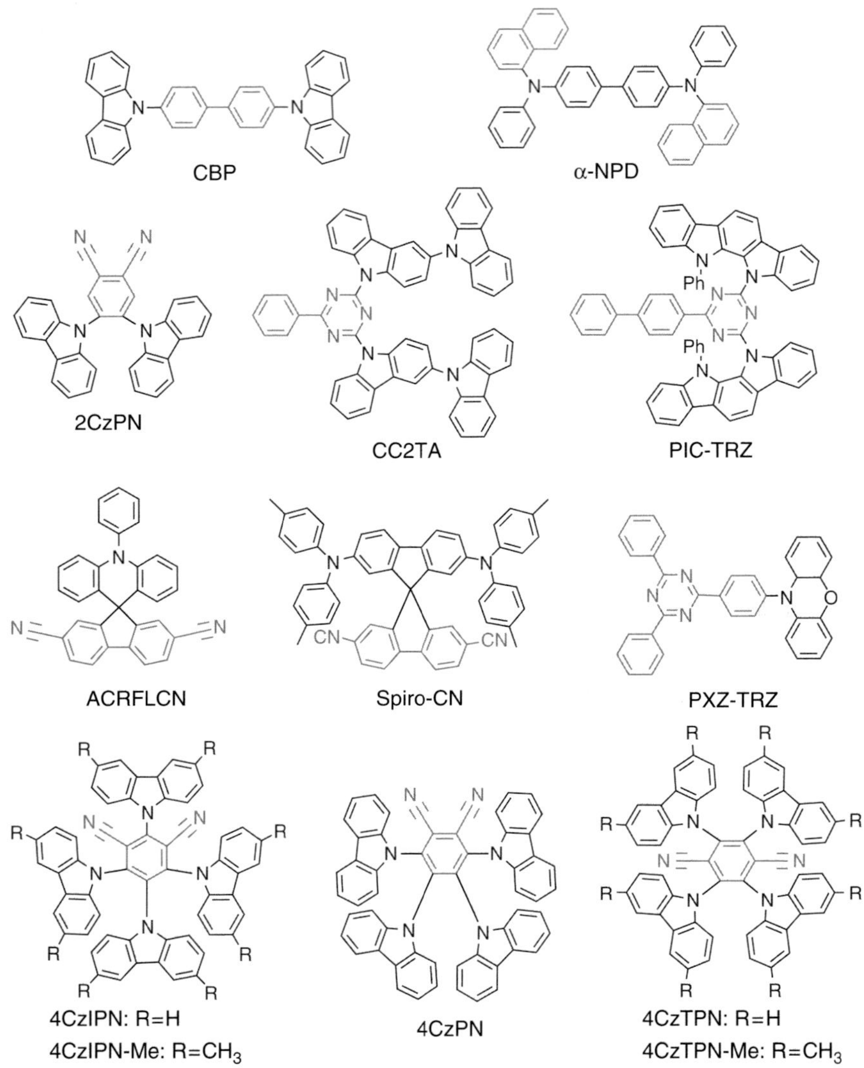

Figure 6.13 The chemical structures of the molecules studied by Bredas and coworkers. (Bredas *et al.* 2017 [6]. Reproduced with permission of American Chemical Society.)

These twisted structures were achieved through steric repulsions between the hydrogen atoms of the acridan unit and the adjacent phenylene spacer. The authors also conclude that the pyrimidine moiety can be used as a universal acceptor in designing deep-blue TADF materials as this group has very weak electron-accepting ability because strong acceptors like triazine can increase the ΔE_{ST}. The authors also reported that pyrimidine moiety can bind to several electron donors and its electronic and photophysical properties can be tuned easily through simple structural modifications.

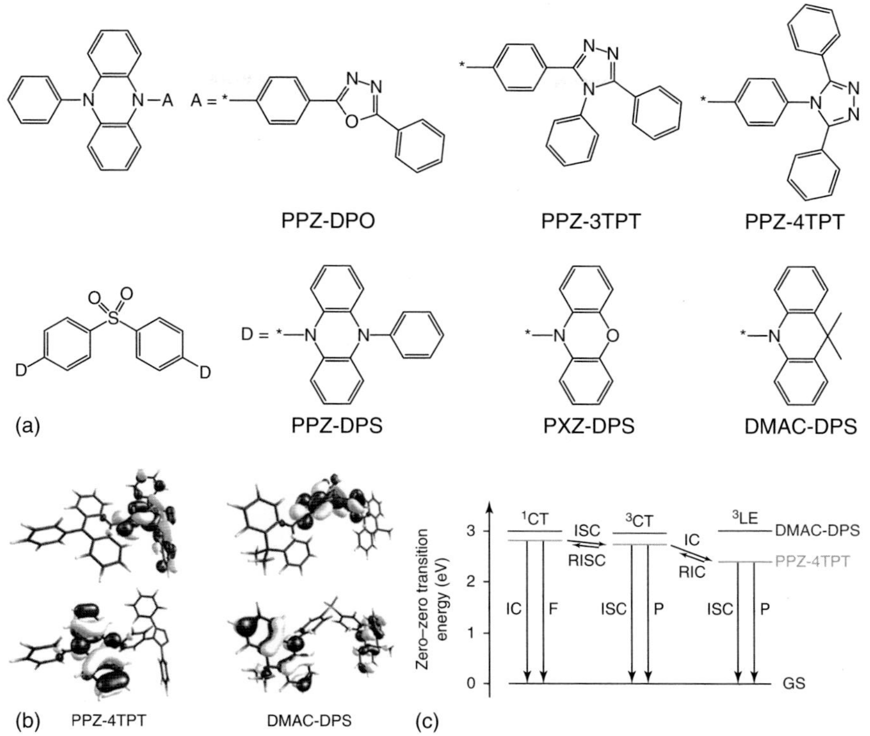

Figure 6.14 (a) Chemical structures of the molecules studied by Adachi and coworkers. (b) The HOMO (lower images) and LUMO pictures of two representative molecules (PPZ-4TPT and DMAC-DPS). (c) Jablonski diagram of the energy levels of PPZ-4TPT and DMAC-DPS calculated in toluene. In the figure, F, P, and GS stand for fluorescence, phosphorescence, and ground state, respectively, while IC, RIC, and ICT denote internal conversion, reverse internal conversion, and intersystem crossing, respectively. The term RISC stands for the reverse intersystem crossing in the system. (Adachi *et al.* 2014 [32]. Reproduced with permission of Nature Publishing Group.)

For white color displays, one needs to prepare OLEDs with red, green, and blue emissions with excellent emission efficiency and high stability. In addition to this, tuning of emission color over the entire visible range is also an important area of research [34, 35]. Several small organic molecules and conjugated polymers have been screened for possible application in OLED devices for full color display. These include polythiophenes, poly(*p*-phenylene vinylene) derivatives, and so on.

Promarak and coworkers [36] studied the photophysical, thermal, electrochemical, and electroluminescent properties of oligoarylenes endcapped with carbazol-*N*-yl-carbazole (CAr) (Figure 6.16) for exploring the possibility of designing color-tunable light-emitting materials. The authors argued that the emission wavelength of these materials can be tuned through structural modifications of the oligoarylene fluorescent cores by varying the degree of π-electronic conjugation as well as through enhancement of the electron-withdrawing ability of the aryl cores.

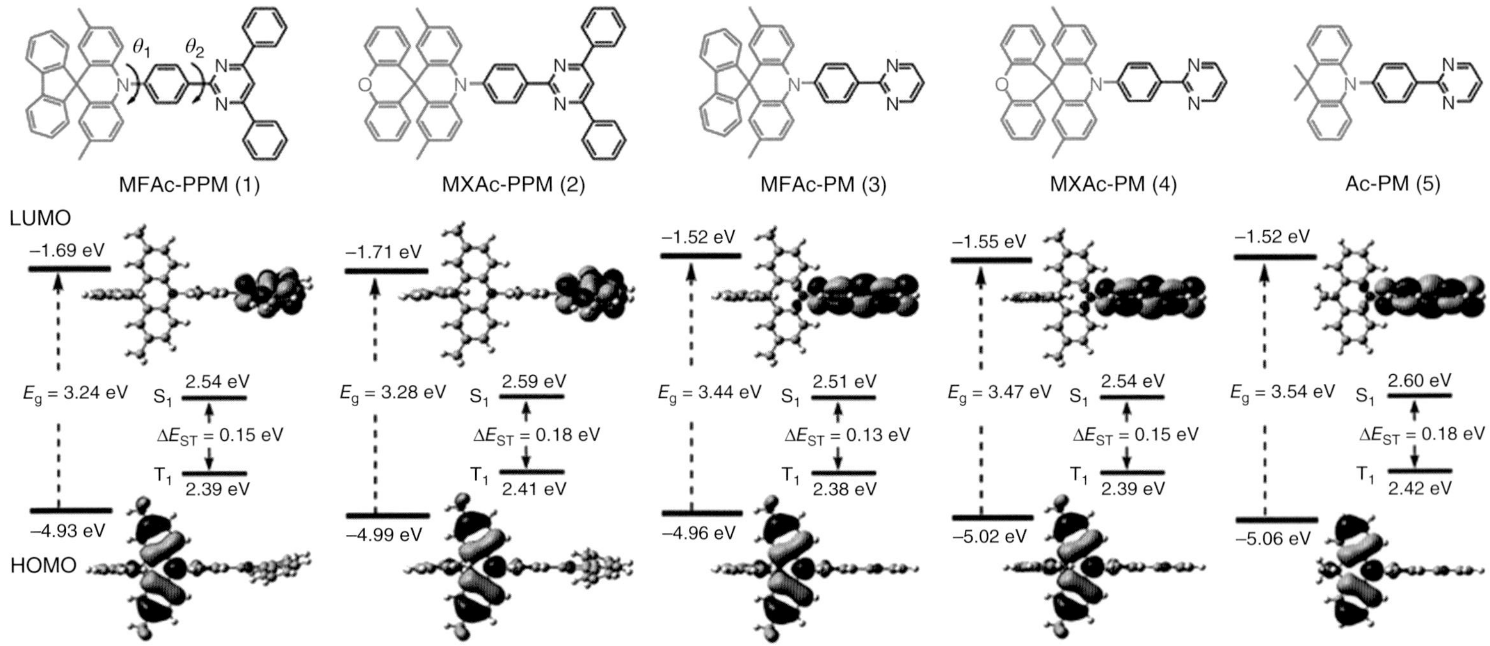

Figure 6.15 Chemical structures of the molecules studied by Yasuda and coworkers. The HOMO and LUMO pictures, and calculated (TDDFT at the PBE1PBE/6-32G(d) level) first singlet and triplet state energies for these molecules are also shown. (Park *et al.* 2017 [33]. Reproduced with permission of Royal Society of Chemistry.)

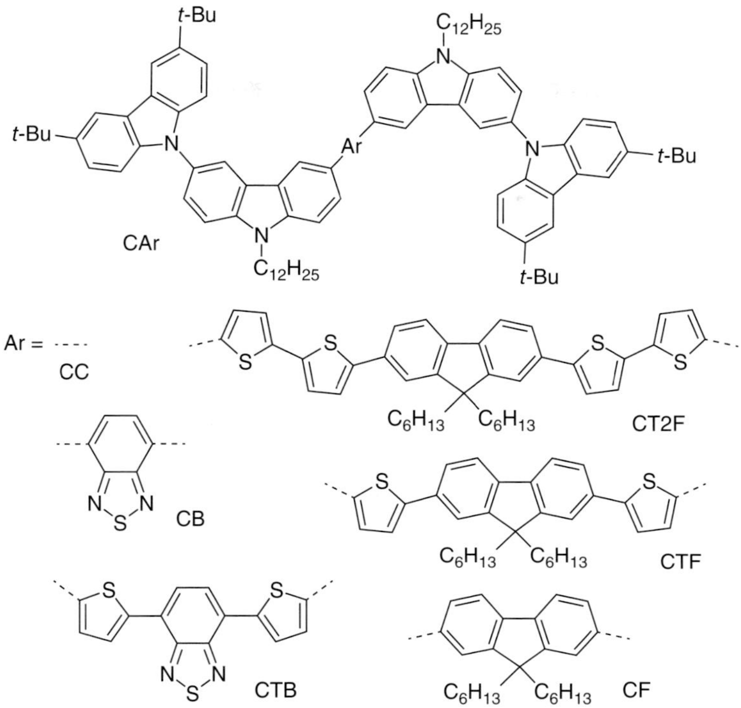

Figure 6.16 Molecular structures of the oligoarylenes CAr, designed by Promarak and coworkers. (Copyright 2015. Reproduced with permission of Wiley.)

6.2.3 In Aggregation-Induced Emission

Some fluorophores show weak emission when dissolved in solvent but show strong emission when clustered or are in solid state. This process is called aggregation-induced emission (AIE) process. Many ICT-based molecules show AIE and have been under intense investigation for possible optoelectronic and bioimaging applications. Some of the aryl-substituted BODIPY dyes are weakly fluorescent/non-fluorescent in polar solvents due to formation of ICT states that are responsible for decrease in the quantum yield of these molecules [37]. It has been reported that these dyes (Figure 6.17) recovered their fluorescence intensities upon aggregation [37]. The red emission of the TICT species boosted significantly upon aggregation through the AIE process. The authors found that the emission intensity can be enhanced by increasing the viscosity of the medium as well through cooling, indicating that restriction of intramolecular rotations result in AIE emission in these molecules.

From the preceding discussion, it is apparent that ICT-based molecules are being used for several technological applications while some studies are still going on to utilize them for future applications. ICT probes are used to quantitative sensing of many chemical species that, in general, are done in solution. However, for some applications, one needs to prepare solid materials with enough strength and durability. This led the scientific community to find possible ways to create

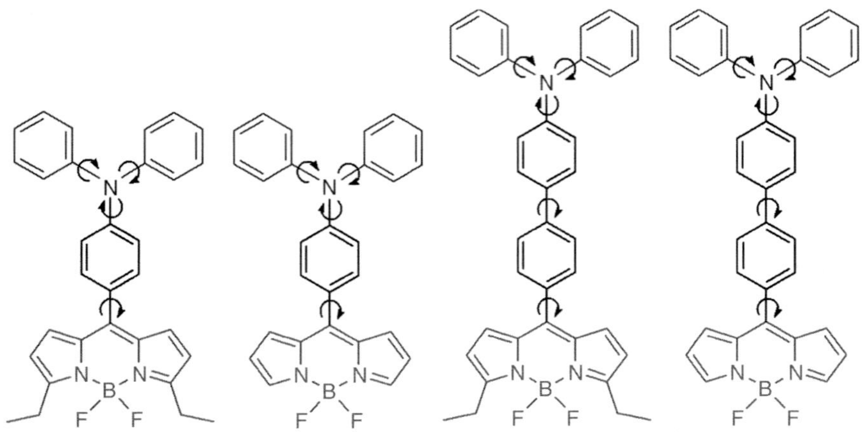

Figure 6.17 The chemical structures of the molecules studied by Hu *et al.* The possible intramolecular rotations are also shown. (Hu *et al.* 2009 [37]. Reproduced with permission of American Chemical Society.)

ICT-based materials that can be suited for possible technological applications. Li *et al.* [38] reviewed some of the methodologies presently being used to prepare self-assembled molecular architectures. Using the phase-transfer methodologies and solvent-vapor technique, in conjunction with chemical modifications of the donor/acceptor unit and conjugation strengths, the authors have been successful in preparing several organic nanostructures, including hollow nanospheres, tubes, and wires with tailored structures and morphologies.

Singh and coworker [39] studied the AIE in two newly developed ICT materials (they called them A1 and A2, Figure 6.18). The authors studied the crystallographic, thermal, and photophysical properties of those compounds. They found that A1 emitted strong green light, while A2 showed relatively weaker blue emission upon photoexcitation. The crystal structure analysis of A1 showed that both the donor and acceptor groups are involved in the intermolecular interactions that in turn restricted the intramolecular rotations of both the groups. As A1 showed strong green light emission and formed good quality crystal, the authors hoped that the compound could be useful in opto-electric devices.

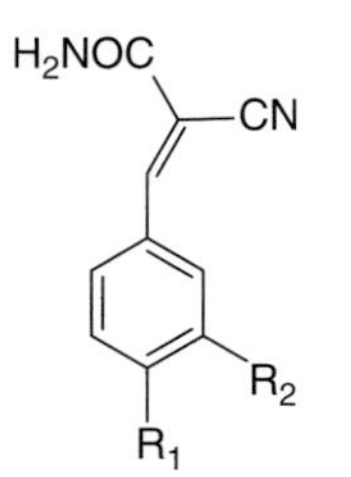

Figure 6.18 Chemical structure of A1 (R_1, R_2 = —OMe) and A2 (R_1 = —OMe, R_2 = —H).

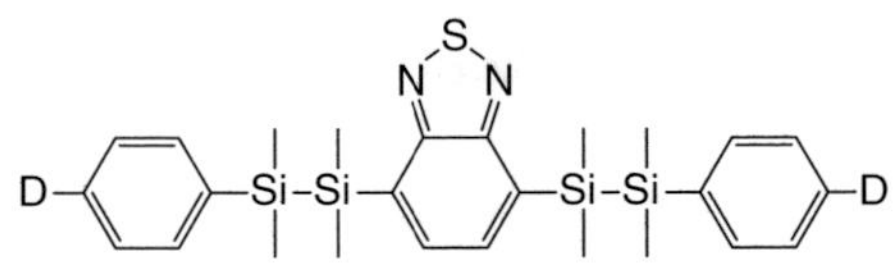

1: OMe
D = **2**: NMe_2
3: OPh

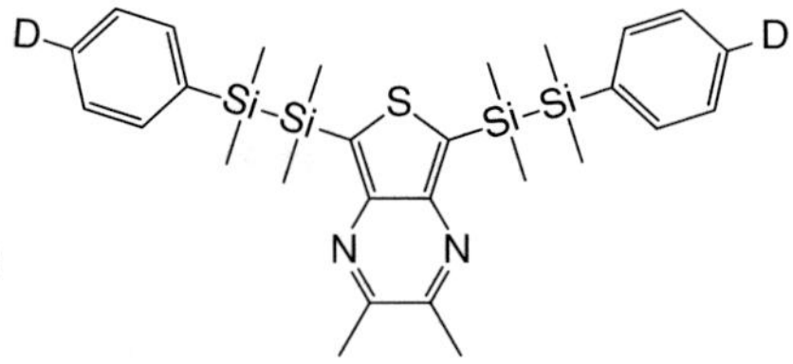

4: OMe
D = **5**: NMe_2
6: OPh

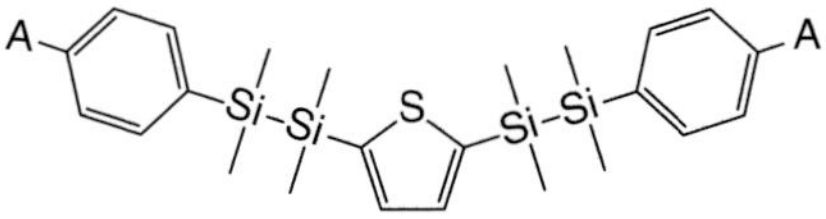

A = **7**: CN
8: CO_2Et

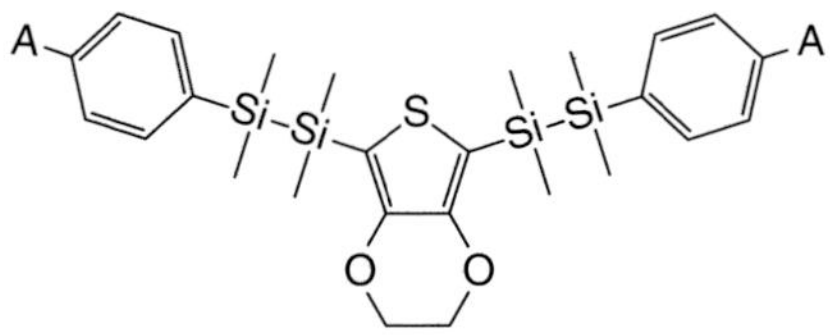

A = **9**: CN
10: CO_2Et

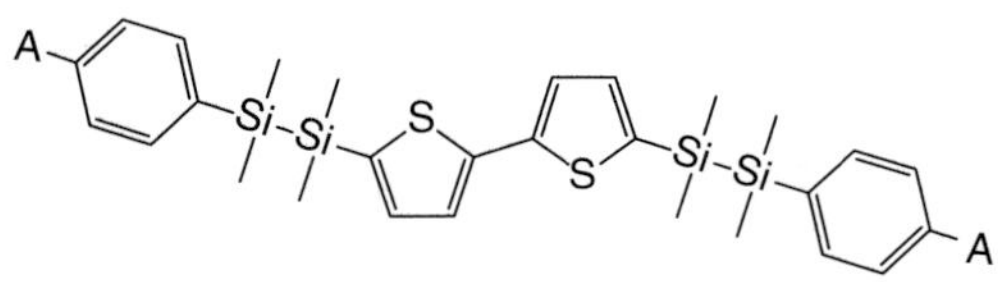

A = **11**: CN
12: CO_2Et

Figure 6.19 Chemical structures of D–A–D (**1–6**) and A–D–A (**7–12**) derivatives mentioned by Shimada *et al*. (Shimada *et al*. 2016 [40]. Reproduced with permission of John Wiley & Sons.)

Shimada *et al.* [40] reported bright solid-state emission from disilane-bridged donor–acceptor–donor (D—Si—Si—A—Si—Si—D) and acceptor–donor–acceptor (A—Si—Si—D—Si—Si—A) compounds (Figure 6.19). They also reported both these compounds showed strong emissions with peaks at around 500 and 400 nm, respectively, in the solid state with quantum yield up to 0.85.

Several pyrazine derivatives have been reported to show ICT upon photoexcitation [41, 42]. Chen *et al.* [43] reported stimuli-dependent transition between the LE to TICT state of pyrazine derivatives in aqueous media. Bhalla and coworkers [44] reported aggregation-induced emission enhancement (AIEE) in a TICT-based pyrazine derivative (Figure 6.20). The TICT molecule consists of a pyrazine scaffold as electron acceptor, while two amino groups act as donor units that are connected to the acceptor unit through rotatable phenyl rings. They also showed that the TICT probe formed aggregates in aqueous media and the aggregates showed copper-induced restriction to intramolecular rotation. The aforesaid restriction led the molecule to act as a "not quenched" probe for detection of copper(II) ions. During the process, the aggregates of the TICT molecule acted as both reactors and stabilizers for generation of copper(II) oxide (CuO) nanoparticles. In the course of the reaction, the molecules in aggregation become oxidized to form polyamine derivative. The authors found

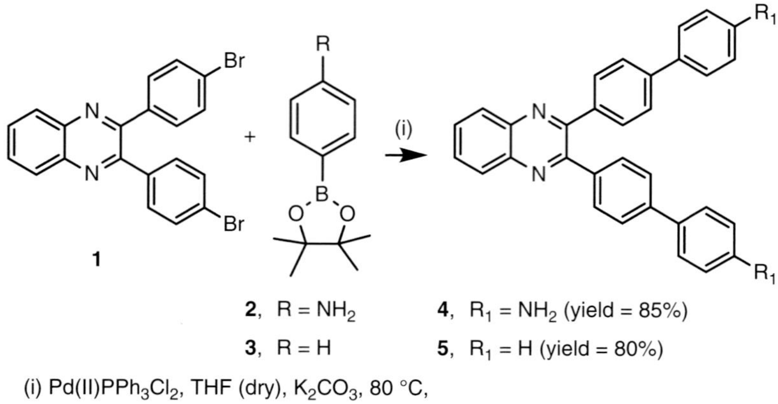

Figure 6.20 The structure and synthetic route of pyrazine-based TICT probe (**4**) and the model compound (**5**) that did not show any TICT behavior. (Deol *et al.* 2016 [44]. Reproduced with permission of American Chemical Society.)

that the oxidized derivative in combination with CuO nanoparticles served as light-harvesting antennas and showed remarkable photocatalytic efficiency in Sonogashira coupling in aqueous medium at room temperature under visible light irradiation. While designing the TICT-AIEE supramolecular assemblies, the authors took advantage of metal-induced restriction of intramolecular rotation of the pyrazine-based TICT probe. They incorporated the amino groups as donor units in the aforesaid molecule as this group is known to interact with soft metal ions. The UV–vis absorption spectrum of the probe in water showed two absorption peaks at 340 and 410 nm, respectively. The authors assigned the high energy peak to $n\pi^*$ transition of the pyrazine moiety, while the low energy peak was assigned to the ICT transition from the donor to the acceptor. The 410-nm peak showed about 6 nm bathochromic shift in tetrahydrofuran (THF) solvent, compared to that in water, while the 340-nm peak remained unaffected due to the change in the medium. The emission spectrum of the probe in 4-(2-hydroxyethyl)-1-piperazineethanesulfonic acid (HEPES) buffer showed two emission bands. The emission band with maximum at 445 nm was assigned to the LE species, while the 555-nm peak was assigned to the ICT species of the probe molecule. The molecule showed only one emission peak at 555 nm in THF upon photoexcitation at 360 nm. The authors carried out temperature-dependent emission measurements of the probe that showed a blueshift in emission maximum of the 555-nm band with an increase in emission intensity as they went from 25 to 75 °C. From the aforesaid thermal behavior of fluorescent species, they concluded that it had to have originated from the TICT state of the molecule. The strong polarity dependence on the emission spectral behavior indicated the formation of ICT state of the molecule in the excited state. The enhanced emission intensity of the probe in glycerol–ethanol mixture with enhanced viscosity also supported the formation of the TICT state of the molecule. To explore the aggregation behavior of the molecule, the authors studied the absorption and emission profiles of the probe in THF–water mixtures with varying compositions. The absorption intensity of the probe was decreased slightly with change in solvent composition in THF–water mixture, while no change in the absorption maxima was observed. The temperature-dependent studies on the absorption spectra of the probe in THF–water mixture showed an increase in the 340-nm band with no significant change in the absorption maximum as the temperature was increased from 25 to 70 °C. From these results, the authors concluded that in the aggregation state the molecules were not arranged in a J/H manner. They also noted that with increase in water content in the THF–water mixture, the emission intensity of the band corresponding to TICT emission was decreased, while a new band at 445 nm (LE band) appeared. To confirm that weak AIEE characteristics of the probe mainly originated due to the incomplete suppression of TICT, the authors synthesized a model derivative which lacked the amino donor groups of the TICT probe (Figure 6.20). The model compound did not show the existence of TICT state, but exhibited strong AIEE behavior in aqueous media. These results indicated that the probe molecule could act as a TICT-AIEE-based dual emission compound.

The authors evaluated the molecular recognition of the metal ions such as Ag^+, Hg^{2+}, Au^{3+}, Cu^{2+}, Zn^{2+}, Fe^{2+}, Pd^{2+}, and so on through the aforesaid TICT-AIEE system by adding their perchlorate/chloride salts using UV–vis absorption and emission spectroscopy. Their investigations revealed that the absorption/emission profile of the TICT-AIEE system was affected only in the presence of Cu^{2+} ions, while other ions did not induce any notable change in its absorption and emission spectra. The authors found that the intensity of absorption bands at 410 and 340 nm gradually decreased due to gradual addition of copper(II) ions (0–2 equiv.) to the aqueous solution of the probe molecule (5 μM). During this process, a new band at 310 nm appeared with two isobestic points at 430 and 320 nm. They assigned that spectral behavior to the formation of a complex of the probe with Cu^{2+} ions at low concentration of the metal ion. Addition of Cu^{2+} ion at higher concentration (3–50 equiv.) led to broadening of the absorption bands as well as formation of two new bands at 280 and 750 nm within 30 min. The intensity of the 280 and 750 nm bands got increased with time to about 60 min. The authors argued that formation of CuO nanoparticles was responsible for the observed spectral change. Upon addition of Cu^{2+} ions (0–50 equiv.) to the solution of the probe molecule (5 μM) in HEPES buffer, the emission intensity of the 555 nm band, that is, the TICT band was decreased with a large increase of the LE band at 445 nm. The authors measured the emission spectrum of this solution after 1 h and found that the intensity of the 445-nm band decreased about 42%. The time-resolved fluorescence studies of the probe in HEPES buffer showed a mono-exponential decay behavior of the probe in the absence of Cu^{2+} ions. However, in the presence of Cu^{2+} ions, the decay curve of the probe was found to be tri-exponential, indicating the presence of three distinct species in the excited state. The time-resolved studies of the probe led the authors to conclude that the presence of Cu^{2+} ions facilitated the decrease in non-radiative rate constant of the LE emission, while the increase in the non-radiative rate constant of the TICT emission factor resulted in enhancement of the emission intensity of the 445-nm band. The authors further showed that the in situ generated oxidized species of TICT-AIEE system with CuO nanoparticles could catalyze Sonogashira couplings under mild conditions with a variety of substrates such as aryl halides with excellent efficiency.

6.2.4 Solar Energy Conversion

To collect and store solar energy, the scientific community is carrying out studies on dye-sensitized solar cells (DSCs). Efficient conversion and storage of solar energy requires collection of charges generated through photoexcitation at the electrodes or storage through formation of chemical bonds. The long lifetime of the photoexcited states due to charge separation or through the formation of triplet state is reported to be crucial for conversion of solar energy. Several donor–acceptor-based ICT-based molecules have been screened for use in DSCs. Some representative examples are discussed in this section. Hagfeldt *et al.* [45] reviewed the recent developments of nanostructured dye-sensitized solar cells (nDSCs), which are believed to be suitable for efficient conversion of solar energy to electricity at low cost. Although the DSCs based on bi- and polypyridyl ruthenium complexes have reported to achieve solar to electrical energy conversion efficiency between 10% and 11% under AM 1.5 irradiation [46], the scientific community is trying to replace these materials with fully organic photosensitizers to reduce the cost as well as to curb environmental hazards. The use of fully organic materials may also facilitate these photosensitizers with high extinction coefficients that could open up the possibility to use them in thinner solar cells such as solid-state nDSC [47]. Several dyes, including coumarins, porphyrins, xanthenes, and polyenes have been examined for possible use in the DSCs with conversion efficiencies between 5% and 9 % with the traditionally used iodide/triiodide redox systems. Edvinsson and coworkers [48] reported the ICT process of some perylene dyes (Figure 6.21) and their performances in DSCs. The authors found that introduction of stronger donor groups to the perylene core at the point which is farthest away from the anchor group pushes up the LUMO of these molecules. This leads to more negative potential with respect to the normal hydrogen electrode (NHE). They also found improved photocurrent with increase in the ICT character of the dyes. They have reported external quantum efficiency more than 70% and solar to electrical conversion energy of about 4% using the dye with highest ICT character, even without using any UV treatment. They have also tested the stability of the dyes used and reported the highest performing dye to be spectrally stable even after 2000 h of irradiation.

6.2.5 ICT- Based Molecules for Sensing Temperature

Measuring temperature with high precision is important for several applications [49, 50]. For example, *in situ* large-area or gradient temperature measurements with high spatial resolution are frequently required in automobile and aircraft industries, in marine research, and for exploring underground geochemistry, to name a few. Fluorescent thermometers have been studied widely for estimating the temperature of microenvironments, like cells and micro-fluidic devices

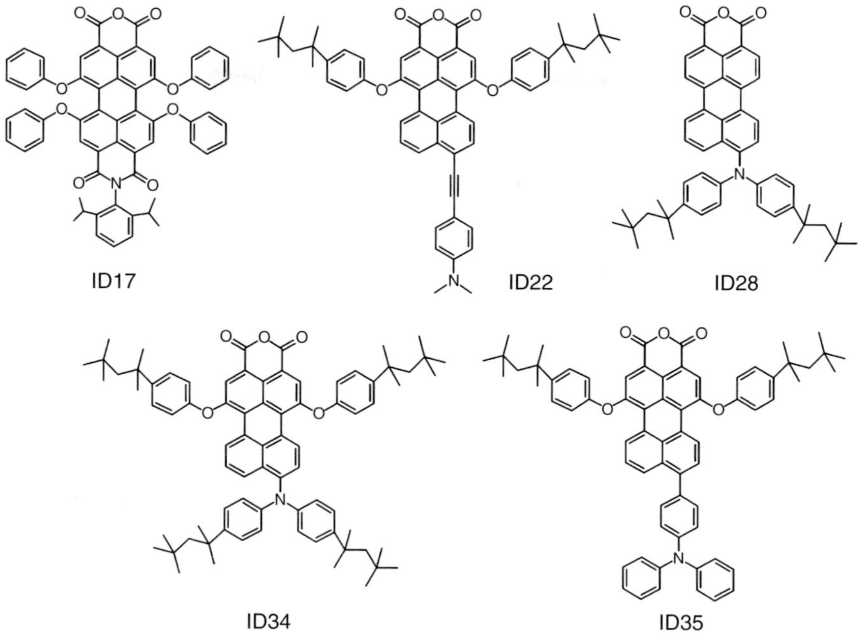

Figure 6.21 Chemical structures of new perylene dyes reported by Edvinsson and coworkers. (Edvinsson *et al.* 2007 [48]. Reproduced with permission of American Chemical Society.)

with high accuracy. In most of the cases, fluorescent organic molecules or inorganic complexes are used whose emission intensities decrease with increase in temperature due to activation of non-radiative deactivation pathways with increasing temperature. Cadmium selenide (CdSe) quantum dots and zinc oxide (ZnO) microcrystals have also been studied for use as "nanothermometers" [51, 52]. In spite of several advantages, these fluorescent thermometers suffer from limitations due to their negative temperature coefficients. On the other hand, a fluorescent probe with positive temperature coefficient, whose emission intensity increases with increasing temperature could be useful in suppressing the background interference at high temperatures. In addition, these probes can be used in conjunction with a negative temperature coefficient probe in which the ratiometric measurements of emission intensities could provide a built-in correction for quantitative detection of temperature of a system. Cao and coworkers [53] have developed an ICT-based ratiometric fluorescent thermometer with a positive temperature coefficient. The ICT probe, namely, *N*,*N*-dimethyl-4-((2-methylquinolin-6-yl)ethynyl)aniline (Figure 6.22) showed increase in fluorescence intensity with increase in temperature in dimethyl sulfoxide (DMSO). The designing principle proposed by the authors to prepare the ICT-based fluorescence thermometer lies in the fact that the molecules that show TICT in the excited state are usually weakly emitting due to the fact that the emission from the zero vibration level of the TICT state is forbidden. On the other hand, as the authors mentioned, the radiative transitions from the TICT state to the ground state are allowed through the assistance of higher and

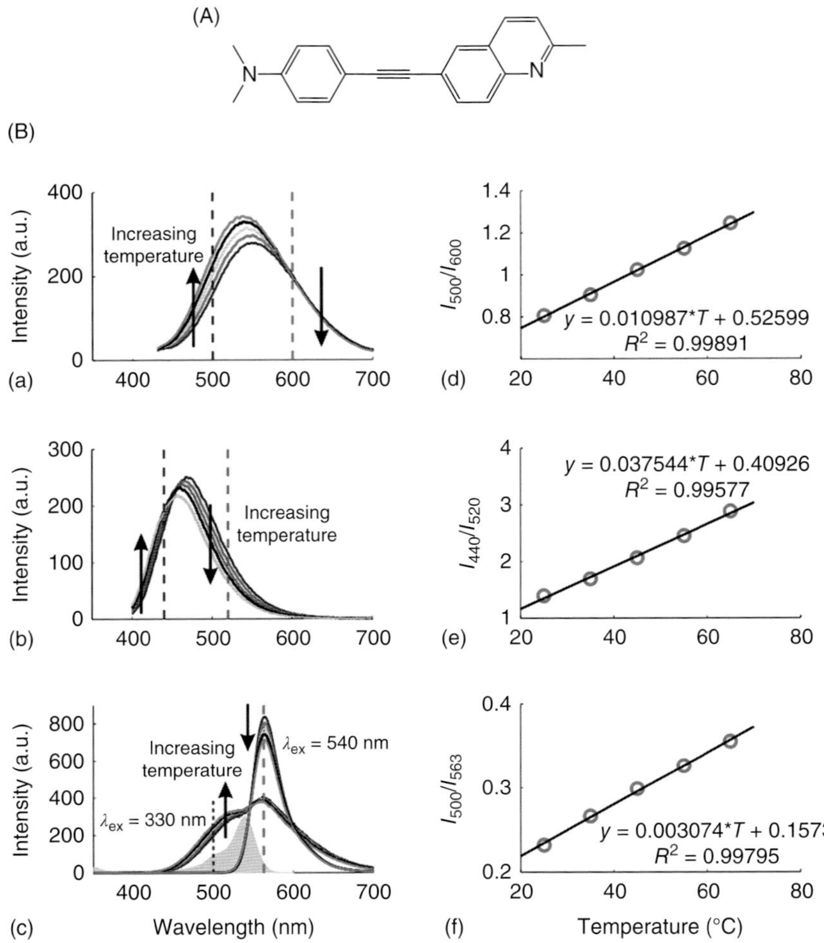

Figure 6.22 Panel (A) shows the structure of the TICT probe used as Cao *et al.* Panel (B) depicts the temperature-dependence fluorescence spectra of the TICT probe (a) in DMSO; (b) in ethyl acetate (EA) (λ_{ex} = 360 nm); (c) the mixture of the probe and rhodamine 6G in DMSO with λ_{ex} 330 and 540 nm, respectively. The temperature was increased from 25 to 65 °C with a step change of 10 °C. The corresponding temperature dependence of emission intensity ratios and the associated best-fit equations of the probe (d) in DMSO at 500 and 600 nm; (e) in EA at 440 and 520 nm; (f) mixture of the probe and rhodamine 6G in DMSO at 500 and 563 nm. (Cao *et al.* 2014 [53]. Reproduced with permission of Royal Society of Chemistry.)

non-totally symmetrical vibrational bands of the TICT state. They also argued that the aforesaid vibrational bands can be activated upon heating, resulting in higher emission intensity with increase in temperature of the medium if the increase in TICT emission can surpass the non-radiative deactivation pathways. The authors confirmed the TICT process in the probe molecule by studying the viscosity-dependent emission measurements in ethanol–glycerol mixtures with varying compositions as well as using DFT calculations. Their quantum chemical calculations indicated an ICT from the phenyl ring to the quinolone ring of the molecule upon photoexcitation. The absorption maxima of the probe does not change much with increase in solvent polarity, while the emission maxima got redshifted substantially with increase in the polarity of the medium. The large Stokes shift of about 210 nm in DMSO of the probe could be useful in minimizing the reabsorption of the emitted photons that in turn may improve the signal-to-noise ratio of fluorescence imaging. The authors found that in DMSO the probe showed unusually positive temperature coefficients (about 0.5% per °C) at around 450 and 600 nm. The fluorescence quantum yield of the probe increases accordingly with increasing temperature. The authors argued that with increase in temperature the vibrational bands at higher energy levels in the TICT state became active, resulting in increased emission intensity. They also assigned the blueshift of emission maximum from 551 nm at 25 °C to 538 nm at 65 °C to increase in contribution of higher vibrational bands with increasing temperatures. The authors pointed out that the fluorescence intensity of a TICT chromophore depends on two competing factors, namely, the rate of radiative TICT emission and non-radiative de-excitations. While both these factors are expected to increase with increasing temperature, the former becomes more relevant for the ICT probe in DMSO solvent, leading to positive temperature coefficient for the probe. A situation may arise when the effect of temperature on the non-radiative de-excitation may get affected more than the TICT emission, and a negative temperature coefficient can be expected. In fact, the authors observed a negative temperature coefficient for the probe in other solvents such as ethyl acetate in which the non-radiative processes leapfrog the TICT fluorescence with increasing temperature. The authors found that the temperature dependence of the aforesaid ICT probe was varied with varying wavelengths, which they claimed to use for ratiometric temperature measurements. They showed that the ratios of emission intensities of the probe at 500 nm to that of 600 nm in DMSO bore a linear relationship with change in temperature, making it a potential candidate for use as a thermometer. The authors also showed that the ICT probe can be used in conjunction with other temperature-sensitive molecules such as rhodamine 6G, which showed a negative temperature coefficient for ratiometric detection of temperature. They also checked the reversibility of the ICT fluorophore by heating and cooling it between 25 and 65 °C and found it to be reversible. As the thermometers based on fluorescence lifetime measurements take relatively longer times and need sophisticated instruments, the emission-based thermometers could be more convenient in detecting temperature with low cost and ease. However, as fluorescence of many TICT molecules are dependent on both viscosity and temperature of the medium, caution must be taken while using these fluorescence-based thermometers in measuring temperatures of systems whose viscosity changes with temperature are significantly high.

Yang and coworkers [54] designed a triarylboron-based fluorescence thermometer for detection over a wide temperature range. It is now well known that like other fluorophores the emission intensity of the TICT-based emitters also decreases with increasing temperature. Another problem that limits the use of TICT probes as fluorescent sensors is that these molecules, in general, show very weak emission intensity. Some TICT molecules show enhanced emission

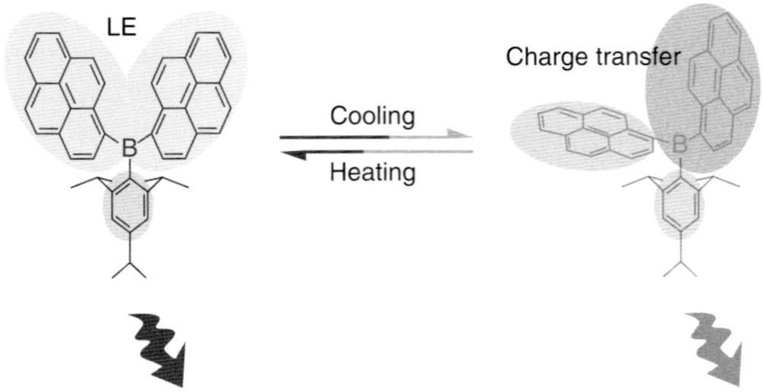

Figure 6.23 A feasible luminescence sensing mechanism of DPTB with changing temperature. (Feng *et al.* 2011 [54]. Reproduced with permission of John Wiley & Sons.)

intensity with increase in temperature of the medium [55]. A colorimetric change can be observed in some cases due to shift of thermal equilibrium between the LE and ICT states [56]. The authors designed a thermosensitive molecule (Figure 6.23), namely, dipyren-1-yl-(2,4,6-triisopropylphenyl)borane (acronymed DPTB) by utilizing the advantages of highly fluorescent arylboron compounds and the TICT process with two reverse-luminescence intensity changes. The authors designed the luminescent pyrene-containing triarylboron molecule, DPTB, considering the following ideas. To facilitate ICT in the aforesaid molecule in the excited state, two electron donor pyrene groups and electron-deficient boron atom with an empty p-orbital were introduced in the molecule. The triisopropylphenyl group was introduced as a stabilizer for the boron compound. The rigid pyrene groups could facilitate the molecule with high emission quantum yield. The authors mentioned that the contribution from the two pyrene moieties were probably not the same, owing to high steric hindrance in the molecule. DPTB showed temperature-dependent luminescence with a very high quantum yield (highest being 0.83 at 50 °C) over a wide range, namely, −50 °C to +100 °C. The luminescence turns from green to blue as the temperature increases. The high quantum yield of the molecule in the studied temperature range also endows the molecule with the ability to avoid the low signal-to-noise ratio generally induced at high temperature. The authors found the absorption spectra of the molecule were not affected much due to change in temperature and only a few changes over the whole temperature range were observed. In contrast to the absorption behavior, the emission profile of DPTB was highly temperature dependent. At room temperature, the molecule showed a wide luminescent band, which the authors identified as dual emission through the decay lifetime measurements of the molecule. They assigned these shorter and longer wavelength species responsible for dual emission to the LE and ICT states of the molecule, respectively. The temperature of the medium affects the dynamic equilibrium between the LE and ICT states of many dual-emitting molecules [57, 58]. The authors observed a similar behavior for DPTB with changing temperature. The luminescence color of the molecule was determined by the population ratio of the two distinct species (LE and ICT). The TICT state that has relatively lower energy was preferentially populated as the temperature is lowered and a bathochromic shift of the luminescence was observed. Conversely, heating of the system led to increased population of the LE state of the molecule and a hypsochromic luminescence was observed. The authors found that with increasing temperature, the luminescence spectrum gradually shifted to the higher energy. They found a linear relationship between the emission maxima (λ_{emi}) of DPTB and the temperature of the medium (2-methoxyethyl ether or MOE). The authors also evaluated the reversibility of the probe in MOE medium by heating and cooling the molecule between −50 and 100 °C by 30 times, which showed the system to be reversible. To study the temperature gradient of a fluid, the authors placed a solution of DPTB in 2-methoxyethyl ether in a quartz tube and heated it from the top and cooled it from the bottom. They found the colors of the top and bottom of the tube to be blue and green, while the temperature of the middle changed from green to cyan to blue with increasing temperature. Therefore, the authors conclude that the temperature of the medium could be realized through the naked eye or a camera. To utilize the TICT-based thermometer to realize the temperature distribution of a certain area, which is often required in automobile and aircraft industries, they sealed the DPTB-MOE solution in a sandwich structure PVC-porous PP-PVC film of $5 \times 5\,cm^2$ area and $60\,\mu M$ thickness. Using their results, the authors claimed that DPTB could be used as a luminescent colorimetric thermometer for *in situ* large-area or gradient temperature measurements with high sensitivity over a wide range of temperature with high spatial resolution. They also showed that the liquid thermometer could be fabricated in various forms, thereby making it useful in various research areas.

From the discussion in this chapter, we understand that several ICT-based molecules have been studied for potential applications as fluorescence as well as colorimetric sensors to detect metal ions and molecules in solution as well as in biological samples, for sensing temperature and viscosity of a medium, and so on. Use of ICT probes for AIE, OLEDs and DSCs are also being reported in literature. As mentioned in Chapter 5, ICT-based molecules have been used to design NLO-active materials that have several applications in technology. Therefore, we can anticipate that several studies of ICT molecules with application in sensing, optoelectronics, and so on, will be reported in near future.

References

1 Haidekker, M.A. *et al.* (2005) *Bioorg. Chem.*, 415.
2 Haidekker, M.A. *et al.* (2000) *Am. J. Physiol. Heart Circ. Physiol.*, **278**, H1401.
3 Sasaki, S., Drummen, G.P.C., and Konishi, G. (2016) *J. Mater. Chem. C*, **4**, 2731.
4 Uoyama, H., Goushi, K., Shizu, K., Nomura, H., and Adachi, C. (2012) *Nature*, **492**, 234.
5 (a) Adachi, C. *et al.* (2014) *Nat. Mater.*, **14**, 330; (b) Adachi, C. *et al.* (2015) *Nat. Commun.*, **6**, 8476.
6 Bredas, J.L. *et al.* (2017) *J. Am. Chem. Soc.*, **139**, 4042.
7 Marks, T.J. *et al.* (1994) *Chem. Rev.*, **94**, 195.

8 Sunahara, H. *et al.* (2007) *J. Am. Chem. Soc.*, **129**, 5597.
9 Sasaki, S. *et al.* (2014) *Tetrahedron*, **70**, 7551.
10 Shigeta, M. *et al.* (2012) *Molecules*, **17**, 4452.
11 Haidekker, M.A. and Theodorakis, E.A. (2010) *J. Biol. Eng.*, **4**, 11.
12 Stsiapura, V. *et al.* (2008) *J. Phys. Chem. B*, **112**, 15893.
13 Liu, J. *et al.* (2015) *Sens. Actuators, B*, **208**, 538.
14 Hiramaya, T. *et al.* (2013) *Chem. Sci.*, **4**, 1250.
15 Umar, S., Jha, A.K., and Goel, A. (2016) *Asian J. Org. Chem.*, **5**, 187.
16 Reja, S.I., Khan, I.A., Bhalla, V., and Kumar, M. (2016) *Chem. Commun.*, **52**, 1182.
17 Peters, T. (1996) *All About Albumin: Biochemistry, Genetics, and Medical Applications*, Academic Press, San Diego, pp. 234–240.
18 Ren, M., Deng, B., Kong, X., Zhou, K., Liu, K., Xu, G., and Lin, W. (2016) *Chem. Commun.*, **52**, 6415.
19 Yu, S., Yang, X., Shao, Z., Feng, Y., Xi, X., Shao, R., Guo, Q., and Meng, X. (2016) *Sens. Actuators, B*, **235**, 362.
20 Wei, X., Yang, X., Feng, Y., Ning, P., Yu, H., Zhu, M., Xi, X., Guo, Q., and Meng, X. (2016) *Sens. Actuators, B*, **231**, 285.
21 Meng, X.M., Ye, W.P., Wang, S., Feng, Y., Chen, M., Zhu, M.Z., and Guo, Q.X. (2016) *Sens. Actuators, B*, **201**, 520.
22 Ito, A., Ishizaka, S., and Kitamura, N. (2010) *Phys. Chem. Chem. Phys.*, **12**, 6641.
23 Kumari, N., Jha, S., and Bhattacharya, S. (2011) *J. Org. Chem.*, **76**, 8215.
24 Hudnall, T.W., Chiu, C.W., and Gabbai, F.P. (2009) *Acc. Chem. Res.*, **42**, 388.
25 Vennesland, B., Comm, E.F., Knownles, J., Westly, J., and Wissing, F. (eds) (1981) *Cyanide in Biology*, Academic, London.
26 Kovalchuk, A., Bricks, J.L. *et al.* (2004) *Chem. Commun.*, 1946.
27 Bhat, H.R. and Jha, P.C. (2017) *J. Phys. Chem. A*, **121**, 3757.
28 Li, Q., Cai, Y., Yao, H., Lin, Q., Zhu, Y.R., Li, H., Zhang, Y.M., and Wei, T.B. (2015) *Spectrochim. Acta, Part A*, **136**, 1047.
29 Thiagarajan, V., Ramamurthy, P., Thirumalai, D., and Ramakrishnan, V.T. (2005) *Org. Lett.*, **7**, 657.
30 Wu, J., Liu, W., Ge, J., Zhang, H., and Wang, P. (2011) *Chem. Soc. Rev.*, **40**, 3483.
31 Tang, C.W. and VanSlyke, S.A. (1987) *Appl. Phys. Lett.*, **51**, 913.
32 Adachi, C. *et al.* (2014) *Nat. Photonics*, **8**, 326.
33 Park, I.S., Komiyama, H., and Yasuda, T. (2017) *Chem. Sci.*, **8**, 953.
34 Chang, C.-J., Yang, C.-H., Chen, K., Chi, Y., Shu, C.-F., Ho, M.-L., Yeh, Y.-S., and Chou, P.-T. (2007) *Dalton Trans.*, 1881.
35 Kim, E. and Park, S.B. (2009) *Chem. Asian J.*, **4**, 1646.
36 Keywin, T., Sooksai, C., Prachumrak, N., Kewpuang, T., Muenmart, D., Namuangruk, S., Jungsuttiwong, S., Sudyoadsuk, T., and Promarak, V. (2015) *RSC Adv.*, **5**, 16422.
37 Hu, R. *et al.* (2009) *J. Phys. Chem. C*, **113**, 15845.
38 Li, Y. *et al.* (2014) *Acc. Chem. Res.*, **47**, 1186.
39 Gupta, V.K. and Singh, R.A. (2017) *Faraday Discuss.*, **196**, 131.

40 Shimada, M., Tsuchiya, M., Sakamoto, R., Yamanoi, Y., Nishibori, E., Sugimoto, K., and Nishihara, H. (2016) *Angew. Chem. Int. Ed.*, **55**, 3022.
41 Hagfeldt, A. and Gratzel, M. (2000) *Acc. Chem. Res.*, **33**, 269.
42 Lu, X., Fan, S., Wu, J., Jia, X., Wang, Z.S., and Zhao, G. (2014) *J. Org. Chem.*, **79**, 6480.
43 Chen, M., Li, L. *et al.* (2015) *Chem. Commun.*, **51**, 10710.
44 Deol, H., Pramanik, S., Kumar, M., Khan, I.A., and Bhalla, V. (2016) *ACS Catal.*, **6**, 3771.
45 Hagfeldt, A., Boschloo, G., Sun, L., Kloo, L., and Pettersson, H. (2010) *Chem. Rev.*, **110**, 6595.
46 O'Regan, B. and Gratzel, M. (1991) *Nature*, **353**, 737.
47 Gratzel, M. (2004) *J. Photochem. Photobiol.*, **164**, 3.
48 Edvinsson, T., Li, C., Pschirer, N., Schoneboom, J., Eickemeyer, F., Sens, R., Boschloo, G., Herrmann, A., Mullen, K., and Hagfeldt, A. (2007) *J. Phys. Chem. C*, **111**, 15137.
49 (a) Kontis, K. (2007) *Aeronaut. J.*, **111**, 495; (b) Wolfbeis, O.S. (2008) *Adv. Mater.*, **20**, 3759.
50 Demas, J.N., DeGraff, B.A., and Colenam, P.B. (1999) *Anal. Chem.*, **71**, 7931.
51 Maestro, L.M. *et al.* (2010) *Nano Lett.*, **10**, 5109.
52 Shinde, S.L. and Nanda, K.K. (2013) *Angew. Chem. Int. Ed.*, **52**, 11325.
53 Cao, C., Liu, X., Qiao, Q., Zhao, M., Yin, W., Mao, D., Zhang, H., and Xu, Z. (2014) *Chem. Commun.*, **50**, 15811.
54 Feng, J., Tian, K., Hu, D., Wang, S., Li, S., Zeng, Y., Li, Y., and Yang, G. (2011) *Angew. Chem. Int. Ed.*, **50**, 8072.
55 Braun, D. and Rettig, W. (1994) *Chem. Phys.*, **180**, 231.
56 Volchkov, V.V. and Uzhinov, B.M. (2008) *High Energy Chem.*, **42**, 163.
57 Rettig, W. and Chandross, E.A. (1985) *J. Am. Chem. Soc.*, **107**, 5617.
58 Yang, J.S., Lin, C.K., Lahoti, A.M., Tseng, C.K., Liu, Y.H., Lee, G.H., and Peng, S.M. (2009) *J. Phys. Chem. A*, **113**, 4868.

7
Summary and Outlook

7.1 Introduction

From the discussion in the preceding chapters it is apparent that the intramolecular charge transfer (ICT) process constitutes a major area of research in molecular science. Since the discovery of the first ICT molecule 4-*N*,*N*-dimethylaminobenzonitrile (DMABN), several related molecules as well as molecules belonging to other classes (coumarin, rhodamine B, etc.) have been studied to understand the underlying principle in detail. In spite of several studies, two issues regarding the ICT process still remains to be resolved fully. One of these concerns the structure of the ICT state, while the other is the pathway through which the ICT occurs. The properties (polarity and hydrogen bonding ability) of the medium play a crucial role in determining the structure and rate of formation of the ICT state. Several experimental and quantum chemical studies have been undertaken by the scientific community to understand the ICT process. As charge transfer can be considered as one of the simple reactions, several theoretical models have been proposed to calculate the rate of the ICT process. Many ICT-based molecules are being used as sensor of medium polarity and viscosity. Several ICT molecules have been studied for possible use in designing nonlinear optical (NLO) materials, in organic light-emitting diodes (OLEds), and so on. This chapter summarizes some of the studies on ICT, discussed in Chapters 1–6.

7.2 Summary of Studies of ICT

In the preceding chapters, we have discussed several aspects of the ICT process and their possible applications. The charge transfer process is ubiquitous in nature as it is a key step in photosynthesis and metabolism. Several novel ICT-based molecules are designed for possible technological applications in optoelectronics, nonlinear optics, sensing, and so on. In several of these applications, metal-ion-based chromophores are being used presently, although several of them lack the required flexibility and are expensive. To get rid of costly metals and to increase the durability and flexibility of the devices, the scientific community has put their efforts in purely organic molecules with the aim of

Intramolecular Charge Transfer: Theory and Applications,
First Edition. Ramprasad Misra and S. P. Bhattacharyya.

using these organic materials for fabricating inexpensive as well as efficient devices. In spite of a large amount of studies, challenges still remain to utilize the potentials of ICT-based molecules fully for betterment of the human life.

In Chapter 1, the basic aspects of the ICT process was discussed, which gives the readers an overview of the content discussed in the following chapters. As we mentioned earlier, DMABN was the first among the ICT molecules reported. This molecule shows dual emission in its electronic emission spectrum, out of which the band at the blue edge was attributed to the "normal" emission arising from the locally excited (LE) state of the molecule. The redshifted band (lower energy band), sometimes referred to as the *abnormal* band, is assigned to the emission form the ICT species originated in the excited state. This discovery fueled the studies of ICT processes in several congeners of DMABN. Over the years, several new ICT molecules have been reported and their spectral characteristics have been discussed in detail. We must mention here that not all ICT molecules show dual emission due to several factors, including the possible decrease in emission quantum yield of the ICT species in polar media. In spite of a large number of studies, some aspects of the ICT process still remain a matter of lively debate till date for the scientific community. One of these unresolved issues is related to the structure of the ICT state, while the other is the dynamic or the pathway through which the ICT process may occur. Initial experimental studies led to the conclusion that in the excited state the ICT molecule gets twisted, leading to a twisted intramolecular charge transfer (TICT) state that has a higher dipole moment than that of the ground state as well as the LE state of this molecule. The progressively higher stability of the ICT species in solvents with increasing polarity supports this twisted mechanism. Later, the idea of the formation of a TICT state was challenged by several research groups as, in many cases, the normal and pretwisted analog of a molecule can give rise to the ICT process. Several new mechanisms concerning the structure of the ICT state were proposed, including rehybridized intramolecular charge transfer (RICT) and wagging intramolecular charge transfer (WICT). It was later found that all of these models have their own merits and demerits and TICT and planar intramolecular charge-transfer (PICT) mechanisms are believed to be models mostly accepted by the scientific community. Over the years, several experimental and quantum chemical studies are reported to investigate the structure and dynamics of the ICT process in several ICT molecules. The dynamics of the ICT process is controlled by several factors. Not only the radiative processes like fluorescence and phosphorescence but also several non-radiative processes are involved in the excited state photophysics of a molecule. Therefore, several experimental and theoretical studies are devoted to find out the possible mechanism and timescale involved in the ICT processes. Not only the organic molecules but several inorganic complexes and biomolecules also undergo the ICT process. In metal complexes, the process is often referred to as the photoinduced electron transfer (PET) process. Studies on the PET process in several inorganic complexes were fueled with the aim of understanding the underlying principle of the photosynthetic system that in turn will help the scientific community to design artificial photosynthetic systems for converting the solar energy to the chemical energy. All these aspects of the charge-transfer process are discussed in Chapter 2. This chapter also deals with several new organic/inorganic molecules that have been developed recently for possible technological applications.

As mentioned earlier, several experimental and theoretical investigations are reported on the ICT process in several molecules. Initially, semiempirical calculations were used to study the ICT process, but the development of the several new computational techniques have helped the scientific community use many state-of-the-art theoretical methods to explore the ICT process. This development has definitely increased our understanding of the ICT process. The electron transfer process is one of the simple reactions in nature. Therefore, several theories have been proposed to find out the rate of this reaction. Marcus theory, proposed by famous scientist R.A. Marcus in the early 1950s that bears his name, is very popular for studying the charge-transfer process. Later, several other theories were proposed in the study of several aspects of the charge-transfer processes. In Chapter 3, we have discussed some of these theoretical studies and models for determining rate of the electron transfer process. The chapter also deals with the experimental techniques that are being used to study the ICT process. Initially, steady-state absorption and emission spectroscopic techniques were used to investigate the ICT process in a molecule. Later, advent of several state-of-the-art spectroscopic techniques has helped the scientific community get better details of the aforesaid process. Vibrational spectroscopic techniques are being used widely to investigate the ICT process. Picosecond time-resolved resonance Raman spectroscopy and femtosecond stimulated Raman spectroscopy (FSRS) are two techniques that have been used recently to study the excited-state photophysics of ICT molecules. Time-correlated single-photon counting (TCSPC) technique is quite popular to study the lifetimes of the excited-state species, which are generally within the range of 1 ns to several microseconds. The LE to ICT reaction time is faster than the instrument response time of a typical TCSPC instruments. Therefore, to study the early processes that occur within a few picoseconds of photoexcitation, several time-resolved spectroscopic techniques with better time resolutions are being used. Transient absorption spectroscopy with sub-picosecond time resolution is one of those techniques. THz spectroscopy is also being used to study the ICT process. Several examples of experimental studies dealing with early photophysics of the ICT process are also discussed in Chapter 3.

The medium in which a reaction is performed plays an important role in determining the rate of a reaction. In the ICT process, a species with higher dipole moment than the ground state is believed to be formed. Therefore, the charge distribution of the molecule in the ground and excited states are expected to be notably different. This, in turn, makes the solvent molecules interact differently with them. In general, the ICT process is favored in polar medium as the polar excited state is more stabilized by the polar solvents. It was later found that not only the polarity but also the other properties of the solvent like viscosity, hydrogen bonding abilities, and temperature in which the ICT reaction occurs affect the process. Therefore, several ICT-based molecules are being used to sense the local polarity and viscosity of the medium. To use these molecules for sensing applications, one needs to understand the exact behavior of a molecule with the change in solvent parameters. It is worthwhile mentioning that it is almost impossible to separate out the effects of solvent polarity and viscosity on the fluorescence property of a molecule; a better understanding of these effects is useful for designing better sensors. The ICT process is studied mainly in the solution phase, while those in gaseous and solid states are rare. For applications in nonlinear optics, optoelectronic devices, and other technological applications, one may need to prepare solid ICT materials. Therefore, several methodologies that rely on the self-assembly of the donor–acceptor ICT molecules have been proposed. These aspects of the ICT process are discussed in Chapter 4.

As we have mentioned several times in this book, ICT molecules are believed to be potential candidates for several technological applications. Other than sensing local environment and chemical species, these molecules are considered for potential applications in OLEDs and also in designing efficient NLO materials. Due to the vast range of studies on designing NLO response materials from ICT molecules, we have separated this from other applications and discussed it in relatively more detail in Chapter 5. The basic principles of the NLO materials and theoretical developments in this field are also discussed. Several computational and experimental studies are devoted with the aim of predicting new ICT molecules for development of potential NLO materials. Studies of some ICT-based molecules for two-photon absorption (TPA) process are also discussed in this chapter. To produce inexpensive OLEDs for white light displays, efforts are on to replace the precious metal ion-based phosphorescent materials that are now used in OLED devises with purely organic materials. Researchers are using the thermally activated delayed fluorescence (TADF) technique in ICT-based organic molecules for possible applications in OLEDs. The TADF process is believed to have capability to harvest the energy efficiently with an internal efficiency almost reaching 100%. To use in a full color display or for white light emission, one needs to use three primary colors: red, green, and blue. Although several OLED displays with red and green emissions are available, that with the capability of blue light (especially deep-blue) emission is still lacking. Several efforts have been devoted to develop materials for blue OLEDs, yet challenges still remain to resolve this issue. One of the important applications of ICT-based molecules is in sensing the local viscosity of several biological systems. Molecular rotors, a class of TICT molecules whose emission quantum yields get affected due to change in rigidity of the medium are used for this purpose. TICT-based molecules are generally weakly fluorescent. If interaction with the medium can restrict the twisting of the molecule, its emission intensity becomes better. Several research groups have exploited this principle to design TICT-based fluorescence "turn on" sensors for selective detective detection of ions and molecules in solution as well as in biological samples. There are a few reports in literature utilizing ICT molecules to design colorimetric sensors to detect the temperature as well as ions. Aggregation-induced emissions (AIEs) have been observed in several organic ICT-based molecules that show weak emission when dissolved in solvents but show intense emission in the solid state or due to cluster formation in solvents in which their solubilities are poor. Of late, ICT-based molecules are being studied for application in solar cells. The possible applications of ICT-based materials in the aforesaid purposes are discussed in Chapter 6.

So, from the preceding discussion it is apparent that ICT-based materials have the potential to be used in designing materials for several technological applications in OLEDs, NLO materials, in sensors, and so on; but more studies are needed to utilize their potentials fully. Several new findings are being reported on the ICT process almost every day and we hope these studies will enrich this field so that it can lead to development of several next-generation devices for technological applications. It is worthwhile to mention that in spite of the huge progress, there are some issues that need to be resolved. The examples cited in Chapter 2 clearly indicate that the structure of the ICT state is still under debate as some evidences point out the formation of the TICT state, while some other results claim it to be otherwise. Although several studies have been carried out to design ICT-molecule-based NLO materials, challenges still remain to obtain materials with high stability and flexibility. The interested readers may go through several references mentioned at the end of each chapter for studies on different ICT molecules. A better understating of the underlying mechanism of the ICT process could be a key to develop new ICT-based materials for future technological applications.

Further Readings

Kuznetsov, A.M. and Ulstrup, J. (1998) *Electron Transfer in Chemistry and Biology: An Introduction to the Theory*, John Wiley & Sons, Inc.

Balzani, V. (ed.) (2001) *Electron Transfer in Chemistry*, John Wiley & Sons, Inc.

Han, K.-L. and Zhao, G.-J. (eds) (2010) *Hydrogen Bonding and Transfer in the Excited State*, vol. **I & II**, John Wiley & Sons, Inc.

Boyd, R. (2008) *Nonlinear Optics*, 3rd edn, Elsevier.

Buckley, A. (ed.) (2013) *Organic Light Emitting Diodes (OLEDs): Materials, Devices and Applications*, Elsevier.

Shigeta, M., Morita, M., and Konishi, G. (2012) *Molecules*, **17**, 4452.

Zhang, Q., Tian, X., Zhou, H., Wu, J., and Tian, Y. (2017) *Materials*, **10**, 223.

Index